高等学校规划教材

Inorganic Chemistry

无机化学

陆家政　陈　菲　主编
曾宪栋　李雪华　副主编

化学工业出版社

·北京·

本书针对药学类专业无机化学的双语课程,将无机化学课程的基本内容进行精选,结合国内高校教学实际,删除原版教材中大量的复杂、繁琐及较深奥的部分,力求使之既能体现无机化学课程的专业基础课特色,又可提高学生的英语应用能力。

鉴于医药学专业学生理论与实验课程并重的特点,全书包含两大部分:Ⅰ Inorganic Chemistry 与 Ⅱ Inorganic Chemical Experiments。Part Ⅰ includings thirteen chapters:1 Introduction, 2 Structures of Atoms, 3 Chemical bonds, 4 Thermochemistry, 5 Chemical Kinetics, 6 Chemical Equilibrium, 7 Solutions, 8 Solubility Equilibrium, 9 Acid-Base Equilibria, 10 An Introduction to Electrochemistry, 11 Chemistry of Coordination Compounds, 12 Nonmetals and Semimetals, 13 Metals。Part Ⅱ includings three chapters:1 Basic Techniques of Experimental Chemistry, 2 Typical Chemical Laboratory Apparatus, 3 Experiments。

本书可作为高等院校本科药学及化学类相关专业的无机化学双语教材,也可作为化学专业英语课的教材或参考书。

图书在版编目(CIP)数据

无机化学 Inorganic Chemistry/陆家政,陈菲主编. —北京:化学工业出版社,2009.9(2019.9重印)

高等学校规划教材

ISBN 978-7-122-06250-5

Ⅰ.无… Ⅱ.①陆…②陈… Ⅲ.无机化学-高等学校-教材 Ⅳ.O61

中国版本图书馆 CIP 数据核字(2009)第 131386 号

责任编辑:宋湘玲 唐旭华 装帧设计:张 辉
责任校对:徐贞珍

出版发行:化学工业出版社(北京市东城区青年湖南街13号 邮政编码100011)
印　　装:北京虎彩文化传播有限公司
787mm×1092mm 1/16 印张19¾ 彩插1 字数529千字 2019年9月北京第1版第4次印刷

购书咨询:010-64518888　　　　售后服务:010-64518899
网　　址:http://www.cip.com.cn
凡购买本书,如有缺损质量问题,本社销售中心负责调换。

定　价:49.00元　　　　　　　　　　　　　　　　　　版权所有　违者必究

《无机化学 Inorganic Chemistry》编写人员名单

主　编　陆家政　陈　菲
副主编　曾宪栋　李雪华

编写人员（以姓名笔画为序）

　　　　伍小云（南方医科大学）
　　　　李君君（广东药学院）
　　　　李雪华（广西医科大学）
　　　　何丽新（广东药学院）
　　　　陆家政（广东药学院）
　　　　陈　菲（广东药学院）
　　　　陈丽江（浙江理工大学）
　　　　姚秀琼（广东药学院）
　　　　贺丽敏（广东药学院）
　　　　蒋　京（广东药学院）
　　　　曾宪栋（广东药学院）
　　　　曾琦华（广东药学院）
　　　　蒋东丽（桂林医学院）
　　　　赖泽锋（广西医科大学）
　　　　管小艳（广东药学院）
　　　　蔡秀兰（广东药学院）

前 言

自 2001 年国家教育部提倡在高等院校实行双语教学以来，很多高等院校都为化学及相关专业的本科生开设了无机化学双语课程。我们在药学专业无机化学课上也进行了积极尝试，课程受到学生的广泛欢迎，但遗憾的是至今很难找到切合相关专业的英文版《无机化学》教材，国外原版教材虽然可保持语言文字上的原汁原味，但往往内容太多，与国内高等院校现行的教学学时相差甚远；国内学者编著的教材可供选择的也为数不多。而我们经过几年的努力，积累了一定的教学经验，因此，联合相关院校组织无机化学双语教学第一线的教师编写了本书。

在本书编写的过程中，对无机化学课程的基本内容进行精选，加强基础，突出重点。删除了以往教材中存在的大量的复杂公式和繁琐计算的推导以及较深奥的理论分析和阐述，力求做到既言简意赅，又具有较完整的无机化学知识体系。竭力保持语言上与外版教材的一致性，又能使其内容更精练、通俗易懂，且更切合目前国内相关专业的教学实际。同时力求使之既能体现无机化学课程的专业基础课特色，又可提高学生的英语应用能力。希望此书的出版能够起到抛砖引玉的作用，为无机化学双语教学质量的进一步提高做出贡献。

全书由陆家政、陈菲主编并统稿。参加编写理论部分的有：陈丽江（第1章），曾宪栋（第2、3章），陈菲（第3章），蔡秀兰（第4章），管小艳（第5章），蒋京（第6章），蒋东丽（第7章），何丽新（第8章），姚秀琼（第9章），曾琦华（第10章），陆家政（第11章），贺丽敏（第12章），伍小云、李君君（第13章）。实验部分由陈菲、陆家政、李雪华和赖泽锋编写。

本书在编撰过程中得到了多方的支持。其中广东药学院教务处将其立项为特色教材予以支持和帮助，在此表示衷心的感谢。

本书可作为高等院校本科药学及化学类相关专业的无机化学双语教材，也可作为化学专业英语课的教材或参考书。

鉴于编者的水平和能力有限，书中尚有不妥之处，恳请专家以及使用本书的老师和同学批评指正。

编 者
2009 年 6 月

目 录

I Inorganic Chemistry ... 1

Chapter 1 Introduction ... 2
1.1 Observations and Conclusions ... 3
1.2 The Scientific Method ... 3
1.3 Units of Measurement ... 5
1.4 Advice on Studying Chemistry ... 8
Key Words ... 8

Chapter 2 Structures of Atoms ... 9
2.1 Particles in Atoms ... 10
2.2 The Bohr Model ... 12
2.3 The Wave Theory of Electrons ... 13
2.4 Heisenberg's Uncertainty Principle ... 14
2.5 The Schrödinger Equation ... 14
2.6 Quantum Numbers ... 15
2.7 Shapes of Atomic Orbitals ... 17
2.8 Many-Electron Atoms ... 19
2.9 Valence Electrons ... 24
2.10 The Periodic Table and Electronic Configurations of Atoms ... 24
2.11 The Periodicity of the Properties of Elements and Atomic Structure ... 26
Key Words ... 30
Exercises ... 31

Chapter 3 Chemical bonds ... 33
3.1 Ionic Bonds ... 33
3.2 Covalent Bonds ... 35
3.3 Lewis Dot Symbols and Lewis Structure ... 37
3.4 Sigma and Pi Bonds ... 43
3.5 Bond Polarity ... 44
3.6 Bond Energy, Bond Length and Bond Angle ... 45
3.7 Molecular Geometry ... 47
3.8 The VSEPR Model ... 47
3.9 Hybrid Orbitals ... 51
3.10 Molecular Orbitals ... 56
3.11 Intermolecular Force ... 64
Key Words ... 71
Exercises ... 71

Chapter 4 Thermochemistry ... 74
4.1 First Law of Thermodynamics ... 74
4.2 Enthalpy and Enthalpy Change ... 75

4.3	Hess's Law	76
4.4	Entropy and the Second Law of Thermodynamics	79
4.5	Spontaneous Processes and Gibbs Free Energy	80
4.6	Temperature and Direction of Spontaneous Change	81
4.7	Standard Free Energy of Formation and Standard Free Energy of Reaction	82
Key Words		83
Exercises		83

Chapter 5 Chemical Kinetics … 85

5.1	Reaction Rates	85
5.2	Theories of Reaction Rates	87
5.3	Effect of Concentration on Rate of Reaction	89
5.4	Rate and Temperature	92
5.5	Catalysis	94
Key Words		95
Exercises		96

Chapter 6 Chemical Equilibrium … 98

6.1	Concept of Equilibrium	98
6.2	Equilibrium Constant	98
6.3	Le Châtelier's Principle	103
Key Words		106
Exercises		106

Chapter 7 Solutions … 108

7.1	Ways of Expressing Concentration	108
7.2	Colligative Properties	113
Key Words		124
Exercises		125

Chapter 8 Solubility Equilibrium … 127

8.1	Saturated Solutions and Solubility	127
8.2	Solubility Equilibria	127
8.3	Factors Affecting Solubility	130
8.4	Predicting Precipitation Reaction	132
8.5	Selective Precipitation	134
8.6	Application of Precipitate on Medicine	137
Key Words		138
Exercises		139

Chapter 9 Acid-Base Equilibria … 141

9.1	Acids and Bases	141
9.2	Ion Product of Water	143
9.3	Strength of Weak Acids and Bases	144
9.4	The pH of Weak Acids and Bases	146
9.5	The Common-Ion Effect	148
9.6	Buffer Solutions	150
Key Words		153
Exercises		153

Chapter 10 An Introduction to Electrochemistry … 155

10.1	Oxidation-Reduction Reactions	155

10.2	Balancing Oxidation-Reduction Equations	158
10.3	Voltaic Cells	161
10.4	Standard Cell Electromotive Force	164
10.5	Electrochemistry and Thermodynamics	171
10.6	Effect of Concentration on Cell EMF	172
10.7	Reduction Potential Diagrams of Elements (Latimer Diagrams)	174
	Key Words	177
	Exercises	177

Chapter 11 Chemistry of Coordination Compounds ... 180

11.1	The History of Coordination Chemistry	180
11.2	Formation of Coordination Compound	182
11.3	Nomenclature of Coordination Compounds	184
11.4	Structure and Isomerism	185
11.5	Coordination Equilibrium in Solution	187
11.6	Valence Bond Theory of Coordination Compounds	191
11.7	The Crystal Field Theory	194
11.8	The Biological Effects of Coordination Compounds	199
	Key Words	200
	Exercises	201

Chapter 12 Nonmetals and Semimetals ... 203

12.1	General Concepts	203
12.2	Hydrogen	203
12.3	Boron	204
12.4	The Group ⅣA Elements, the Carbon Group Elements	205
12.5	The GroupⅤA Elements, the Nitrogen Group Elements	208
12.6	The Group ⅥA Elements, the Oxygen Group Elements	211
12.7	The Group ⅦA Elements, the Halogens	215
12.8	The Group ⅧA Elements, the Noble Gases	218
	Key Words	218
	Exercises	219

Chapter 13 Metals ... 220

13.1	The Group IA Elements	220
13.2	The Group ⅡA Elements	224
13.3	The Group ⅢA Elements	229
13.4	A Survey of Transition Metals	232
13.5	Chemistry of Some Transition Metals	234
	Key Words	239
	Exercises	239

Ⅱ Inorganic Chemical Experiments ... 241

Chapter 1 Basic Techniques of Experimental Chemistry ... 242

1.1	Safety	242
1.2	Recording Results	243
1.3	Weighing	243
1.4	Concerning Liquids	244

Chapter 2 Typical Chemical Laboratory Apparatus ... 252

Chapter 3 Experiments ... 254
3.1 Some Elementary Operation ... 254
3.2 Acid-Base Titration ... 256
3.3 Electrolyte Solutions ... 259
3.4 Preparation of Officinal Sodium Chloride and Its Purity Examination ... 262
3.5 Oxidation-Reduction Reactions ... 267
3.6 Coordination Compounds ... 271
3.7 The Halogens ... 275
3.8 Chromium, Manganese and Iron ... 278
3.9 Copper, Silver, Zinc and Mercury ... 283
3.10 Determination of the Ionization Constant of HAc by pH Meter ... 287
3.11 Preparation of Ammonium Iron (II) Sulfate ... 291
3.12 Preparation and Content Determination of Zinc Gluconate ... 294
Appendix ... 296
Reference ... 308
元素周期表

I

Inorganic Chemistry

Chapter 1　Introduction

Chemistry is an integral part of the science curriculum both at the high school and the early college level. At these levels, it is an introduction to a wide variety of fundamental concepts enabling the students to acquire tools and skills useful at the advanced levels. If you study further, you will find that chemistry is studied in any of its various sub-disciplines. So what is chemistry?

Chemistry (from Egyptian kēme, meaning "earth") is the science studying the composition, structure, properties of matter, and the changes it undergoes during chemical reactions. Matter is the physical material of the universe, it is anything that has mass and occupies space. All the matter that you are likely to meet in your everyday life is made of chemical. So chemists seek to understand how chemical transformations occur by studying the physical and chemical properties of matter. And it takes a long time for chemistry to develop and mature from its infancy (see Table 1.1).

Table 1.1　The development of chemistry

The Phases of Development	Worthy	Contribution
The genesis of chemistry (From B.C. 3000)	Ancient alchemists and metallurgists	The process for the metal purification
The early chemistry (From A.D. 800)	Medieval muslims: Geber (815), al-Kindi (873), al-Razi (925), al-Biruni (1048)	The introduction of precise observation and controlled experimentation into the field and the discovery of numerous chemical substances
	Indian alchemists and metallurgists	
The emergence of chemistry in Europe (From Dark ages, A.D. 476~1000)	Paracelsus (1493~1541)	Iatrochemistry
	Robert Boyle (1627~1691)	The Boyle's Law
Modern chemistry (From 1773)	Antoine Lavoisier (1743~1794)	The law of conservation of mass
	John Dalton (1766~1844)	The atomic theory
	Lavoisier ("Father of Modern Chemistry", 1743~1794)	1. The oxygen theory of combustion 2. Fit all experiments into the framework of a single theory 3. The consistent use of the chemical balance 4. Use oxygen to overthrow the phlogiston theory 5. A new system of chemical nomenclature 6. Contribution to the modern metric system
	Friedrich Wöhler (1880~1882)	The discovery that many natural substances and organic compounds can be synthesized in a chemistry laboratory
	Dmitri Mendeleev (1834~1907)	The periodic table of the chemical elements

Chemistry is largely an experimental science, and a great deal of knowledge comes from laboratory research. Knowledge of chemistry is needed in every modern science from astronomy to zoology. Chemists participate in many fields, such as the development of new drugs, and the problem of environmental pollution. And most industries have a basis in chemistry.

Today computer and sophisticate electronic equipment may be used to study the microscopic structure and chemical properties of substances or to analyze toxic substances in the soil or other samples. Indeed, because chemistry has diverse applications and connects the other natural sciences, such as astronomy, physics, material science, biology, and geology, it is often called the "central science."

The disciplines within chemistry are traditionally grouped by the type of matter being studied or the kind of study. However, there are four basic disciplines within chemistry, which are the required courses for any learners on the fields of natural science. These include inorganic chemistry, the study of the properties and behavior of inorganic matter; organic chemistry, the study of the structure, properties, composition, reactions, and preparation of organic matter; physical chemistry, the energy related studies of chemical systems at macromolecular and submolecular scales within the field of chemistry traditionally using the principles and practices of thermodynamics, quantum chemistry, statistical mechanics and kinetics; analytical chemistry, the study of the chemical composition and structure of natural and artificial materials.

Inorganic chemistry is a creative field that has applications in every aspect of the chemical industry-including catalysis, materials science, pigments, surfactants, coatings, medicine, fuel, and agriculture. It covers all chemical compounds except the myriad organic compounds (compounds containing C—H bonds), which are the subjects of organic chemistry. Inorganic chemistry, like many scientific fields, becomes more and more interdisciplinary. But chemists in the fields such as materials science or polymer science still strongly recommend getting a degree in inorganic chemistry. A degree in the basic discipline will give a better understanding of bonding, valence, and orbital theory. In addition, it is also important to learn inorganic chemistry and see how it applies in other areas.

1.1 Observations and Conclusions

Chemistry, like all sciences, is based on observations. It is a challenge to grasp the connotation within the information obtained and make a conclusion. Repeating experiments is an effective way to test your thought. If observations are made carefully, you would make the same conclusion. Other people in other places or at other times would also make the same observations if they did the same experiments. But they might explain their observations in different point. For example, before the seventeenth century, people usually accounted for their observations in terms of religion, while scientists today usually interpret their observations in terms of atoms and molecules.

1.2 The Scientific Method

Scientific methodology has been practiced in some forms for at least one thousand years. The development of the scientific method is inseparable from the history of science itself. Now it is acceptable to use experiment to understand nature. But the ancient Greeks did not rely on experiments to test their ideas. One way is to just talk about it. This method is unreliable. It requires proofs to determine whether a statement is correct. Since Ibn al-Haytham (Alhazen, 965~1039), the emphasis has been on seeking truth. However there are difficulties in a formulaic statement of method. As William Whewell (1794~1866) noted in his *History of Inductive Science* (1837) and in *Philosophy of Inductive Science* (1840), multiple steps are needed in scientific method.

Now all sciences, including the social sciences, employ variations of the scientific method. It is the process for experimentation to explore observations, answer questions and consequently construct an accurate representation of the world. The scientific method will help you to focus on your science fair project question, construct a hypothesis, design, execute, and evaluate your experiment. It requires intelligence, imagination, and creativity. Through the use of standard procedures and criteria we hope to minimize the influences of bias or prejudice in the experiment when developing a theory.

Now there is a hypothetico-deductive model for scientific method.
(1) Define the problem carefully
(2) Gather informations and resources (background research)
(3) Form hypothesis
(4) Perform experiment and collect information or data about the system
(5) Analyze data
(6) Interpret data and deduce a prediction that serve as a starting point for new hypothesis
(7) Publish results
(8) Retest (frequently done by other independent scientists)

Note that this method can never absolutely verify 3. It can only falsify 3. And the iterative cycle inherent in this step-by-step methodology goes from point 3 to 6 back to 3 again.

In all science disciplines, the words "hypothesis," "model," "theory" and "law" possess different connotations in relation to the stage of acceptance or knowledge about a series of phenomena.

Based on the informations that was gathered, the researcher formulated a hypothesis, a tentative explanation for a set of observations, or alternately a reasoned proposal suggesting a possible correlation among a set of phenomena. Scientists are free to use whatever resources they have, such as their own creativity, ideas from other fields, induction, Bayesian inference, and so on, to imagine possible explanations. The history of science is filled with stories of scientists claiming a "flash of inspiration", or a hunch. Further experiments are then devised to test the validity of the hypothesis in as many ways as possible, and the process begins anew.

Any useful hypothesis will enable predictions, by reasoning deductive reasons. It might predict the outcome of an experiment in a laboratory setting or the observation of a phenomenon in nature. The prediction can also be statistical and only talk about probabilities. If the predictions are not proved by observation or experience, the hypothesis is not yet useful for the method, and must wait for others to rekindle its line of reasoning.

Model is reserved for situations when it is known that the hypothesis has at least limited validity. An often-cited example is the Bohr model of the atom. In an analogy to the solar system, this model describes the electrons as moving in circular orbits around the nucleus. It does not depict accurately what an atom "looks like," but the model succeeds in mathematically representing the energies of the quantum states of the electron in the hydrogen atom.

If the experiments bear out the hypothesis it may come to a theory or law of nature. If the experiments do not bear out the hypothesis, it must be rejected or modified.

In science, a scientific law is a concise verbal statement or a mathematical equation that summarizes the relationship of phenomena that is always the same under the same conditions. We tend to think of the laws of nature as the basic rules under which nature operates. That is, the laws of natured describe the behaviour of matter. For example, Sir Isaac Newton's second law of motion ($F=ma$) means that the mass or in the acceleration of an object is always proportional to the object's force.

In science, a theory is a unifying explanation of the general principles of certain phenomena with considerable evidence or facts to support it. Hypotheses that survive many experimental tests of their validity may evolve into theories.

A scientific theory or law represents a hypothesis, or a group of related hypotheses confirmed through repeated experimental tests. Science progresses by cycles of suggested theories and tests by experiment. No matter how elegant a theory is, its predictions must agree with experimental results if we are to believe that it is a valid description of nature. So as soon as a new theory has been suggested, new experiments should be launched to test it. If the experimental results support the theory, it is accepted. If they do not, it must be modified or a new theory invented. Accepted scientific theories and laws become a part of our understanding of the universe and the basis for exploring less well-understood areas of knowledge. In part because of the unavailable necessary technology, probing or disproving a theory can take years, even centuries. For example, it took more than 2000 years to work out atomic theory proposed by Democritus, and ancient Greek philosopher.

A theory can be proved wrong, but it can never be proved right. And theories are not easily discarded; new discoveries are first assumed to fit into the existing theoretical framework. It is only when, after repeated experimental tests, the new phenomenon cannot be accommodated; scientists seriously question the theory and attempt to modify it.

There is always the possibility that a new observation or a new experiment will conflict with a long-standing theory. The overturning of an important theory opens new frontiers in science. For example, before the sixteenth century, people thought that Earth was the centre of the universe. Through many years of observation, Copernicus published his observations of the planets. He found that Earth and the other planets revolve around the sun. This was the beginning of the scientific revolution.

1.3 Units of Measurement

Measurement is also central to the sciences and engineering; modern technology and development are based on measurements. Generally, there are two kinds of measurements: quantitative measurements and qualitative measurements.

Qualitative measurements do not involve numbers. It involves an indepth understanding of human behaviour and the reasons that govern human behaviour. Qualitative research bases on the reasons behind various aspects of behaviour. Simply put, according to what, where, and when of quantitative research, it investigates the why and how of decision making.

Quantitative measurements, over qualitative observations, are numerical observations of a value, or quantity, such as volume, length etc. It is the systematic scientific investigation of quantitative properties, phenomena and their relationships. Quantitative research is widely used in both the natural and social sciences, from physics and biology to sociology and journalism. Quantitative measurements are made with instruments. The application of instruments extends the human senses of sight, smell, taste, touch, and hearing. For example, you can measure your weight with a scale, time with a watch or clock and the speed of a car with a speedometer. To simplify the process of keeping records, standardized symbols and equations are used by chemists in recording their measurements and observations. This form of representation provides a common basis for communications with other chemists.

Every measurement has two parts, a number and a unit. In science, units are essential to state measurements correctly. To say that the length of a pen is 12.5 is meaningless. We must specify that the length is 12.5 centimeters (cm).

The definition, agreement and practical use of units of measurement have played a crucial role in human endeavour from early ages up to this day. A unit of measurement is a standardized quantity of a physical property, used as a factor to express occurring quantities of that property. Units of measurement were among the earliest tools invented by humans. Disparate systems of measurement used to be very common. Now a global standard, the International System of units (the modern form of the metric system) has been or is in the process of being adopted throughout the world.

Traditional Systems

Prior to the global adoption of the metric system many different systems of measurement had been in use. The earliest known uniform systems of weights and measures seem to have all been created sometime in the 4th and 3rd millennia BC among the ancient peoples of Mesopotamia, Egypt and the Indus Valley, and perhaps also Elam in Persia as well. Many of these were related to some extent or other. Often they were based on the dimensions of parts of the body and the natural surroundings as measuring instruments. As a result, units of measure could vary not only from location to location, but from person to person.

Metric Systems and SI Units

A unit of measurement is an agreed-on standard with which other values are compared. Agreement on such standards is not easy to achieve, because different standards have been used within the scientific community. Five thousand years ago in Egypt, people use the cubit, the length of a person's arm from the elbow to the tip of the middle finger, as the unit for the measurement of length. Different sized people had different rulers. Now people in the United States use the English system of units-inches, feet, ounces, pounds, and so forth, which are based on various familiar human lengths. For example, the foot is the length of a typical human foot; the yard is the distance from the tip of the nose to the end of an extended finger. Scientists, on the other hand, use the meter, an international standard unit that was originally defined as 10^{-7} times the length of a line from the North Pole to the equator. In everyday life, volume can be measured in pints, quarts, and gallons, but the scientific standards are the cubic meter and the liter ($10^{-3}\,m^3$). For people to be able to reproduce each other's measurements, they must agree on standard units. The units used for scientific measurements are those of the metric system.

The metric system was first developed in France in 1791. After that a number of metric systems of units have evolved and are used as the system of measurement in most countries throughout the world. For many years scientists recorded measurements in metric units, which are related decimally, that is, by powers of 10. In 1960, the General Conference of Weights and Measures, the international authority on units, has attempted to work exclusively with a single set of units. It proposed a revised metric system called the International System of Units (abbreviated SI Units, from the French Système International d'Unités) for use in scientific measurements. The SI system has seven base units from which all other units are derived. Table 1.2 lists these base units and their symbols. An important feature of modern systems is standardization. The size of each SI base unit is defined exactly and universally recognized. For example, the standard of mass is the mass of a platinum-iridium cylinder kept in a vault near Paris. All other SI units of measurement can be derived from these base units. For example, volume is not a base unit in the SI system because it can be obtained from the base unit for length. A cube with 1 m on a side has a volume of 1 m^3. In chemistry the common volume is smaller than a cubic meter, so volumes are often expressed using the liter, which is defined to be exactly 10^{-3} m^3.

Scientists all over the world use SI units. The purpose of adopting SI units was to make communication easier between scientists and engineers working in different scientific subjects and in different countries.

Table 1.2 SI base units

Physical quantity	Name of unit	Abbreviation	Physical quantity	Name of unit	Abbreviation
Mass	Kilogram[1]	kg	Temperature	Kelvin	K
Length	Meter	m	Luminous intensity	Candela	cd
Time	Second	s^2	Amount of substance	Mole	mol
Electric current	Ampere	A			

1. The official spelling is kilogramme.
2. The abbreviation sec is frequently used.

Base and Derived Units

Different systems of units are based on different choices of a set of fundamental units. Using physical laws, units of quantities can be expressed as combinations of units of other quantities. Thus only a small set of units is required. These units are taken as the base units. Other units are derived units. Which units are considered base units is a matter of choice. The most widely used system of units is the International System of Units, or SI. There are seven SI base units (See Table 1.3). All other SI units can be derived from these base units.

Table 1.3 Prefixes used with SI units

Prefix	Symbol	Meaning	Example	Prefix	Symbol	Meaning	Example
yotta	Y	10^{24}	1 yottameter (Ym)=1×10^{24} m	deci-	d	10^{-1}	1 decimeter (dm) =0.1m
zetta	Z	10^{21}	1 zettameter (Zm)=1×10^{21} m	centi-	c	10^{-2}	1 centimeter (cm) =0.01m
exa-	E	10^{18}	1 exameter(Em)=1×10^{18} m	milli-	m	10^{-3}	1 millimeter (mm) =0.001m
peta	P	10^{15}	1 petameter (Pm)=1×10^{15} m	micro-	μ[①]	10^{-6}	1 micrometer (μm) =1×10^{-6} m
tera-	T	10^{12}	1 terameter (Tm)=1×10^{12} m	nano-	n	10^{-9}	1 nanometer(nm)=1×10^{-9} m
giga-	G	10^{9}	1 gigameter (Gm)=1×10^{9} m	pico-	p	10^{-12}	1 picometer (pm)=1×10^{-12} m
mega-	M	10^{6}	1 megameter(Mm)=1×10^{6} m	femto-	f	10^{-15}	1 femtometer (fm) =1×10^{-15} m
kilo-	k	10^{3}	1 kilometer (km)=1×10^{3} m	atto-	a	10^{-18}	1 attometer (am) =1×10^{-18} m
hecto-	h	10^{2}	1 hectometer(hm)=1×10^{2} m	zepto-	z	10^{-21}	1 zeptometer (zm) =1×10^{-21} m
deka-	da	10^{1}	1 dekameter(dam)=1m	yocto-	y	10^{-24}	1 yoctometer(ym)=1×10^{-24} m

① This is the Greek letter μ (pronounced "mew").

The sizes of the SI base units are not always convenient. For example, a meterstick is a little longer than a yardstick. Using a meterstick to measure the diameter of a penny or the distance between two cities would be awkward. Like metric units, SI units are modified in decimal fashion by a series of prefixes to make larger and smaller units from the base units. For example, one centimetre is 10^{-2} meter, or one-hundredth (0.01) of meter. Table 1.3 shows the SI prefixes with their symbols.

Sample How many micrograms (μg) in a milligram (mg)?
Solution: 1000 micrograms=1 milligram, and 1000 milligrams=1 gram.

1.4 Advice on Studying Chemistry

Here are some useful advices for learners.

(1) Find other students who are in your class to study with.

(2) Schedule a regular time to study each day. Don't wait until the week before an exam to begin studying. Don't stay up all night the night before the big game. In Chemistry, explanations are usually based on concepts and skills learned previously. Don't let yourself get behind.

(3) Do the assignments before coming to class. Read slowly and carefully. Try to understand, not memorize. (However, some memorization is necessary, you cannot learn to think without knowing something to think about.) Make sure you understand the worked-out examples, and do the chapter problems and check your answers to them in (bracket) parentheses. Write down any questions you have so don't forget them when you get to class.

(4) After each class, review what you have just learned before beginning the next assignment.

(5) Pay careful attention to vocabulary. An important part of a first course in any subject is learning the meaning of the terms used in the field (which, unfortunately, do not always mean the same thing in other fields). If you don't remember the meaning of a term when you meet it again later, use the index to find the definition. If you come across unfamiliar words, look up their meaning in a chemical dictionary. Different people have different learning styles so choose the one that suit you.

Key Words

chemistry ['kemistri] 化学
matter ['mætə] 物质
inorganic chemistry [ˌinɔːgænik 'kemistri] 无机化学
organic chemistry [ɔːˈgænik 'kemistri] 有机化学
physical chemistry ['fizikəl 'kemistri] 物理化学
analytical chemistry [ˌænəˈlitikəl 'kemistri] 分析化学

hypothesis [haiˈpɔθisis] 假设
prediction [priˈdikʃən] 预言
model ['mɔdl] 模型
law [lɔː] 定律
theory ['θiəri] 理论
quantitative ['kwɔntitətiv] 定量的
qualitative ['kwɔlitətiv] 定性的
metric system ['metrik 'sistəm] 国际公制,米制
unit ['juːnit] 单位

Chapter 2 Structures of Atoms

The fundamental unit of a chemical substance is atom, which is derived from the Greek atomos, meaning "uncuttable". Atomic theory owes its origin to two ancient Greek philosophers, Leucippus (ca. 450 BC) and Democritus (ca. 470~380 BC). They believed that matter contains indivisible parts, atomos. Unfortunately, they lacked the necessary scientific evidence to support their assertions, so their ideas were not widely accepted.

To a chemist, every sample of matter is made up of immense numbers of atoms. For example, a chemist views a glass of water as a swarm of water molecules, each composed of two atoms of hydrogen and one atom of oxygen (Figure 2.1).

Modern atomic theory began with the English chemist John Dalton (1766~1844), who developed and published an atomic description supported by experimental data. Dalton's experiments show that elements always combine in fixed proportions. For example, when oxygen and hydrogen combine to make water, the ratio of masses is always 8 : 1; 8 g of oxygen combine with 1 g of hydrogen to make 9 g of water.

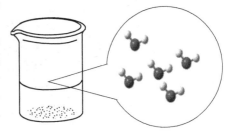

Figure 2.1 A glass of water and water molecules

Over time, chemists and physicists have refined Dalton's theory as they have learned more about the details of atomic structure.

The essential features of our modern understanding of atoms can be summarized as follows.

1. All matter is composed of atoms.
2. All atoms of a given element have identical chemical properties.
3. Atoms of each different element have distinct properties. Symbols are used to designate each different type of atom.
4. Atoms form chemical compounds by combining in whole number ratios. All samples of a pure compound have the same combination of atoms.
5. In chemical reactions, atoms change the ways they are combined, but they are neither created nor destroyed.

Consequently, chemists have created a shorthand language of symbols, formulas, and equations that convey information about atoms and molecules in a simple, straightforward manner. Symbols are used to designate each different type of atoms. These symbols in turn are combined into formulas that describe the compositions of more complicated chemical substances. Formulas can then be used to write chemical equations that describe how molecules are changed in a chemical reaction. An equation in which atoms of each type are neither created nor destroyed is said to be a balanced chemical equation. In accord with feature 5 of atomic theory, all proper chemical equations must be balanced. We will describe how to balance chemical equations in Chapter 10.

2.1 Particles in Atoms

Electrons

Important clues about the structure of atoms came from experiments that used electrical force. The simplest of these experiments used two metal plates called electrodes sealed inside a glass tube along with a sample of a gas. Figure 2.2 shows a schematic drawing of this type of apparatus. One of the metal plates was given a large positive electrical charge, but the other was given a large negative charge. When the charges became large enough, electrical forces caused an electrical discharge (similar to a lightning bolt) to leap across the space between the plates.

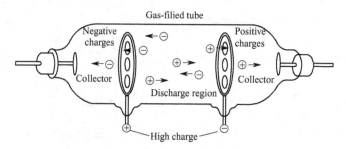

Figure 2.2 Gas discharge tube in operation

This high-energy discharge caused some of the atoms of the gas to fall apart. The pieces of the broken atoms turned out to be charged particles. Consistent with coulomb's law, particles with positive charges moved toward tile negative electrode, while particles with negative charges moved toward the positive electrode. When the electrodes had holes in them, some of these charged particles passed through the electrodes and were captured by detectors at the ends of the tube.

This experiment showed that atoms are made up of a collection of smaller fragments that possess positive and negative charges. The energy supplied by the discharge between the electrodes was enough to break atoms into these smaller fragments.

The experiment was repeated using different types of gases. Different gas in the tube changed the behavior of the positively charged particles, but the negatively charged particles always acted the same. These negatively charged fragments are common to all atoms. They are called electrons.

The discovery of the electron prompted a series of more sophisticated experiments designed to reveal even more about its nature. J. J. Thomson performed one such experiment with a device called a cathode ray tube, which is illustrated in Figure 2.3. A cathode ray is a beam of electrons generated by all electrical discharge. Because an electron beam is a collection of moving electrical charges, it is affected by electrical and magnetic forces. When either type of force is applied at right angles to the direction of motion of an electron beam, it causes the beam to bend. The amount of bending depends on the electrons' velocity, charge, and mass. In Thomson's experiment a cathode ray was subjected simultaneously to electrical and magnetic forces, as indicated in Figure 2.3. By measuring the amount of magnetic force required to exactly counterbalance the deflection of the beam by a known electrical force, Thomson was able to calculate the ratio of the electron's charge to its mass:

$$\text{Charge/Mass} = e/m = 1.76 \times 10^{11} \text{ C/kg}$$

Combining this value with Thomson's measurement of charge/mass ratio, Millikan computed the mass of a single electron.

$$m_{electron} = e/(e/m) = -1.6 \times 10^{-19} \text{ C}/(-1.76 \times 10^{11} \text{ C/kg}) = 9.1 \times 10^{-31} \text{ kg}$$

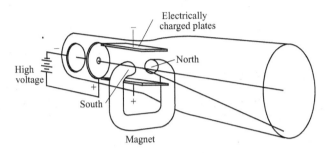

Figure 2.3　A cathode ray tube

Nucleus

The nature of electrons had been elucidated by the experiments of Thomson and Millikan, but the nature of the positive particles was still unknown. Also, it was not sure how the particles fit together to make an atom. Based on the Thomson's experiments, scientists hypothesized that the atom was similar to a chocolate chip cookie, with negative electrons (the "chips") embedded in the positive "dough" of the atom. The problem of "now to balance repulsion between like charges and attraction between unlike charges in an atom" was explained by this way. Such experiments and hypothesis led to the "cookie" model.

The definitive experiment was carried out in 1909 by Ernest Rutherford (see Figure 2.4), which showed how charges and masses are distributed in an atom.

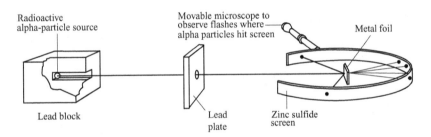

Figure 2.4　Rutherford's α-particle scattering experimental installation

In Rutherford's experiment, alpha particles were projected to a thin gold foil, which can be considered as one layer aurum atoms. Alpha particles are high-energy, positively charged fragments of helium atoms emitted during radioactive decay of unstable elements such as uranium. According to the "cookie" model, the mass of each aurum atom in the foil should have been spread evenly over the whole atom. The particles should slow down and change direction only by a small amount as they passed through the foil. The results of the experiment were quite unexpected. Most alpha particles passed directly through the gold foil, some were deflected slightly, and only a few particles bounced back.

To explain his experiment, Rutherford proposed a new model for atomic structure. He suggested that every atom contains a tiny central core where all the positive charge and most of the mass are concentrated. This central core is surrounded by the electrons. Electrons occupy a huge volume, which is compared with the size of the nucleus. The relatively heavy alpha particles passed through the gold atoms in the foil, were almost unaffected by electrons. Each electron has such a small mass that this volume appears as empty space to an alpha particle. An alpha particle was deflected only when it passed very near a nucleus. Only a few alpha particles bounced back when they hit on the nucleus, as shown schematically in Figure 2.5.

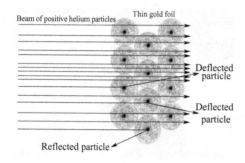

Figure 2.5 Schematic view of Rutherford's scattering experiment

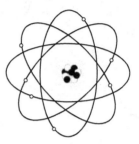

Figure 2.6 A planetary model of the atom

A new model of the atom, named as planetary model is represented in Figure 2.6, which proposed that the atom has a positively charged care called the nucleus, is what finally emerged.

Experiments on nuclei showed that the nucleus contains two types of subatomic fragments called protons and neutrons. A proton's positive charge is equal in magnitude to the negative charge of an electron. The mass of a proton is almost 2000 times greater than the mass of an electron. On the other hand, neutrons contribute mass while they are electrically neutral. The mass of a neutron is almost the same as the proton.

Three particles-electrons, protons, and neutrons, combine in various numbers to make the different atoms of all the elements of the periodic table. However, the chemical properties of atoms and molecules stem mainly from their electrons.

2.2 The Bohr Model

The most important properties of atomic and molecular structure can be exemplified by using an even close to reality atom model that is called the Bohr Model. This model was proposed by Niels Bohr in 1913, it is not completely correct, but it has many features that are approximately correct and it is sufficient for much of our discussion. The correct theory of the atom is called quantum mechanics; the Bohr Model is an approximation to quantum mechanics that has the virtue of being much simpler.

The Bohr Model is similar to the planetary model of the atom. In the Bohr Model, the neutrons and protons occupy a dense central region called the nucleus, and the electrons orbit the nucleus much like planets orbiting the Sun. The difference is that the orbits are quantized in the Bohr Model. It means that only certain orbits with certain radii and energy are allowed.

Bohr's postulates are as follows.

(1) Bohr first modified Rutherford's planetary electrons by putting a restriction on their

orbits. As long as the planetary electron stays in its prescribed orbit, it will net absorb or radiate energy. This was the introduction of the concept of the energy level, or stationary state.

(2) The spectral lines arose as a result of a transition from one such energy level, or state, to another. The lowest energy state is generally termed the ground state. The states with successively more energy than the ground state are called the first excited state, the second excited state, and so on. Differences in energy levels were to be in terms of Plank's equation.

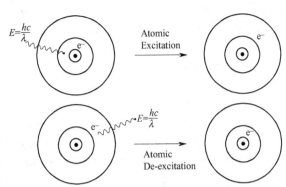

Figure 2.7 Excitations by absorption of light and de-excitation by emission of light

$$\Delta E = E_{final} - E_{initial} = h\nu \tag{2.1}$$

In each case the wavelength of the emitted or absorbed light, the energy is exactly the same as the difference between the two orbits. This energy may be calculated by dividing the product of the Planck constant and the speed of light hc by the wavelength of the light λ, see Figure 2.7. Thus, an atom can absorb or emit only certain discrete wavelengths (frequencies or energies).

2.3 The Wave Theory of Electrons

Depending on the experimental circumstances, the radiation appears to have either a wave like or particle-like character. To answer the question of why the energy of the electron should be quantized in Bohr's atom model for the hydrogen atom, the French physicist Louis de Broglie (1892~1987), extended this idea. If radiant energy could behave as though it were stream of particles, could matter possibly show the properties of a wave? De Broglie suggested that the electron orbiting the nucleus of a hydrogen atom could be thought of as a wave, with a particular wavelength, which is associated with the electron's movement.

Light possesses momentum that can be determined by measuring the pressure that an intense light beam exerts on an object. This momentum (p) is given by the following equation.

$$\text{Momentum} = \text{Energy/Velocity} \quad \text{or} \quad p = E/c \tag{2.2}$$

If we replace E with Einstein's expression for the energy of a photon, and convert frequency into wavelength, we obtain an equation that can be applied to either light or electrons.

$$E = h\nu = hc/\lambda \quad \text{so} \quad p = hc/\lambda c = h/\lambda \tag{2.3}$$

According to classical physics, the momentum of an electron is equal to its mass (m) times its velocity (ν). Combining these equations gives the de Broglie equation, which links the mass of an object with its wavelength.

$$m\nu = h/\lambda \quad \text{or} \quad \lambda = h/m\nu \tag{2.4}$$

Because de Broglie's hypothesis is applicable to all matter, any object of mass m and velocity ν would give rise to a characteristic matter wave.

Sample 2.1 What wavelengths are associated with (1) an electron traveling at 1.00×10^5 m/s and (2) a Ping-Pong ball of mass 2.7 g and diameter 40 mm traveling at 30 m/s?
Solution: This problem deals with de Broglie equation.

$\lambda_{electron} = h/mv = (6.626 \times 10^{-34}$ kg·m²/s$) / (9.109 \times 10^{-31}$ kg$)(1.00 \times 10^5$ m/s$)$
$\qquad = 7.27 \times 10^{-9}$ m
$\lambda_{ball} = h/mv = (6.626 \times 10^{-34}$ kg·m²/s$)/(2.7 \times 10^{-3}$ kg$)(3 \times 10^1$ m/s$) = 8.2 \times 10^{-33}$ m

The wavelength of the Ping-Pong ball is even shorter than the radius of a single nucleus, but the wavelength of the electron is the same size as atomic radii. Thus Ping-Pong ball wavelengths are inconsequential, but electron wavelengths have a significant effect on electron-atom interactions.

Within a few years after de Broglie published his theory, the wave properties of the electron were demonstrated experimentally. The diffraction patterns were observed after passing electron beams through metals. This established the wave nature of electrons.

2.4 Heisenberg's Uncertainty Principle

The discovery of the wave properties of matter raised some new and interesting questions about classical physics. For an electron, can we calculate exactly its position, its direction of motion, and its speed of motion at the same time?

Instead of things being exactly specified in space, microcosmic particles are distributed over some volume. Werner Heisenberg, a German physicist, showed this theory in the 1920s. According to Heisenberg, particles have the curious property that their velocity and position cannot be "pinned down" at the same time. Thus if a particle can be pinpointed in a specific location, its velocity must be uncertainty. Conversely, if the velocity of a particle is known precisely, its location must be uncertainty. Heisenberg summarized this uncertainty, which is known as the uncertainty principle. The more accurately we know position, the more uncertain we are about velocity, and vice versa.

De Broglie's wave theory and Heisenberg's uncertainty principle set a new stage for the theory of atomic structure. It is not appropriate to imagine that the electrons moving on well-defined circular orbits around the nucleus.

In order to get over this problem, we use the probability of finding the electron in a given volume of space. The probability of finding an electron at a given point in space is determined from the function Ψ^2. Ψ is the wave function, which is a mathematical function which can be used to describe the behavior of an electron-wave.

2.5 The Schrödinger Equation

In 1926 an Austrian physicist, Erwin Schrödinger introduced his wave equation. It was the basis for the quantum theory that underlies our current understanding of atomic and molecular structure.

The Schrödinger wave equation may be represented in several forms. Equation (2.5) gives the form of the Schrödinger wave equation that is appropriate for motion, which can be applied to the motion of a particle in three-dimensional space.

$$\frac{\partial^2 \Psi}{\partial x^2} + \frac{\partial^2 \Psi}{\partial y^2} + \frac{\partial^2 \Psi}{\partial z^2} + \frac{8\pi^2 m}{h^2}(E-V)\Psi = 0 \qquad (2.5)$$

Ψ is the Greek symbol representing the wave function; E is the total energy of the system; V is the potential energy of the system; m is the mass of the electron; h is Planck's constant; x, y, and z are the space coordinates in three dimensions.

It is not important to remember the details of this equation now. Without going into the mathematical sense of the equation, let us note that acceptable solutions can only be obtained with quite definite, discrete values of the electron's energy. The various functions Ψ_1, Ψ_2, $\Psi_3 \cdots \Psi_n$ that are solutions of the wave equation correspond to energies E_1, E_2, $\cdots E_n$, respectively. Quantization of the energy of a microsystem thus stems directly from solution of the wave function, Ψ has no apparent physical sense, but its square Ψ^2 characterizes the probability of the electron concerned being at a given point in space that is the probability density.

The mathematics of quantum theory is too complicated for our purposes. Nevertheless, by making various approximations in the equations, it is possible to describe the quantum features of atoms and molecules. These features agree with the observations of experimental chemistry and physics. Because the theory describes experimental observations accurately, quantum mechanics is firmly established as a reasonable description of matter at the atomic and molecular level.

2.6 Quantum Numbers

Recall from section 2.1 that every electron has the same charge, and the same mass. However, electrons in atoms display four other characteristic properties. These properties are energy, orbital shape, orbital orientation, and spin orientation. If you want to describe an atom, then you have to list the properties of all its electrons. Unfortunately, this list will be very tedious to construct. Another way to describe an atom is to solve the Schrödinger equation and determine the wave function (Ψ) for each of its electrons. However, these wave equations are too abstract to most chemists. Instead, chemists use sets of integers and half-integers called quantum numbers to specify the values of the electron's quantized properties. Each electron in an atom has a set of four quantum numbers, which cab is used to describe an atomic electron completely, we need only to specify a value for each of its four quantum numbers.

An orbital can be described definitely by a set of three integral numbers known as quantum numbers (Table 2.1): principal quantum number (n), angular momentum quantum number (l), and magnetic quantum number (m). Electron's spin orientation can be described by The forth quantum number (m_s).

Principle Quantum Number (n)

We have already encountered the principal quantum numbers, n, in the Bohr model of the hydrogen atom. As the name suggests, this is the most important of the four quantum numbers specifying the major energy level to which an electron can be assigned. It is a positive integer with values lying between the limits $1 \leqslant n < \infty$, and is related to the size of the shell: $n=1, 2, 3, 4, \cdots$ (corresponding to K, L, M, N, $\cdots$ shell of Bohr theory). Allowed values arise when the radial part of the Schrödinger equation is solved. All of the electrons in an atom possessing one principal quantum number exist in the same shell, or energy level. Electrons in $n=1$ shell are all located in the first major energy level; $n=2$ electrons are located in the second major energy level. The number of electrons in a particular energy level is restricted to $2n^2$ electrons or less.

Two more quantum numbers, l and m, appear when the angular part of the Schrödinger equation is solved.

Angular Momentum Quantum Number (l)

The value of l determines the shape of the atomic orbital, and the orbital angular momentum of the electron. Among the electrons occupying any particular shell, those present in sublevels of identical, values are located in the same subshell. For each value of n (the principal quantum number), the l quantum number has allowed values.

$$l=0,1,2,3,\cdots,n-1$$

If $n=1$, there can only be one possible value for l. The value for l may be any integral number from 0 to $n-1$. Thus, for the case in question, $n=1$, $l=0$. These are the subshells of the principal shell, differing slightly in energy from each other but each nearly the same as the principal shell. The number of subshell within any energy level or shell turns out to be equal to the quantum number n. The fine structure in atomic spectra is accounted for by the existence of the subshells with their slightly different energies described by the l quantum number. Historically, orbital shapes have been identified with letters rather than numbers. These letter designations correspond to the values of l as follows.

Value of l	0	1	2	3	4
Orbital letter	s	p	d	f	g

An orbital is named by listing the numerical value for n, followed by the letter that corresponds to the numerical value for l. Thus an electron with quantum numbers $n=3$, $l=0$ is described as a 3s orbital. A 5f orbital has $n=5$, $l=3$. Notice that the restrictions on l mean that many orbital names do not correspond to actual orbitals. For example, when $n=1$, l can only be zero. Thus 1s orbitals exist, but them are no 1p, 1d, or 1f orbitals. Similarly, there are 2s and 2p orbitals but no 2d, 2f or 2g orbitals. Remember that n restricts l but l does not restrict n. Thus a 10d orbital ($n=10$, $l=2$) is rather unstable but perfectly legitimate, whereas there is no orbital with $n=2$, $l=10$.

For a given atom, a series of orbitals with different values of n but the same values of l (e.g. 1s, 2s, 3s, 4s, ⋯) differ in their relative size (spatial extent). The larger the value of n is, the larger the orbital is. An increase in size also corresponds to an orbital being more diffuse and higher energy. $E_{1s}<E_{2s}<E_{3s}<E_{4s}<\cdots$

For a given atom, a series of orbitals with same values of n but the different values of l (e.g. 3s, 3p, 3d) differ in their shapes. The larger the value of l is, the higher energy the orbital is.

$$E_{3s}<E_{3p}<E_{3d}$$

Magnetic Quantum Number (m)

The value of the magnetic quantum numbers gives information about the directionality of an atomic orbital. The angular momentum of the electron is responsible for inducing a magnetic field does an electric current passed through a loop of wire. The effect is to split the sublevel into still finer levels, each of slightly different energies, out each nearly the same as the sublevel energy. The magnetic quantum numbers, m, has integral values between $+l$ and $-l$, including zero.

$$m=-l,\cdots,0,\cdots,+l$$

or putting it another way.
$$m = 0, \pm 1, \pm 2, \pm 3, \cdots, \pm l$$
There are, thus $2l+1$ values of m for any l, expressing the orientation of the subshells in space.

Before we place electrons into atomic orbitals, we must define one more quantum number m_s.

Magnetic Spin Quantum Number (m_s)

In addition to the magnetic effect due to the angular momentum of the electron spinning about the nucleus, the electron spins about its own axis. The effect shows up as a modification of the angular momentum. It can have two possible values, $+1/2$ or $-1/2$, as the electron's spin orientation about its axis lies in one direction or the other. Whereas an atomic orbital is defined by a unique set of three quantum numbers, an electron in an atomic orbital is defined by a unique set of four quantum numbers: n, l, m and m_s. As there are only two values of m_s, an orbital can accommodate only two electrons. An orbital is fully occupied when it contains two electrons which are spin-paired; one electron has a value of $m_s = +1/2$ and the other, $m_s = -1/2$. Thus, the six possible p-type electrons will distribute themselves into three orbitals, in pairs with opposite spins. The ten d-type electrons show up in five orbitals, in pairs with opposite spins.

An atomic orbital is defined by a unique set of three quantum numbers, n, l and m. The first four shells (levels) in the hydrogen atom are listed with their quantum number in Table 2.1.

Table 2.1 Electronic quantum numbers and atomic orbitals in the hydrogen atom

Shell	n	l	Orbital	m	Degeneracy	
K	1	0	1s	0	1	1
L	2	0	2s	0	1	4
	2	1	2p	$-1, 0, 1$	3	
M	3	0	3s	0	1	9
	3	1	3p	$-1, 0, 1$	3	
	3	2	3d	$-2, -1, 0, 1, 2$	5	
N	4	0	4s	0	1	16
	4	1	4p	$-1, 0, 1$	3	
	4	2	4d	$-2, -1, 0, 1, 2$	5	
	4	3	4f	$-3, -2, -1, 0, 1, 2, 3$	7	

Each electron in an atom is unique, no two electrons can have the same four quantum numbers in the same atom.

2.7 Shapes of Atomic Orbitals

The chemical properties of atoms are determined by the behavior of their electrons. Because atomic electrons are described by orbitals, these electron interactions can be described in terms of orbital interactions. The two characteristics of orbitals that determine how electrons interact are orbital shapes and energies. Orbital shapes describe the distribution of electrons in three-dimensional space. Orbital energies will be discussed in Section 2.8.

The size and shape of an orbital are determined by the quantum numbers n and l. As n increases, the size of the orbital increases, and as l increases, the shape of the orbital becomes more elaborate.

The shapes of orbitals strongly influence chemical interactions. So we need to have detailed pictures of orbital shapes to understand the chemistry of the elements. The quantum number $l=0$ corresponds to an s orbital. According to the restrictions on quantum numbers, there is only one s orbital for each value of the principal quantum number. All s orbitals are spherical, with radii increase as n increases.

And recall from Section 2.5 that an electron is a particle-wave delocalized in three-dimensional space in a way described by a wave equation (Ψ). An orbital depiction gives a spatial view of the wave equation and provides a "map" of how the electron wave is distributed in space. There are several ways to draw representations of these three-dimensional maps. Each drawing shows some important orbital features, but none shows all of them. We use four different representations: plots of the wave equation (Ψ) vs. distance from the nucleus (r), plots of the square of the wave equation (Ψ^2) vs. r, pictures of electron density, and pictures of electron contour surfaces.

A plot of Ψ vs. r is the most mathematically direct way of depicting a wave equation. A plot of the 2s orbital is shown in Figure 2.8(a) Shows the distribution of the electron in space depends on Ψ^2, the square of the wave equation. Thus a plot of Ψ^2 vs. r provides more information about electron distributions. The plot of Ψ^2 vs. r for the 2s orbital, shown in Figure 2.8(b) describes how electron density varies with distance from the nucleus.

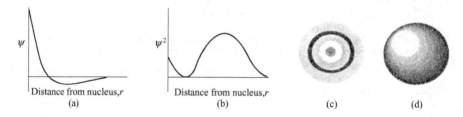

Figure 2.8   The maps of electron particle-wave distributing in space

One type of orbital picture is a two-dimensional dot pattern in which the density of dots represents electron density. Such an orbital density picture of the 2s orbital is shown in Figure 2.8(c) note that this two dimensional pattern of dots shows a cross-sectional slice through the middle of the orbital.

Orbital density pictures are probably the most comprehensive views we can draw, but they require much time and care. A simplified orbital picture is provided by an electron contour drawing. In this representation, we draw a contour surface that encloses almost all the electron density. Commonly, "almost all" means 95%. Thus the electron density is high inside the contour surface but very low outside the surface, Figure 2.8(d) shows the contour surface depiction of the 2s orbital.

The quantum number $l=1$ corresponds to a p orbital. Figure 2.9 shows contour diagrams of the three 2p orbitals in three-dimensional. Because a p electron can have any of three values ($+1$, 0, -1) for m, there are three similarly shaped p orbitals for each value of m. All p orbitals are spindly, with one preferred axis. Notice that each has its electron density concentrated along its preferred axis on both sides of the nucleus. Each p orbital is oriented along one cartesian axis, perpendicular to the other two, with the nucleus at the center of the system. To recognize these orientations, we can subscript the orbitals accordingly: p_x, p_y and p_z.

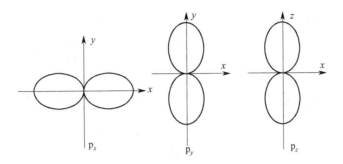

Figure 2.9 Diagrams of the three p orbitals

As n increases, the shapes of the p orbitals become more complicated, but their outermost features do not change. An electron in a 3p orbital presents the same orbital characteristics as one in a 2p orbital, except that the 3p orbital is bigger.

The quantum number $l = 2$ corresponds to a d orbital. A d electron can have any of five values for m (-2, -1, 0, $+1$, or $+2$), so them are five different orbitals in each set. The shapes of the d orbitals are showed by the contour drawings in Figure 2.10, three of them look like three-dimensional "cloverleafs" lying in a cartesian plane with their lobes pointed between the axes. A subscript is used to identify the plane in which they lie: d_{xy}, d_{xz}, and d_{yz}. A fourth orbital is also a clover-leaf in the xy plane, but its lobes point along the x and y ax-

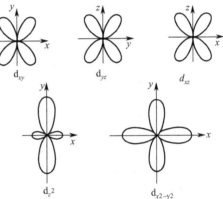

Figure 2.10 Diagrams of the five d orbitals

es. This orbital is designated $d_{x^2-y^2}$. The fifth orbital looks quite different. Its major lobes point along the z axis, but there is also a "doughnut" of electron density in the xy plane. This orbital is designated d_{z^2}.

The chemistry of all the common elements can be described completely by using s, p, and d orbitals, so we need not extend our catalog of orbital shapes to the f orbitals and beyond.

2.8 Many-Electron Atoms

The Helium Atom

The next simplest atom is He, and for its two electrons $e_{(1)}$ and $e_{(2)}$, three electrostatic interactions as below must be considered.

(ⅰ) attraction between electron $e_{(1)}$ and the nucleus;
(ⅱ) attraction between electron $e_{(2)}$ and the nucleus;
(ⅲ) repulsion between electrons $e_{(1)}$ and $e_{(2)}$.

The net interaction will determine the energy of the system. In atoms with more than one electron, the orbital energies are assumed to be similar to the orbital energy levels in the hydrogen atom, except subshells of the same shell are different.

Electron Configurations

The 1s orbital is the lowest energy orbital and closest to the nucleus. In the ground state of the hydrogen atom, the lowest energy state the electron is in the 1s orbital. In orbital energy diagrams, electrons are represented by arrows. The first arrow in an orbital is usually drawn pointing up representing an electron with spin up. Figure 2.11 shows the orbital energy diagram for the hydrogen atom in its ground state and excited state.

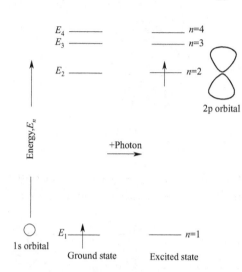

Figure 2.11 The orbital energy diagram for the hydrogen atom

Drawing an orbital energy diagram is time-consuming. The electrons among an atom can be shown more easily by writing the electron configuration of the atom. To write an electron configuration, write the principal quantum number of the main shell and the letter for the subshell. Show the number of electrons in the subshell by a right superscript. For example, the electron configuration for the hydrogen atom in its ground state is $1s^1$.

$$\text{number of shell} \leftarrow 1s^1 \rightarrow \text{subshell}$$

with number of electrons in subshell indicated by the superscript.

This electron configuration means that the hydrogen atom's one electron is in the 1s orbital.

Any other electron configuration of the hydrogen atom is an excited state. For example, the electron configuration $2s^1$ (meaning that the hydrogen atom's one electron is in the 2s orbital) represents an excited state of the hydrogen atom. An atom has infinite number of excited states.

Now consider the ground state electronic configurations of isolated atoms of all the elements (Table 2.2). These are experimental data, and are nearly always obtained by analyzing atomic spectra. Most atomic spectra are too complex for discussion here and we take their interpretation on trust.

The ground state electronic configurations of He can be write as $1s^2$ respectively. The 1s atomic orbital is fully occupied as He, which is often written as [He] for simply. In the next two elements, Li and Be, the electrons go into the 2s orbital, and then from B to Ne, the 2p orbitals are occupied to give the electronic configurations as $[He]2s^2 2p^x$ ($x = 1-6$). When $x = 6$, the energy level (or shell) with $n = 2$ is fully occupied, and the configuration for Ne can be written as [Ne]. The filling of the 3s and 3p atomic orbitals takes place in an analogous sequence from Na to Ar. The last element in the series, Ar has the electronic configuration $[Ne]3s^2 3p^6$.

A detailed inspection of Table 2.2 makes it obvious that it is difficult to recite every atom's ground state electronic configuration. And there is no one sequence that represents accurately the occupation of different sets of orbitals with increasing atomic number. The following rules can be used to forecast the ground state electronic configurations for most of *neutral atoms*.

Table 2.2 Ground state electronic configurations of atoms

Period	Atomic Number	Element	Ground state electronic configuration	Period	Atomic Number	Element	Ground state electronic configuration
1	1	H	$1s^1$	6	55	Cs	$[Xe]6s^1$
	2	He	$1s^2$		56	Ba	$[Xe]6s^2$
2	3	Li	$[He]2s^1$		57	La	$[Xe]5d^1 6s^2$
	4	Be	$[He]2s^2$		58	Ce	$[Xe]4f^1 5d^1 6s^2$
	5	B	$[He]2s^2 2p^1$		59	Pr	$[Xe]4f^3 6s^2$
	6	C	$[He]2s^2 2p^2$		60	Nd	$[Xe]4f^4 6s^2$
	7	N	$[He]2s^2 2p^3$		61	Pm	$[Xe]4f^5 6s^2$
	8	O	$[He]2s^2 2p^4$		62	Sm	$[Xe]4f^6 6s^2$
	9	F	$[He]2s^2 2p^5$		63	Eu	$[Xe]4f^7 6s^2$
	10	Ne	$[He]2s^2 2p^6$		64	Gd	$[Xe]4f^7 5d^1 6s^2$
	11	Na	$[Ne]3s^1$		65	Tb	$[Xe]4f^9 6s^2$
	12	Mg	$[Ne]3s^2$		66	Dy	$[Xe]4f^{10} 6s^2$
	13	Al	$[Ne]3s^2 3p^1$		67	Ho	$[Xe]4f^{11} 6s^2$
3	14	Si	$[Ne]3s^2 3p^2$		68	Er	$[Xe]4f^{12} 6s^2$
	15	P	$[Ne]3s^2 3p^3$		69	Tm	$[Xe]4f^{13} 6s^2$
	16	S	$[Ne]3s^2 3p^4$		70	Yb	$[Xe]4f^{14} 6s^2$
	17	Cl	$[Ne]3s^2 3p^5$		71	Lu	$[Xe]4f^{14} 5d^1 6s^2$
	18	Ar	$[Ne]3s^2 3p^6$		72	Hf	$[Xe]4f^{14} 5d^2 6s^2$
4	19	K	$[Ar]4s^1$		73	Ta	$[Xe]4f^{14} 5d^3 6s^2$
	20	Ca	$[Ar]4s^2$		74	W	$[Xe]4f^{14} 5d^4 6s^2$
	21	Sc	$[Ar]3d^1 4s^2$		75	Re	$[Xe]4f^{14} 5d^5 6s^2$
	22	Ti	$[Ar]3d^2 4s^2$		76	Os	$[Xe]4f^{14} 5d^6 6s^2$
	23	V	$[Ar]3d^3 4s^2$		77	Ir	$[Xe]4f^{14} 5d^7 6s^2$
	24	Cr	$[Ar]3d^5 4s^1$		78	Pt	$[Xe]4f^{14} 5d^9 6s^1$
	25	Mn	$[Ar]3d^5 4s^2$		79	Au	$[Xe]4f^{14} 5d^{10} 6s^1$
	26	Fe	$[Ar]3d^6 4s^2$		80	Hg	$[Xe]4f^{14} 5d^{10} 6s^2$
	27	Co	$[Ar]3d^7 4s^2$		81	Tl	$[Xe]4f^{14} 5d^{10} 6s^2 6p^1$
	28	Ni	$[Ar]3d^8 4s^2$		82	Pb	$[Xe]4f^{14} 5d^{10} 6s^2 6p^2$
	29	Cu	$[Ar]3d^{10} 4s^1$		83	Bi	$[Xe]4f^{14} 5d^{10} 6s^2 6p^3$
	30	Zn	$[Ar]3d^{10} 4s^2$		84	Po	$[Xe]4f^{14} 5d^{10} 6s^2 6p^4$
	31	Ga	$[Ar]3d^{10} 4s^2 4p^1$		85	At	$[Xe]4f^{14} 5d^{10} 6s^2 6p^5$
	32	Ge	$[Ar]3d^{10} 4s^2 4p^2$		86	Rn	$[Xe]4f^{14} 5d^{10} 6s^2 6p^6$
	33	As	$[Ar]3d^{10} 4s^2 4p^3$	7	87	Fr	$[Rn]7s^1$
	34	Se	$[Ar]3d^{10} 4s^2 4p^4$		88	Ra	$[Rn]7s^2$
	35	Br	$[Ar]3d^{10} 4s^2 4p^5$		89	Ac	$[Rn]6d^1 7s^2$
	36	Kr	$[Ar]3d^{10} 4s^2 4p^6$		90	Th	$[Rn]6d^2 7s^2$
5	37	Rb	$[Kr]5s^1$		91	Pa	$[Rn]5f^2 6d^1 7s^2$
	38	Sr	$[Kr]5s^2$		92	U	$[Rn]5f^3 6d^1 7s^2$
	39	Y	$[Kr]4d^1 5s^2$		93	Np	$[Rn]5f^4 6d^1 7s^2$
	40	Zr	$[Kr]4d^2 5s^2$		94	Pu	$[Rn]5f^6 7s^2$
	41	Nb	$[Kr]4d^4 5s^1$		95	Am	$[Rn]5f^7 7s^2$
	42	Mo	$[Kr]4d^5 5s^1$		96	Cm	$[Rn]5f^7 6d^1 7s^2$
	43	Tc	$[Kr]4d^5 5s^2$		97	Bk	$[Rn]5f^9 7s^2$
	44	Ru	$[Kr]4d^7 5s^1$		98	Cf	$[Rn]5f^{10} 7s^2$
	45	Rh	$[Kr]4d^8 5s^1$		99	Es	$[Rn]5f^{11} 7s^2$
	46	Pd	$[Kr]4d^{10}$		100	Fm	$[Rn]5f^{12} 7s^2$
	47	Ag	$[Kr]4d^{10} 5s^1$		101	Md	$[Rn]5f^{13} 7s^2$
	48	Cd	$[Kr]4d^{10} 5s^2$		102	No	$[Rn]5f^{14} 7s^2$
	49	In	$[Kr]4d^{10} 5s^2 5p^1$		103	Lr	$[Rn]5f^{14} 6d^1 7s^2$
	50	Sn	$[Kr]4d^{10} 5s^2 5p^2$		104	Rf	$[Rn]5f^{14} 6d^2 7s^2$
	51	Sb	$[Kr]4d^{10} 5s^2 5p^3$		105	Ha	$[Rn]5f^{14} 6d^3 7s^2$
	52	Te	$[Kr]4d^{10} 5s^2 5p^4$		106	Sg	$[Rn]5f^{14} 6d^4 7s^2$
	53	I	$[Kr]4d^{10} 5s^2 5p^5$		107	Bh	$[Rn]5f^{14} 6d^5 7s^2$
	54	Xe	$[Kr]4d^{10} 5s^2 5p^6$		108	Hs	$[Rn]5f^{14} 6d^6 7s^2$
					109	Mt	$[Rn]5f^{14} 6d^7 7s^2$

(1) Aufbau principle (Building-up principle)

The atomic orbitals are filled in the order of increasing energy. The orbital energy levels differ from element to element. The sequence of energy levels determines the sequence of quantum states for the electrons, as they fill the subshells and shells in building up the atoms from their nuclear cores.

The energy spacing between sets of orbitals gets smaller as the principal quantum number increases. For instance, the energy gap between the $n=1$ and $n=2$ orbital is greater than the gap between the $n=2$ and $n=3$ orbitals (see Figure 2.11, 2.12). At the same time, orbital splitting between different l values increases as nuclear charge increases (see Figure 2.12). For example, the energy difference between the 3s and 3p orbitals is greater than the energy difference between the 2s and 2p orbitals. This is because screening becomes increasingly important for large nuclear charges. As Z increases, the stability of the s orbitals increases more rapidly than the stability of the d orbitals.

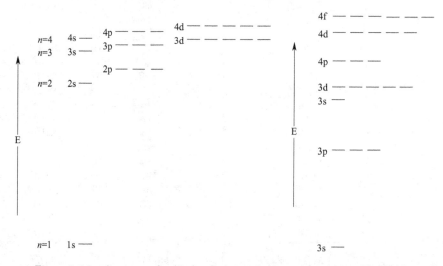

Figure 2.12 diagrams of orbital splitting and energy increasing with n and l

As a consequence of these two trends in energy spacing, the 4s and 3d orbitals of the elements from $Z=19$ to $Z=30$ have nearly the same energy. Figure 2.12 also shows the order of orbital stability for the $n=3$ and $n=4$ orbitals for these elements. Notice that the most stable $n=4$ orbital (4s) is more stable than the least stable $n=3$ orbitals (3d). Experiments confirm that potassium ($Z=19$) has the ground state configuration [Ar] $4s^1$ rather than [Ar] $3d^1$. In potassium and other neutral atoms in this row of the periodic table, the 4s orbital is slightly more stable than 3d orbitals. In calcium ($Z=20$) the 4s orbital is filled. Now the 3d set fills, starting with scandium ($Z=21$) and finishing with zinc ($Z=30$).

This pattern repeats for the 5s and 4d pair of orbitals for neutral atoms between $Z=37$ and $Z=48$. Here, the 5s orbital is more stable than the 4d orbital. Furthermore the 4f orbital is very effectively screened by $n=3$ electrons, so this orbital does not fill until after the 5s, 4d, 5p, and 6s orbitals. Thus Rb has the configuration [Kr]$5s^1$, and Sr has the configuration [Kr] $5s^2$. The (n)s orbital is always more stable than its $(n-1)$d.

Pauli summarizes the sequence of orbital filling without referring to an energy level diagram, show as Figure 2.13. The orbitals are listed in a table in which all orbitals with the same principal quantum number n lie in the same row and all orbitals with the same angular momentum quantum number l values lie in the same column. The order of orbital filling follows the 45-degree diagonal lines that move up and across the table from right to left. In constructing a configuration this way, we must remember the capacity of each set of orbitals. This is shown at the bottom of the columns. Using this mnemonic, we can quickly write the ground state configuration of element with many electrons.

(2) Pauli Exclusion Principle

Each electron in the atom has a unique set of the four quantum numbers n, l, m, and m_s. No two electrons in the same atom can have all four quantum numbers identical. All electron configurations must obey this fundamental law of quantum mechanics, which is known as the Pauli Exclusion Principle. This principle named from the Austrian physicist Wolfgang Pauli, who won the Nobel Prize in Physics in 1945.

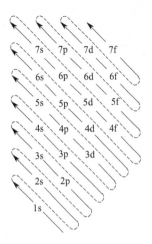

Figure 2.13 Pauli's approximate energy level order

This principle can be derived from the mathematics of quantum mechanics, but it cannot be rationalized in a simple manner. Nevertheless, all experimental evidence upholds the idea. As a result of the Pauli Exclusion Principle, an orbital can hold only two electrons. the two electrons must have opposite spins, since there are only two possible values for m_s ($+1/2$ and $-1/2$). The exclusion principle is a statement of an experimental fact with no explanation according to the Schrödinger model of the atom, but with very important effects in systems that have more than one electron. All electrons tend to avoid each other because all electrons are negatively charged and like charges repel each other.

(3) Hund's Rule

If more than one atomic orbital of the same energy is available, electrons may not be spin-paired in an orbital until each orbital in the set contains one electron with same spin; electrons will occupy different atomic orbitals, as far as possible. Hund's rule is based on the results of measurements of magnetic properties.

For a given value of the angular momentum quantum number l, m will split into a number of orbitals equal to all the integral numbers from $-l$ to $+l$, including zero. A p subshell, having an l value of 1, will subdivide the subshell into three orbitals of equal energy according to $m=+1, 0, -1$ Where one spectral line was before, three now exist. A 5-fold subdivision will take place for a d subshell: $l=2$ and $m=+2, +1, 0, -1, -2$, producing five orbitals of equal energy. Such states are said to be degenerated which means they have equal energy: s-type orbitals cannot be degenerate; p-type orbitals are triply degenerate; dtype orbitals exhibit 5-fold degeneracy. Note that since these orbitals axe of equal energy, they will not normally show up as separate spectral lines.

Carbon atoms, for example, have six electrons. For the Aufbau principle, the first four fill the 1s and 2s energy levels without doubt, but how are the final two electrons arranged in the 2p orbitals? Three different arrangements of these electrons appear to be consistent with the Aufbau principle.

① The electrons could be paired in the same 2p orbital (same m value but different m_s values).

② The electrons could be placed in different 2p orbitals with the same spin orientation (different m values but the same m_s value).

③ The electrons could be placed in different 2p orbitals with opposite spin orientations (different m values and different m_s values).

These three arrangements have different energies because when electrons are close together, they repel one another more than when they are far apart. Thus given a choice between otherwise equal-energy orbitals, an electron will occupy the orbital which keeps it farthest from other electrons in the system. Placing two electrons in different p orbitals keeps them relatively far apart, so an atom is more stable with the two electrons in different p orbitals. Thus arrangements ② and ③ are more stable than arrangement ①.

Arrangement ② and ③ look spatially equivalent, but theory and experiment show that a configuration that gives unpaired electrons the same spin orientation is always more stable than one that gives them opposite orientations. This fundamental feature of ground state configurations is called Hund's rule.

According to Hand's rule, for an atom in its ground state configuration, all unpaired electrons have the same spin orientation.

2.9 Valence Electrons

When chemical reactions take place, bonds break in the reactants and the new bonds reform in the products. It is said in another word, electrons rearrange around the nuclei. However, not all the electrons are influenced in. The outer electrons of an atom always show a significant activity. The outer electrons are called valence electrons, and the outer shell is called the valence shell. The electron configuration can be used to estimate the valence electrons or valence shell of an element, chlorine, for example. The electron configuration of chlorine is $[Ne]3s^2 3p^5$, then the valence electrons are those in shell 3s and 3p. In chemistry, valence electrons are important in determining how an element reacts with other elements. The number of electrons in an atom's valence shell governs its bonding behavior. Therefore, elements with the same number of valence electrons are grouped together in the periodic table.

2.10 The Periodic Table and Electronic Configurations of Atoms

Periods

The configuration of the electron shell of an unexcited atom depends on the charge of its nucleus. Elements with the same principal quantum number n form a quantum shell of clouds of about the same size. Shells with $n=1, 2, 3, 4, \cdots$, are designated respectively by the letters K, L, M, N, $\cdots$, P. A horizontal row of the periodic table that terminates in one of the rare gases is called a period. A period is a series of elements whose atoms have the same number of electron shell. The number of a period coincides with the principal quantum number n of the outer shell.

Groups

In accordance with the maximum number of electrons in the outer shells of ground state atoms, all elements of the periodic system are divided into eight groups. The position of s and p elements in the groups is determined by the total number of electrons in the outer shell. For example, phosphorus ([Ne] $3s^2 3p^3$), with five electrons in the outer shell, belongs to group V, argon ([Ne] $3s^2 3p^6$) to group Ⅷ, calcium ([Ar] $4s^2$) to group Ⅱ, and so on.

The elements of groups are divided into subgroups. The s and p elements constitute what are called the main subgroups (subgroup A), and the d elements a secondary subgroup (subgroup B).

Transition Elements

All the elements in the first transition series tend to lose two electrons to become positive ions, and the remaining electrons are always 3d electrons. When an atom is ionized, the electrons highest in energy are the most easily removed. These ions do not have rare gas electronic configurations, and most of them are colored. In aqueous solution, Cu^{2+} is light blue, Ni^{2+} is bright apple green, Mn^{2+} is pale pink, and Co^{2+} is rosy pink, for example. The Zn^{2+} ion, which has completely filled subshells, is colorless.

Because it is not possible for the transition metals to achieve a rare gas configuration by a simple loss or gain of one or two electrons, most of these elements can form more than one type of cation. You are already familiar with Fe^{2+} and Fe^{3+}, Cr^{2+} and Cr^{3+} ions, among others.

Using the Periodic Table to Write Electron Configurations

If we had to draw the orbital energy diagram for an atom to write its electron configuration, writing electron configurations would be a tiring work. Fortunately, the electron configurations for most atoms can be written from the periodic table. In Periodic Table, the valence electron configurations for all the elements are shown in their grids in the periodic table. Because the core is the same for all the elements in a period, the symbol for the core is shown at the end of previous row. Most of the electron configurations were found by both experiment and calculation.

You may wonder why potassium and calcium have electrons in the 4s orbital in stead of in the 3d orbital. Remember that the higher the principal quantum number, the more closely spaced the energies of the orbitals. Compared to the difference in energy between the first and second shell, the difference in energy between the third and fourth shells is small. In potassium, calcium, and scandium atoms, the 4s orbital is lower in energy than the 3d orbital because the 4s orbital penetrates to the nucleus more than the 3d orbital.

Electron configurations make clear the reason for the structure of the periodic table. Hydrogen, helium, and the elements in group Ⅰ A and Ⅱ A are called s-block elements; the outer electrons of these elements are all in s subshells. The elements in group Ⅲ A-0 are called p-block elements. Going across a row of the periodic table, the difference between p-block elements lies in the number of p electrons in the outer shell. As atomic number increases by one, the number of p electrons in the outer shell also increases by one. All p-block elements have filled s subshells in their outer shells. Some also have filled d and f subshells outside the core. The s-block and p-block elements are the elements that we have previously called main group elements.

The elements in group Ⅲ B-Ⅱ B are called d-block elements. The d-block elements are the elements that we have been referring to as transition elements. In most cases, going across a row in the periodic table, the difference between d-block elements lies in the number of d electrons in the next-to-the-outermost or $(n-1)$ shell. As atomic number increases by one, the number of d electrons in the $(n-1)$ shell also increases by one. The d-block elements in the fourth period (Sc-Zn) are called the first transition series. The d-block elements in the fifth period (Y-Cd) are referred to as the second transition series and the d-block, elements in period (Lu-Hg) as the third transition series.

If we consider the elements in order of increasing atomic number, the first element that does not have the electron configuration expected from its place in the periodic table is chromium (atomic number 24). From its position between vanadium [Ar] $3d^3 4s^2$ and manganese [Ar] $3d^5 4s^2$, chromium would be expected to have the electron configuration [Ar] $3d^4 4s^2$. Instead, chromium has the electron configuration [Ar] $3d^5 4s^1$. At the present time, chemists do not agree on a reason for the unexpected electron configuration of chromium. However, note that both the 3d and the 4s subshells in the chromium atom are half-filled. Copper, silver, molybdenum and gold, which have unexpected electron configurations, also have half-filled subshells.

The elements in the two rows at the bottom of the periodic table, those in the lanthanide and actinide series, are called f-block elements. In most cases, the difference between f-block elements going across a row in the periodic table lies in the number of f electrons in the next-to-the-next-to-the-outermost or $(n-2)$ shell. All f-block elements in a period have the same outer electron configuration, that is $6s^2$ or $7s^2$. The f-block elements are sometimes called inner-transition elements because the difference between elements in the same period is one more shell in and from the outside of the atom than the difference of transition elements.

Figure 2.14 shows the orbitals that are being filled for elements in different parts of the periodic table. Figure 2.14 shows the periodic table with the lanthanides and actinides or f-block elements in their proper place between the s-block elements and the d-block elements. We do not ordinarily use the expanded form of the periodic table because it is too wide.

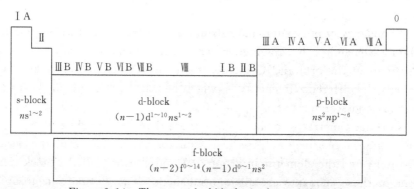

Figure 2.14 The s, p, d, f-blocks in the periodic table

2.11 The Periodicity of the Properties of Elements and Atomic Structure

Atomic Radii

Real atoms and ions are not hard balls. There is no sharp boundary that defines the outside of an atom or ion. However, charge density diagrams and radial distribution curves similar to those in Figures 2.8(c) and (d) suggest that the charge clouds of atoms occupy fairly well-defined volumes. The sizes of the circles in Figure 2.15 are proportional to the best available set of estimates of radii of atoms in metals, in molecules of elements, and in molecular compounds.

As you can see from Figure 2.15 in general, the atomic radii of main group elements decrease from left to right across rows in the periodic table. The principal quantum number of the outer electrons is the same all across a period, the outer electrons are in the same shell. The outer electrons do not shield each other very well. On the other hand, the charge on the nucleus increases from left to right across a row in the periodic table. As a result, the outer electrons are attracted more strongly and pulled in tow nucleus, making the atom smaller.

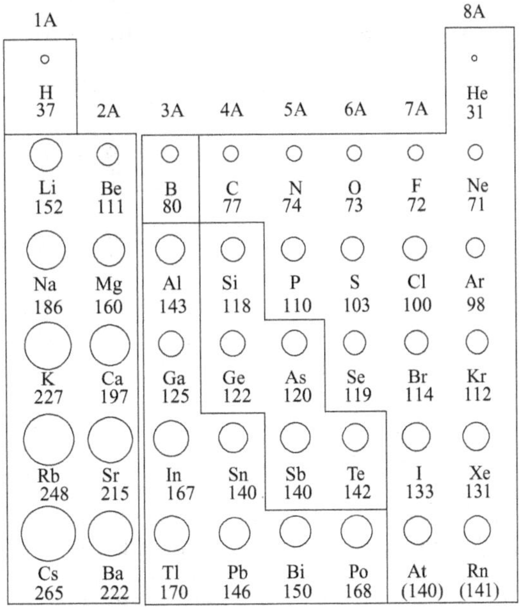

Going down main groups, atoms get bigger. The number of shells of electron increases from top to bottom of a group. The higher the principal quantum number of outer shell is, the larger the radius of atom is. Although the charge on the nucleus creases going down a group, the effective nuclear charge (the charge that is outer electrons actually feel) does not increase as much.

1A							8A
H 37	2A	3A	4A	5A	6A	7A	He 31
Li 152	Be 111	B 80	C 77	N 74	O 73	F 72	Ne 71
Na 186	Mg 160	Al 143	Si 118	P 110	S 103	Cl 100	Ar 98
K 227	Ca 197	Ga 125	Ge 122	As 120	Se 119	Br 114	Kr 112
Rb 248	Sr 215	In 167	Sn 140	Sb 140	Te 142	I 133	Xe 131
Cs 265	Ba 222	Tl 170	Pb 146	Bi 150	Po 168	At (140)	Rn (141)

Figure 2.15 The atomic radii of main group elements

The radii of transition metals generally first decrease and then increase each series. Changes in radius are small compared to the changes in size of atoms of main group elements across a period. The smallest main group atom in a period is only about half as big as the biggest main group atom in the period. The smallest transition metal atom in a period is three quarters or more the size of the largest transition metal atom in the period. The small change in size across a series of transition metal is explained by the fact that electrons are being added to an inner shell. These additional inner electrons screen the outer electrons. Effective nuclear charge does not increase very much, so size does not decrease very much. Toward the end of each series, repulsion the increasing number of electrons makes the atomic radius increase.

Ionic Radii

Many of the properties of ions are explained by their charges and sizes. In Figure 2.16, the sizes of some monatomic ions formed from main group elements are compared with the sizes of the atoms from which the ions are formed. As you can see, cations (positive ions) are smaller than the atoms from which they are formed. When positive ions are formed from main group elements, the outer shell of electrons is removed. It is not surprising that the cations are smaller than the atoms from which they are formed. Anions (negative ions) are larger than the atoms from which they are formed. Adding electrons to the outer shell results the increasing of repulsion between electrons. Therefore, the electron cloud spreads out and the anions are bigger than the atoms.

Species that have the same electron configurations are called isoelectronic. For example,

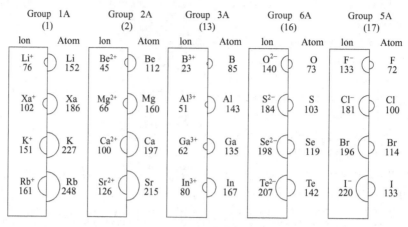

Figure 2.16 Some ionic radii compared with the atomic radii

O^{2-}, F^-, Ne, Na^+, Mg^{2+}, and Al^{3+} are isoelectronic because all have the electron configuration $1s^2 2s^2 2p^6$. For isoelectronic species, the higher the nuclear charge, the smaller the species are. The radii sequence of these isoelectronic is: $O^{2-} > F^- > Ne > Na^+ > Mg^{2+} > Al^{3+}$.

The physical and chemical properties of the elements and their compounds depend on the outer electron configuration. Then members of a group have similar properties because they have the same outer electron configuration. The chemical nature of an element depends on the capacity of its atoms to donate or accept electrons. All of the properties of atoms that measure the ease of loss or gain of electrons are related to their size. Two such properties are ionization energy and electron affinity.

Ionization Energy

It always requires energy to remove an electron from a neutral atom, that is to produce a singly positive ion. However, the amount of energy required to remove one electron varies from atom to atom. If we examine the experimental values of this ionization energy, we learn a great deal about the stabilities of different electronic configurations.

We define the ionization energy (I) as the energy that be expended to remove an electron from a single, isolated gas phase atom, that is, to carry out the reactions.

$$X(g) \longrightarrow X^+(g) + e^-$$

The X in the reaction above represents any element. We start with a gaseous atom to avoid any extra energy be expended to break chemical bonds, or so. For atoms with many electrons, we can remove a second electron, a third, and so on. The second ionization energy is the energy required for the reaction.

$$X^+(g) \longrightarrow X^{2+}(g) + e^-$$

Ionization energies are usually reported either in electron volts per atom or kilojoules per mole.

Table 2.3 lists the first ionization energies of the elements of the second period from Li to Ne. As we look the list of values, we are easy to find the fact that the first ionization energy generally increases as we go across the period. It is because the nuclear charge is increasing, but the orbitals being filled are all $n=2$ orbitals, which are approximate distance from the nucleus. Thus as Z increases, the outer electrons are held more and more tightly and it requires more energy to remove one electron.

Table 2.3 First ionization energies of the elements of the second period

Z	Atom	Electronic configuration	I_1/eV	I_1/(kJ/mol)
3	Li	$1s^2 2s^1$	3.39	5.20×10^2
4	Be	$1s^2 2s^2$	9.32	8.99×10^2
5	B	$1s^2 2s^2 2p^1$	8.30	8.00×10^2
6	C	$1s^2 2s^2 2p^2$	11.3	1.09×10^3
7	N	$1s^2 2s^2 2p^3$	14.5	1.40×10^3
8	O	$1s^2 2s^2 2p^4$	13.6	1.31×10^3
9	F	$1s^2 2s^2 2p^5$	17.4	1.68×10^3
10	Ne	$1s^2 2s^2 2p^6$	21.6	2.08×10^3

The alkali metal, Li, has the lowest first ionization energy of the elements in this period, as it is relatively easy to remove the single s electron and form an ion with rare gas configuration. The rare gas, Ne, has the highest ionization energy of the elements of this period because it is difficult to remove an electron from this very stable configuration. However, there are two exceptions to the general trend of increasing ionization energy with increasing atomic number.

Electron Affinity

Almost all atoms can accept one more electron than is present in the neutral atom and become an anion. For many atoms this process is accompanied by a release of energy. The released energy is defined as the electron affinity (abbreviated E). To be precise, what we have just defined is the first electron affinity when the following reaction occurs.

$$X(g) + e^- \longrightarrow X^-(g)$$

Second and third electron affinities can also be defined when add the second or third electron to an ion. It is quite difficult to measure electron affinities experimentally. And you may see different values in different references for the experimental uncertainty. The main reason is atoms that do not readily accept an electron. The most accurately known values are those for the halogens. Values of the electron affinity for the elements of groups ⅤA, ⅥA, and ⅦA are given in Table 2.4. Electron affinity is also a periodic property on the whole. As a general trend we note that the electron affinity increases as we go across a period from left to right. With some notable exceptions, the electron affinity decreases as we go down a group, that is, as the electron being added is further from the nucleus. The electron affinities of N, O, and F are significantly less than the electron affinities of the elements directly beneath each of them.

Table 2.4 Electron affinities of groups ⅤA, ⅥA, and ⅦA

Atom	E/eV	Atom	E/eV	Atom	E/eV
N	0.00	O	1.47	F	3.45
P	0.77	S	2.08	Cl	3.61
As	0.80	Se	2.02	Br	3.36
Sb	1.05	Te	1.97	I	3.06
Bi	1.05	Po[1]	(1.8)	At[1]	(2.8)

[1] The values for Po and At are estimated.

The usual convention in discussing energy changes is to give the positive energy absorbed when a reaction takes place. As the electron affinity is the energy released when reaction takes place, it is defined in contravention to the customary procedure.

Electronegativity

The concept of electronegativity was introduced by Linus Pauling in 1932. Electronegativity is a measure of the ability of a bonded atom to attract the electrons in the bond from the other atom or atoms to which it is bonded. An atom with a high electronegativity is able to pull electrons toward itself and away from an atom with a lower electronegativity. In these qualitative terms it is easy to understand the concept of electronegativity, but it is more difficult to obtain a quantitative value for the electronegativity of an atom. The principal reason for this is that the electronegativity is a property of the bonded atom (not the isolated gaseous atom) and therefore is not a constant for a particular atom but varies somewhat as the nature of the bonding differs in different molecules. When Pauling introduced the concept of electronegativity, he devised a scale based on the energy required to break a bond between two atoms. It is Pauling's scale that is most often used, and a set of values is given in Table 2.5.

Values of the electronegativities are very helpful for understanding certain properties of chemical bonds, but they should be used to make quantitative predictions cautiously. General features of the electronegativity scale which you should be familiar are the following.

Table 2.5 Electronegativity of the atoms (Pauling's scale)

H 2.18																	He
Li 0.98	Be 1.57											B 2.04	C 2.55	N 3.04	O 3.44	F 3.98	Ne
Na 0.93	Mg 1.31											Al 1.61	Si 1.90	P 2.19	S 2.58	Cl 3.16	Ar
K 0.82	Ca 1.00	Sc 1.36	Ti 1.54	V 1.63	Cr 1.66	Mn 1.55	Fe 1.80	Co 1.88	Ni 1.91	Cu 1.90	Zn 1.65	Ga 1.81	Ge 2.01	As 2.18	Se 2.55	Br 2.96	Kr
Rb 0.82	Sr 0.95	Y 1.22	Zr 1.33	Nb 1.60	Mo 2.16	Tc 1.90	Ru 2.28	Rh 2.20	Pd 2.20	Ag 1.93	Cd 1.69	In 1.73	Sn 1.96	Sb 2.05	Te 2.10	I 2.66	Xe
Cs 0.79	Ba 0.89	La 1.10	Hf 1.30	Ta 1.50	W 2.36	Re 1.90	Os 2.20	Ir 2.20	Pt 2.28	Au 2.54	Hg 2.00	Tl 2.04	Pb 2.33	Bi 2.02	Po 2.00	At 2.20	Rn

(1) Fluorine is the most electronegative element and its electronegativity is set at 3.98. The least electronegative element is cesium with an electronegativity of 0.79.

(2) Generally, electronegativities increase as we go across a period from left to right, and decrease as we go down a group in the table. The electronegativities of all the elements in a series of transition metals are very similar.

(3) Very roughly, an electronegativity value of 2 divides the metals and the nonmetals. Most metals have an electronegativity value less than 2, most nonmetals have a value greater than 2.

Key Words

atom [ˈætəm] 原子
electron [iˈlektrɔn] 电子
nucleus [ˈnjuːkliəs] 原子核（nuclear 的复数）
proton [ˈprəutɔn] 质子
neutron [ˈnjuːtrɔn] 中子
uncertainty principle [ʌnˈsəːtnti ˈprinsəpl] 测不准原理
probability density [ˌprɔbəˈbiliti ˈdensiti] 概率密度
quantum number [ˈkwɔntəm ˈnʌmbə] 量子数
principle [ˈprinsəpl] 原理，主要的
momentum [məuˈmentəm] 动量
magnetic [mægˈnetik] 磁的

spin [spin] 自旋
degeneracy [di'dʒenərəsi] 简并
atomic orbital [ə'tɔmik 'ɔːbitl] 原子轨道
electron configuration [iˈlektrɔn kənˌfiɡjuˈreiʃən] 电子组态
ground state [ɡraʊnd steit] 基态
neutral atom [ˈnjuːtrəl ˈætəm] 中性原子
energy level [ˈenədʒi ˈlev(ə)l] 能级
exclusion [iksˈkluːʒən] 互斥，不相容
hund's rule [ˈhʌnds ruːl] 洪特规则
valence electron [ˈveiləns iˈlektrɔn] 价电子
periodic table [ˌpiəriˈɔdikˈteibl] 周期表
period [ˈpiəriəd] 周期

group [ɡruːp] 族
transition element [trænˈziʒən ˈelimənt] 过渡元素
block [blɔk] 区
periodicity [ˌpiəriəˈdisiti] 周期
cation (positive ion) [ˈkætaiən] 阳离子
anion (negative ion) [ˈænaiən] 阴离子
ionization energy [ˌaiənaiˈzeiʃən ˈenədʒi] 电离能
electron affinity [iˈlektrɔn əˈfiniti] 电子亲合能
electronegativity [iˌlektrɔˈneɡəˈtiviti] 电负性

Exercises

1. Describe what atoms "look like", what they are composed of, and how they behave.

2. Write a short description of (1) Uncertainty Principle; (2) Schrödinger Equation; (3) Aufbau Principle; (4) Pauli Exclusion Principle; (5) valence electrons.

3. What is the wavelength of a neutron traveling at a speed of 3.60 km/s? Neutrons of these speeds are obtained from a nuclear pile.

Answer: $\lambda_{neutron} = 1.0 \times 10^{-10}$ m.

4. How many subshells are there in the N shell? How many orbitals are there in the f subshell?

Answer: Four subshells in the N shell. Seven orbitals in the f subshell.

5. Write the orbitals with following quantum numbers:
(1) $n=2, l=1$ (2) $n=3, l=2$ (3) $n=4, l=0$
(4) $n=2, l=1, m=-1$ (5) $n=4, l=0, m=0$

Answer: (1) 2p   (2) 3d   (3) 4s   (4) $2p_z$ (5) 4s

6. Which of the following electron configurations are possible? Explain why the others are not.
(1) $1s^2 2s^2 2p^6$ (2) $1s^2 2s^2 2p^6 3s^1 3p^6$ (3) $1s^2 2s^2 2p^8$ (4) $1s^2 2s^1 2p_x^2 2p_y^2 2p_z^2 3s^2 3p_x^2$

Answer: (1) $1s^2 2s^2 2p^6$ is possible, and (2), (3), (4) are impossible.

(2) $1s^2 2s^2 2p^6 3s^1 3p^6$: 3s is not full, which disagree with the *Aufbau* principle;

(3) $1s^2 2s^2 2p^8$: 2p has more than 6 electrons, which disagree with the Pauli Exclusion Principle;

(4) $1s^2 2s^1 2p_x^2 2p_y^2 2p_z^2 2s^3 3p_x^2$: $3p_x$ has 2 electrons but $3p_y$, $3p_z$ are empty, which disagree with the Hund's Rule.

7. Fill in the table below:

Atomic Number	Electron Configurations	Electron Configurations of Valence Electrons	Period	Group
49				
	$1s^2 2s^2 2p^6$			
		$3d^5 4s^1$		
			6	ⅡB

Answer:

Atomic Number	Electron Configurations	Electron Configurations of Valence Electrons	Period	Group
	$[Kr]4d^{10}5s^25p^1$	$5s^25p^1$	5	ⅢA
10		$2s^22p^6$	2	0
24	$[Ar]3d^54s^1$		5	ⅥB
80	$[Xe]4f^{14}5d^{10}6s^2$	$5d^{10}6s^2$		

8. Copper has the ground state configuration $[Ar]3d^{10}4s^1$. Give the group and period for this element. Classify it a main-group, a d-transition, or an f-transition element.

   Answer: Copper is a d-transition element in the ⅠB group and the forth period.

9. With the aid of a periodic table, arrange the following in order of increasing electrnegativity:

   Cs, S, N, P, F, Ca, Zn.

   Answer: Cs(0.79)<Ca(1.00)<Zn(1.65)<P(2.19)<S(2.58)<N(3.04)<F(3.98).

Chapter 3 Chemical bonds

Only the noble gases occur naturally as separate atoms with full electronic shells showing very stable chemical properties. Chemical compounds are formed by the joining of two or more atoms and the total energy of the combination is lower than the separated atoms. The bound state implies a net attractive force between the atoms. So chemical bonds are the forces that hold atoms together in molecules of elements such as O_2 and N_2, in compounds and in metals. During the formation of chemical bonds every atom becomes much more stable, or less reactive with a full valence shell. This can be achieved by different ways causing three major types of bonds exist.

1. Ionic bonds
2. Covalent bonds
3. Metallic bonds

The types of bonds in a substance are largely responsible for not only the physical and chemical properties of the substance but also the attraction one substance has for another. For example, sodium chloride dissolves in water much better than in oil because of the differences in bonding; aqueous solution of salt conducts electricity whereas aqueous solution of sugar doesn't. In this chapter, we will discuss the bonds that hold atoms together in ionic compounds and in molecules, that is, ionic bonds and covalent bonds.

The geometry of a molecule or polyatomic ion is one of its most important characteristics. Without information of their geometry you cannot understand their properties or the reactions that they undergo. Reactions that take place in living systems are especially sensitive to the geometry of the species involved. In this chapter, we will learn how to predict molecular geometry qualitatively.

In fact, virtually all substances that are liquids at room temperature are molecular substances. Their molecular geometry, bond energies, and many aspects of chemical behaviour are influenced by covalent bonds, forces within molecules. The physical properties of molecular liquids and solids, however, are due largely to intermolecular forces, which are between molecules. By understanding the nature and strength of intermolecular forces, we can relate the composition and structure of molecules to their physical properties.

3.1 Ionic Bonds

Formation of ionic bonds

The chemical bond between positively charged ions and negatively charged ions through electrostatic attraction is called an ionic bond. Normally, atoms are neutral and have no charge. In order to gain stable electron configurations, atoms will sacrifice their neutrality by either transferring or sharing their valence electrons. Elements that are described as "metallic" tend to lose electrons and elements that are described as "non-metallic" tend to gain electrons. In the extreme case one or more atoms lose one or more of its outermost electrons thus becoming a positive ion (cation) or they will gain one or more electrons thus becoming a negative ion (anion). The oppositely charged ions will attract each other causing them to come together and form a bond. Most salts are ionic. Any metal will combine chemically with any non-metal to form ionic bonds. Typical of ionic bonds are those in the alkali halides.

Take sodium chloride, NaCl as an example. When ionic compound NaCl is formed from a sodium atom and a chloride atom, one electron is transferred from the sodium atom (Na) to the chlorine atom (Cl) forming a cation (Na^+) and an anion (Cl^-). These ions are then attracted to each other in a 1∶1 ratio to form NaCl.

$$Na([Ne]3s^1) + Cl([Ne]3s^2 3p^5) \rightarrow Na^+([Ne]) + Cl^-([Ar]) \rightarrow NaCl$$

Sodium loses its outer electron to give it the electron configuration of the noble gas neon. Chlorine atom accepts the electron from sodium atom to form a chloride ion and give it the electron configuration of the noble gas argon. Once this has happened, each ion acquires the noble gas configuration. The oppositely charged ions are then attracted to each other, and their bonding releases energy. The bonding energy from the electrostatic attraction of the two oppositely-charged ions has a large enough negative value and the overall bonded state energy is lower than the unbonded state. So the reaction is able to take place. The larger the resulting energy change the stronger the bond. When metals with low electronegativity combine with non-metals with high electronegativity, the reactions are prone to happen because the energy change of the reaction is most favorable.

Pure ionic bonding is not known to exist. Large amount of experiments indicate that all ionic bonds have a degree of covalent bond or metallic bond. The degree of ionic bond increases with the increasing difference in electronegativity between two atoms.

Sample 3.1 Use electron configurations to describe the formation of each ionic compound: (a) aluminum oxide from aluminum atoms and oxygen atoms, (b) lithium hydride from lithium atoms and hydrogen atoms, (c) magnesium nitride from magnesium atoms and nitrogen atoms. Write formulas for each of the compounds formed.

Solution:

(a) $2Al([Ne]3s^2 3p^1) + 3O([He]2s^2 2p^4) \rightarrow 2Al^{3+}([Ne]) + 3O^{2-}([Ne]) \rightarrow Al_2O_3$

(b) $Li[He]2s^1 + H(1s^1) \rightarrow Li^+([He]) + H^-([He]) \rightarrow LiH$

(c) $3Mg([Ne]3s^2) + 2N([He]2s^2 2p^3) \rightarrow 3Mg^{2+}([Ne]) + 2N^{3-}([Ne]) \rightarrow Mg_3N_2$

Ionic compounds

Ionic compounds in the solid state form a continuous ionic lattice structure in an ionic crystal. They generally have higher melting points and boiling points compared to covalent compounds. Because of the extremely polar bonds in ionic compounds, most of them are soluble in water but insoluble in nonpolar solvents. When ionic compounds dissolved in water, they conduct electricity because in solution the ions are free to move and carry the electrical charge from the anode to the cathode. Ionic substances also conduct electricity when molten because atoms are mobilized and electrons can flow directly through the ionic substance in a molten state. Table 3.1 lists some common substances in ionic form.

Table 3.1 Substances in ionic form

Simple Cations		Simple Anions		Anions from Organic Acids	
aluminum	Al^{3+}	arsenide	As^{3-}	acetate	$C_2H_3O_2^-$
barium	Ba^{2+}	azide	N_3^-	formate	HCO_2^-
beryllium	Be^{2+}	bromide	Br^-	oxalate	$C_2O_4^{2-}$
cesium	Cs^+	chloride	Cl^-	hydrogen oxalate	$HC_2O_4^-$
calcium	Ca^{2+}	fluoride	F^-	Other Anions	
chromium(II)	Cr^{2+}	hydride	H^-	hydrogen sulfide	HS^-
chromium(III)	Cr^{3+}	iodide	I^-	telluride	Te^{2-}
chromium(VI)	Cr^{6+}	nitride	N^{3-}	amide	NH_2^-
cobalt(II)	Co^{2+}	oxide	O^{2-}	cyanate	OCN^-
cobalt(III)	Co^{3+}	phosphide	P^{3-}	thiocyanate	SCN^-
copper(I)	Cu^+	sulfide	S^{2-}	cyanide	CN^-
copper(II)	Cu^{2+}	peroxide	O_2^{2-}		
copper(III)	Cu^{3+}	Oxoanions			
gallium	Ga^{3+}	arsenate	AsO_4^{3-}		
gold(I)	Au^+	arsenite	AsO_3^{3-}		
gold(III)	Au^{3+}	borate	BO_3^{3-}		
helium	He^{2+}	bromate	BrO_3^-		
hydrogen	H^+	hypobromite	BrO^-		
iron(II)	Fe^{2+}	carbonate	CO_3^{2-}		
iron(III)	Fe^{3+}	hydrogen carbonate	HCO_3^-		
lead(II)	Pb^{2+}	chlorate	ClO_3^-		
lead(IV)	Pb^{4+}	perchlorate	ClO_4^-		
lithium	Li^+	chlorite	ClO_2^-		
magnesium	Mg^{2+}	hypochlorite	ClO^-		
manganese(II)	Mn^{2+}	chromate	CrO_4^{2-}		
manganese(III)	Mn^{3+}	dichromate	$Cr_2O_7^{2-}$		
manganese(IV)	Mn^{4+}	iodate	IO_3^-		
manganese(VII)	Mn^{7+}	nitrate	NO_3^-		
mercury(II)	Hg^{2+}	nitrite	NO_2^-		
nickel(II)	Ni^{2+}	phosphate	PO_4^{3-}		
nickel(III)	Ni^{3+}	hydrogen phosphate	HPO_4^{2-}		
potassium	K^+	dihydrogen phosphate	$H_2PO_4^-$		
silver	Ag^+	permanganate	MnO_4^-		
sodium	Na^+	phosphite	PO_3^{3-}		
strontium	Sr^{2+}	sulfate	SO_4^{2-}		
tin(II)	Sn^{2+}	thiosulfate	$S_2O_3^{2-}$		
tin(IV)	Sn^{4+}	hydrogen sulfate	HSO_4^-		
zinc	Zn^{2+}	sulfite	SO_3^{2-}		
Polyatomic Cations		hydrogen sulfite	HSO_3^-		
ammonium	NH_4^+				
hydronium	H_3O^+				
nitronium	NO_2^+				
mercury(I)	Hg_2^{2+}				

3.2 Covalent Bonds

Although the concept of molecules goes back to the seventeenth century, it was until early in the twentieth that chemists began to understand how and why molecules form. In

1916 Gilbert Lewis described the sharing of electron pairs between atoms. The term "covalence" in regard to bonding was first used in 1919 by Irving Langmuir in a *Journal of American Chemical Society* article entitled The Arrangement of Electrons in Atoms and Molecules.

(p. 926) ···we shall denote by the term covalence the number of pairs of electrons which a given atom shares with its neighbors.

The formation of covalent bonds also can be represented using Lewis dot symbols (see Section 3.3). For example, hydrogen gas forms the simplest covalent bond in the diatomic hydrogen molecule.

$$H\cdot + \cdot H \longrightarrow H:H \qquad (3.1)$$

This type of electron pairing is an example of a covalent bond, a bond in which two valence electrons are shared by two atoms, in contrast to the transfer of electrons in ionic bonds. Covalent compounds are compounds that contain only covalent bonds. For the sake of simplicity, the shared pair of electrons is often represented by a single line. Thus the covalent bond in the hydrogen molecule can be written as H-H. In a covalent bond, each electron in a shared pair is attracted to the nuclei of both atoms. This attraction holds the two atoms in H_2 together and is responsible for the formation of covalent bonds in other molecules.

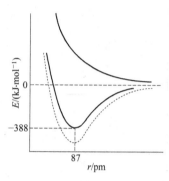

Figure 3.1 The potential energy-nuclear distance curve of two H atoms

Valence bond theory describes a chemical bond as the overlap of atomic orbitals. In the case of the hydrogen molecule (Figure 3.1), the 1s orbital of one hydrogen atom overlaps with the 1s orbital of the second hydrogen atom. Attraction increases as the distance between the atoms gets closer. When the maximum overlap of orbitals occurs, the H_2 molecule forms. But nuclear-nuclear repulsion becomes important if the atoms approach too close.

Covalent bonds can be single, double, or triple. In a single bond, two atoms are held together by one electron pair, such as H-H, H-Cl, and F-F. Many compounds are held together by multiple bonds, that is, bonds formed when two atoms share two or more pairs of electrons.

If two atoms share two pairs of electrons, the covalent bond is called a double bond. Double bonds are found in molecules of oxygen (O_2), carbon dioxide (CO_2), and ethylene (C_2H_4).

$$O=O \qquad O=C=O \qquad H_2C=CH_2 \qquad (3.2)$$

A triple bond arises when two atoms share three pairs of electrons, as in the nitrogen molecule (N_2) and the acetylene molecule (C_2H_2).

$$N\equiv N \qquad HC\equiv CH \qquad (3.3)$$

Double and triple bonds occur frequently, atoms can achieve a complete octet (see Section 3.3) by sharing more than one pair of electrons between them. Multiple bonds increase electron density between two nuclei, which decrease nuclear repulsion. As a result, the nuclei move closer together, so double bonds are shorter than single bonds and triple bonds are shortest of all.

3.3 Lewis Dot Symbols and Lewis Structure

Although some ionic compounds are common in our daily life such as NaCl, most compounds are molecular, such as water (H_2O), sulphuric acid (H_2SO_4), methane (CH_4), and ammonia (NH_3). The formation of ionic bond had been successfully explained after the discovery of the electron, the nuclear atom, the development of the periodic table and concept of electron configuration. Chemists also try a lot to develop an explanation of the chemical bonding in molecules based on electrons. In 1916, the American chemist Gilbert Newton Lewis (1875~1946) proposed that chemical bonds in molecules are formed by atoms sharing pairs of outer electrons in order to achieve a more stable electron configuration. He assumed that unshared electrons are also paired. He also suggested that groups of eight electrons (octets) around atoms, a structure of noble gases, have a special stability. Lewis's idea gave chemists a rationale for molecule and compound formation. It described the covalent bond as a shared pair of electrons. Although Lewis's idea was very useful, it did not explain why or how the electrons were shared. A genuine theory of the covalent bond was not possible until after the development of the quantum theory.

When atoms interact to form a chemical bond, only their valence electrons are concerned. To keep track of valence electrons in a chemical reaction, and to make sure that the total number of electrons does not change, Lewis invented a simple and convenient way of showing the valence electrons-Lewis dot symbols.

Lewis dot symbols

A Lewis dot symbol consists of the symbol of an element and one dot for each valence electron in an atom of the element. Lewis symbols are used mainly for s- and p-block elements. For these elements, except for helium, the number of valence electrons each atom has is equal to the group number of the element. For example, Li is a group I A element and has one dot for one valence electron; Be, a group II A element, has two valence electrons (two dots); and so on. The Lewis symbols for the atoms in the second period are

$$Li\cdot \quad \cdot Be\cdot \quad \cdot \dot{B}\cdot \quad \cdot \dot{C}\cdot \quad :\dot{N}\cdot \quad :\dot{O}: \quad :\ddot{F}: \quad :\ddot{Ne}:$$

In Lewis dot symbols, the dots are placed on the four sides of the atomic symbol, the top, the bottom, the left and right sides. Each side can accommodate up to two electrons and all four sides of the symbol are equivalent. Before 1924, in the Lewis dot symbols for beryllium, boron, and carbon, the dots were placed on each of the four sides of the symbol. This description indicated that the number of the unpaired electrons would be equal to the number of bonds that an element usually forms in compounds. Beryllium usually forms two bonds; boron forms three, and so on. Neon, a noble gas, does not form any bonds. In 1924 the idea that two of the electrons of beryllium, boron, and carbon are paired was developed.

Elements in the same group have similar outer electron configurations and hence similar Lewis dot symbols. Because the transition metal, lanthanide, and actinides all have incompletely filled inner shells, in general we cannot write simple Lewis dot symbols for them.

Sample 3.2 Write the Lewis dot symbols for the elements of s- and p-block in the third period.

Solution:

$$\text{Na}\cdot \quad \cdot\text{Mg}\cdot \quad \cdot\dot{\text{Al}}\cdot \quad \cdot\dot{\text{Si}}\cdot \quad :\dot{\text{P}}\cdot \quad :\dot{\text{S}}: \quad :\dot{\text{Cl}}: \quad :\ddot{\text{Ar}}:$$

Covalent bonding between many-electron atoms involves only the valence electrons. Consider the fluorine molecule, F_2. The electron configuration of F is $1s^2 2s^2 2p^5$. The 1s electrons are low in energy and do not participate in bond formation. Thus each F atom has seven valence electrons in the 2s and 2p subshell. According to the Lewis dot symbol of F atom, there is only one unpaired electron on F, so the formation of the F_2 molecule can be represented as follows.

$$:\ddot{\text{F}}\cdot + \cdot\ddot{\text{F}}: \longrightarrow :\ddot{\text{F}}:\ddot{\text{F}}: \qquad (3.4)$$

Only two valence electrons participate in the formation of F_2. The other, nonbonding electrons, are called lone pairs-pairs of valence electrons that are not involved in covalent bond formation, that is, valence electron pairs without bonding or sharing with other atoms. Thus, the number of lone electrons plus the number of bonding electrons equal the total number of valence electrons from a compound.

In certain circumstances, lone pairs will participate in the formation of coordinate covalent bonds (see Chapter 11) in which they provide both of the paired electrons.

In the formation of covalent bonds, each atom donates half of the electrons to be shared. If the electronegativity of the two bonded atoms are either equal or the difference of their electron-egativity is no greater than 1.7, the bond would be covalent and the atoms can only share the bonding electrons. If the electronegativity difference is greater than 1.7, the atom with higher electronegative would attract the electrons greatly and cause the transfer of electrons from the less electronegative atom. This would be an ionic bond. Theoretically, even ionic bonds have some covalent character. Thus, the boundary between ionic and covalent bonds is vague.

Lewis Structure

The structure used to represent covalent compounds, such as H_2 and F_2, are called Lewis structure. a representation of covalent bond in which shared electron pairs are shown either as lines or as pairs of dots between two atoms, and lone pairs are shown as pairs of dots on individual atoms. Only valence electrons are shown in a Lewis structure.

Take the Lewis structure of the water molecule as example. O atom has two unpaired electrons, so O might have two covalent bonds. H atom has only one electron, so it can form only one covalent bond. Thus the Lewis structure for water is

$$\text{H}:\ddot{\text{O}}:\text{H} \quad \text{or} \quad \text{H}-\ddot{\text{O}}-\text{H}$$

The Lewis structure shows that the O atom has two lone pairs and the H atom has no lone pair.

Octet Rule

The noble gases have very stable electron arrangements, as evidenced by their high ionization energies, low affinity for additional electrons, and general lack of chemical reactivity. Except for helium (with a filled 1s shell), noble gases have eight electrons in their valence shells. So many atoms undergoing reactions also end up with eight valence electrons forming a full outer shell. This observation has led to a simple chemical rule of thumb known as the octet rule: In molecules or ions atoms tend to gain, lose, or share electrons so as to have eight valence electrons in their outer shells, which match the nearest noble-gas electron configuration ($ns^2 np^6$). This 8-electron configuration is especially stable because with 8 valence electrons, the s- and p-orbitals are completely filled. In simple terms, molecules or ions tend to be most stable when the outermost electron shells of their constituent atoms contain eight electrons. To achieve this state, reaction of atoms occurs primarily in two ways, ionically and covalently.

The octet rule is commonly used in drawing Lewis structures. In terms of Lewis dot symbols, an octet can be thought of as four pairs of valence electrons arranged around the atom. For example, the bonding in carbon dioxide (CO_2) shows that all atoms are surrounded by 8 electrons ($\ddot{\text{O}}::\text{C}::\ddot{\text{O}}$). CO_2 is thus a stable molecule.

The formation of NaCl also illustrates the octet rule. Sodium (Na) with an electron configuration of $1s^2 2s^2 2p^6 3s^1$ sheds its outermost 3s electron and the Na^+ ion has an electron configuration of $1s^2 2s^2 2p^6$ matching the electron configuration of neon. Chlorine (Cl) with an electron configuration of $1s^2 2s^2 2p^6 3s^2 3p^5$ takes on the electron shed by sodium to fill its outermost third shell with eight electrons. Then the Cl^- ion's electron configuration becomes $1s^2 2s^2 2p^6 3s^2 3p^6$ matching the configuration of argon.

$$\text{Na}([\text{Ne}]3s^1) + \text{Cl}([\text{Ne}]3s^2 3p^5) \rightarrow \text{Na}^+([\text{Ne}]) + \text{Cl}^-([\text{Ar}]) \rightarrow \text{NaCl} \quad (3.5)$$

Sample 3.3 Apply the octet rule to the elements below to determine how many electrons each of these atoms would like to gain or lose.

C, O, Na, P, S, Cl, Ca

Solution:

According to the octet rule

Elements	Electrons to be gained or shed	Elements	Electrons to be gained or shed
C	4	S	2
O	2	Cl	1
Na	−1①	Ca	−2
P	3		

① "−" means the number of electrons to be shed.

Sample 3.4 Illustrate the combination of F and Cl giving FCl where both atoms have a complete octet.

Solution:

(seven valence electrons) $:\ddot{\text{F}}\cdot$ $\cdot\ddot{\text{Cl}}:$ (seven valence electrons)

(eight valence electrons) $:\ddot{\text{F}}:\ddot{\text{Cl}}:$ (eight valence electrons)
↑
shared electrons forming a covalent bond

The octet rule is applicable to the main-group elements, especially carbon, nitrogen, oxygen, and the halogens, but also the metals in group I A and group II A, such as sodium or magnesium.

Although the octet rule is a very useful rule, there are many exceptions to it. Hydrogen

(H) and helium (He) don't follow the octet rule, but rather the "duet" rule (2 electrons) because there is no 1p subshell, thus shell 1 can only have at most 2 valence electrons. Lithium needs to lose one electron to attain this stable configuration. In boron compounds, boron often only has 6 electrons in the valence shell, called as electron deficiency. This is because the atom has fewer than eight electrons, and has no unpaired electrons to make more bonds. Atoms with 3 or more electron shells can accommodate more than eight electrons in their outer shell, for example, S in SF_6 and P in PCl_5. And in general, the octet rule is not applicable to the transition elements. These elements seek additional stability by having half-filled or filled d or f subshell orbitals.

Although the octet rule does not accurately predict the electron configurations of all molecules and compounds, it provides a useful framework for introducing many important concepts of bonding. It is the guiding principle for much of what we discuss in this chapter.

Writing Lewis Structure

We can write the Lewis structure of a molecule by the following steps.

Step 1 Choose a central atom. Generally the central atom has lower electronegativity than bonded atoms.

Step 2 Count electrons and write initial Lewis structure.

Note that non-valence electrons are not represented in Lewis structures. The total number of electrons represented in a Lewis structure is equal to the sum of the numbers of valence electrons on each individual atom.

For polyatomic ions, if there is a negative charge, add one electron to the total for every negative charge. If there is a positive charge indicated on the molecular formula, subtract one electron for every positive charge indicated. Then divide the total number of available electrons by 2 to obtain the number of electron pairs available.

Once the total number of available electrons has been determined, electrons must be placed into the structure. The initial Lewis structure should be drawn with each bonded atom connected by a single bond to the central atom, although not all atoms have to be connected to show bonds.

Step 3 Place remaining electrons as lone pairs.

Elements C, N, O and F should always be assumed to obey the octet rule.

Lone pairs should initially be placed on bonded atoms (other than hydrogen) until each bonded atom has eight electrons in bonding pairs and lone pairs; extra lone pairs may then be placed on the central atom. When in doubt, lone pairs should be placed on more electronegative atoms first.

Once all lone pairs are placed, atoms, especially the central atoms, may not have an octet of electrons. In this case, the atoms must form a double bond. a lone pair is moved to form a second bond between the two atoms. As a result all atoms can satisfy the octet rule.

For example, the correct Lewis structure for CO_2 is $\overset{..}{\underset{..}{O}}=C=\overset{..}{\underset{..}{O}}$, not $:\overset{..}{\underset{..}{O}}-C-\overset{..}{\underset{..}{O}}:$.

Sample 3.5 Write the Lewis structure for each using lines to represent shared electron pairs: (1) hydrogen fluoride, HF; (2) ammonia, NH_3; (3) nitrogen, N_2; (4) fluorine, F_2; (5) hydrogen sulphide, H_2S; (6) methane, CH_4.

Solution:

(1) HF: valence electrons: $1+7=8$ (4 pairs) initial Lewis structure: H—F

Lone pairs: 3   final Lewis structure: H—$\ddot{\underset{..}{F}}$:

(2) NH_3: valence electrons: 5+3=8 (4 pairs) initial Lewis structure: H—N—H, H below N

Lone pairs: 1   final Lewis structure: H—$\ddot{N}$—H, H below N

(3) N_2: valence electrons: 5+5=10 (5 pairs) initial Lewis structure: N—N
To satisfy the octet rule for N atom, final Lewis structure should be: :N≡N:

(4) F_2: valence electrons: 7+7=14 (7 pairs) initial Lewis structure: F—F

Lone pairs: 6   final Lewis structure: :$\ddot{\underset{..}{F}}$—$\ddot{\underset{..}{F}}$:

(5) H_2S: valence electrons: 2+6=8 (4 pairs) initial Lewis structure: H—S—H

Lone pairs: 2   final Lewis structure: H—$\ddot{\underset{..}{S}}$—H

(6) CH_4: valence electrons: 4+4=8 (4 pairs) initial Lewis structure: H—C—H with H above and below

Final Lewis structure: H—C—H with H above and below

The formation of these molecules satisfies the octet rule.

Exceptions to the Octet Rule

Only a minority of compounds have an octet of electrons in their valence shell. Incomplete octets are common for compounds of groups II A and III A such as beryllium, boron, and aluminum. Compounds with more than eight electrons in the Lewis representation of the valence shell of an atom are common for elements of groups VI A to 0, such as phosphorus, sulfur, iodine, and xenon. So when writing the Lewis structure, if electrons remain after the octet rule has been satisfied, place them on the elements in the third period or beyond.

Sample 3.6 Write the Lewis structure of BF_3 and SF_6.
Solution:
(1) BF_3:
Sum the valence electron for BF_3: 3+7×3=24 (12 pairs, B as the central atom)
Initial Lewis structure:

F   F
 \ /
  B
  |
  F

Then place remaining electrons (24−6=18, 9 pairs). Note that F always obeys the octet rule. So the Lewis structure for BF_3 is

:$\ddot{\underset{..}{F}}$:   :$\ddot{\underset{..}{F}}$:
  \ /
   B
   |
  :$\ddot{\underset{..}{F}}$:

Boron has only six electrons around it and cannot satisfy the octet rule.
(2) SF_6

Sum the valence electron for SF_6: $6+7\times 6=48$ (24 pairs, F as the central atom)

Initial Lewis structure:

$$\begin{array}{c} F\ \ F\ \ F \\ \diagdown | \diagup \\ S \\ \diagup | \diagdown \\ F\ \ F\ \ F \end{array}$$

Then place remaining electrons ($48-12=36$, 18 pairs). Note that F always obeys the octet rule. So the Lewis structure for SF_6 is

$$\begin{array}{c} :\ddot{\underset{..}{F}}:\ :\ddot{\underset{..}{F}}:\ :\ddot{\underset{..}{F}}: \\ \diagdown | \diagup \\ S \\ \diagup | \diagdown \\ :\ddot{\underset{..}{F}}:\ :\ddot{\underset{..}{F}}:\ :\ddot{\underset{..}{F}}: \end{array}$$

Sulfur has 12 valence electrons around it and exceeds the octet rule.

However, when it is necessary to exceed the octet rule for one of several third-period (or beyond) elements, assume that the extra electrons are placed on the central atom. For example, the triiodide ion, I_3^-. Its Lewis structure is $[\ :\ddot{\underset{..}{I}}\text{-}\ .\ \ \ddot{\underset{..}{I}}\text{-}\ .\ \ \ddot{\underset{..}{I}}:\ ]^-$. There are 10 electrons around the central I.

Resonance

For some molecules and ions, multiple atoms of the same type surround the central atom. It is difficult to determine which lone pairs should be moved to form double or triple bonds. This is especially common for polyatomic ions. The nitrate ion (NO_3^-), for instance, must form a double bond between nitrogen and one of the oxygens to satisfy the octet rule for nitrogen. However, because the molecule is symmetrical, it does not matter which of the oxygens forms the double bond. When this situation occurs, the molecule's Lewis structure is said to be a resonance structure. In the case of NO_3^-, there are three possible resonance structures. Expressing resonance when drawing Lewis structures may be done by drawing each of the possible resonance forms and placing double-headed arrows between them.

Sample 3.7 Write the Lewis structure of the nitrite ion, NO_2^-. Experiments show that the two N-O bonds are equivalent.

Solution:

Step 1 Nitrogen is the least electronegative atom, so it is the central atom.

Step 2 Count valence electrons, $(6\times 2)+5=17$. The ion has a charge of -1, so the total number of electrons is 18.

Step 3 Place electron pairs and write the initial Lewis structure

$$O-N-O$$

14 electrons remain that should initially be placed as 7 lone pairs.

Step 4 Satisfy the octet rule for all atoms and write the Lewis structure.

$$\left[\ :\ddot{\underset{..}{O}}\diagup^{\displaystyle N}\diagdown\ddot{\underset{..}{O}}:\ \right]^-$$

Since experiments show that the two N-O bonds are equivalent, we therefore must have a resonance structure.

Step 5 Write the resonance structure.

Two Lewis structures must be drawn, one with each oxygen atom double-bonded to the nitrogen atom. The second oxygen atom in each structure will be single-bonded to the nitrogen atom. Draw a double-headed arrow between the two resonance forms.

$$\left[\overset{\cdot\cdot}{\underset{\cdot\cdot}{O}}\overset{N}{\diagdown}\overset{\cdot\cdot}{\underset{\cdot\cdot}{O}}\cdot\right]^{-} \longleftrightarrow \left[\cdot\overset{\cdot\cdot}{\underset{\cdot\cdot}{O}}\overset{N}{\diagdown}\overset{\cdot\cdot}{\underset{\cdot\cdot}{O}}\right]^{-}$$

In Lewis resonance structures, the structure is written such that it appears the molecule may switch between multiple forms. However, the molecule itself exists as a hybrid of the forms. For example, in the case of the nitrite ion, there are one single bonds and one double bond in each resonance form. But experiments show that two N—O bonds are equivalent, the length and bond energy of each is somewhere between that of a single bond and a double bond. The resonance structure should not be interpreted to indicate that the molecule switches between forms, but that the molecule acts as the average of multiple forms.

Lewis structures for molecules and polyatomic ions are very useful. But Lewis structures do not concern the geometries of molecules, nor do they explain how and why atoms share electrons. Later we will learn a very simple method for predicting geometry, which gives the correct geometry for most molecules. Then, two theories that explain how and why covalent bonds form will be discussed.

3.4 Sigma and Pi Bonds

Sigma bonds

Bonds formed by endwise (head-on) overlap of orbitals are call sigma (σ) bonds which are symmetrical about the internuclear axis. This means that a σ-bond is symmetrical with respect to rotation about the bond axis. If you cut the bond in half, the cross section is a circle with the line between the centers of the two nuclei in its center, as shown in Figure 3.2. In this diagram, there are quite a few examples of sigma bonds. It does not matter what shapes the orbitals have or what types they are. They can be s orbitals or p orbitals or hybrid orbitals which will be mentioned in Section 3.9.

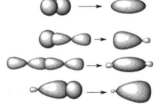

Figure 3.2 Formation of sigma bonds by endwise overlap of atomic orbitals

In chemistry, σ bonds are the strongest type of covalent bond. Single bonds are always sigma bonds. The electrons in these bonds are sometimes referred to as sigma electrons.

Pi bonds

Bonds formed by sidewise overlap are called pi (π) bonds where two lobes of one involved electron orbital overlap two lobes of the other involved electron orbital. Contrast to σ bond, a π bond is not symmetrical about the internuclear axis. In a π bond (Figure 3.3), the electron density is in two separate regions resulting from two areas of overlap of atomic orbitals, one above and the other below the plane of the molecule. Both parts make one π bond.

Electrons in π bonds are sometimes referred to as π electrons. The overlap between the component p-orbitals is significantly less due to their parallel orientation. So the electron density of π bonds is farther from the positive charge of the atomic nucleus, which requires more energy. This indicates that π bonds are usually weaker than sigma bonds.

A double bond consists of one sigma bond and one π bond. The combination of π and σ bond is stronger than either bond by itself. The enhanced strength of a multiple bond *vs.* a

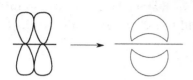

Figure 3.3 Two p orbitals forming a π bond

single (sigma bond) results a contraction in bond lengths. For example, carbon-carbon bond lengths are ethane (154 pm), ethylene (133 pm) and acetylene (120 pm).

Rotation around single bonds is free. Rotation around double bonds is not free because double bonds consists one σ bond and one π bond. A π bond cannot rotate because rotation involves destroying the parallel orientation of the constituent p orbitals and the π bond would be broken. Free rotation about single bonds and lack of free rotation about double bonds are important factors influencing the properties of many molecules, including physiologically important molecules like proteins.

Sample 3.8 How many sigma bonds and how many π bonds does the molecule ethene, C_2H_4, contain?

Solution:

From the Lewis structure for this compound shown below, it can be seen that there are one double bond (between the two carbons) and four single bonds. Each single bond (C—H) is a sigma bond, and the double bond (C=C) is made up of one sigma bond and one pi bond, so there are five sigma bonds and one pi bond in the molecule ethene.

$$\begin{array}{c} H \\ \diagdown \\ C = C \\ \diagup \diagdown \\ H H \end{array}$$

In a triple bond, the first bond to form is a single, sigma bond and the next two to form are both pi, such as the formation of N_2 molecule (Figure 3.4).

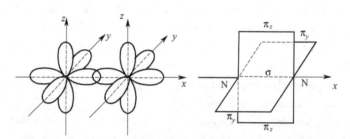

Figure 3.4 The formation of N_2 molecule (triple bond)

3.5 Bond Polarity

A covalent bond, as we have said, is the sharing of an electron pair by two atoms. Two atoms with the same electronegativity will share the bonding electron pairs equally, such as H_2 molecule. The atoms in H_2 are identical, so the electrons spend the same amount of time in the vicinity of each atom, that is, the bonding electrons are evenly distributed between the bonded atoms. Bonds in which electron pairs are shared equally are called nonpolar bonds. Both the bond between two oxygens as in O_2 and that between two nitrogens like N_2 are nonpolar bonds.

However, in the covalently bonded HCl molecule, the electronegativity of Cl atoms is higher than that of H atoms. So Cl atom will attract the bonding electrons closer to itself. The bonding pairs of electrons will spend more time in the vicinity of Cl atom than H atoms. Such bond is called a polar bond, because the electron pair is not shared equally. As a result in HCl molecule the chlorine end will be partially negative (symbol δ^-) since the electrons are closer to the chlorine (Figure 3.5). The hydrogen end will be partially positive (symbol δ^+) since the bonding pair is farther from the hydrogen. This two pole condition is called a dipole and it generates a dipole moment that is a vector force directed toward the higher electronegative atom in the bond. The greater the difference in the electronegativity between the two bonded atoms, the more polar the bonds are.

Figure 3.5 The dipole moment in HCl molecule

The polar covalent bonds can be thought of as being intermediate between a completely nonpolar covalent bond, in which the sharing of electrons is exactly equal, and an ionic bond, in which the transfer of the electron(s) is nearly complete and the bond can be thought as completely polar. We can take the difference between the electronegativity of the atoms to determine the polarity of a covalent bond. If the result is between 0.4 and 1.7, generally, the bond is polar covalent.

3.6 Bond Energy, Bond Length and Bond Angle

Bond Energy

Bond energy (E), a measure of bond strength in a chemical bond, is the amount of energy it takes to break that bond, which is exactly the same as the amount of energy released when the bond is formed. The bond energy is essentially the average enthalpy change for a gas reaction to break all the similar bonds. Take methane as an example, 1662 kJ is required to break all four C—H bonds for a mole of methane, and 435 kJ is required to break a single C—H bond for a mole. So the average C—H bond energy in methane, $E_{C—H}$, is 1662/4 (=416) kJ/mol, not 435 kJ/mol. The latter is called as bond dissociation energy. Bond energy (E) should not be confused with bond dissociation energy. Another example, an O—H bond of a water molecule (H—O—H) has 493.4 kJ/mol of bond dissociation energy, and 424.4 kJ/mol is needed to cleave the remaining O—H bond. But the bond energy of the O—H bonds in water is 458.9 kJ/mol, which is the average of the values.

The magnitude of the bond energy for a particular bond is determined by several contributing factors. But usually the most important is the difference in the electronegativity of the two atoms bonding together. The energies of bonds between atoms of substantially different electronegativities tend to be high, e.g., the 110 kcal of the O—H. The larger the bond energy, the more energy is needed to break the bond and the stronger the bond. Thus bonds between atoms of differing electronegativities are apt to be very strong and stable.

On the other hand, the energies of bonds between atoms of similar electronegativity tend to be smaller. For example, in O—O and C—C, there is no electronegativity difference between the atoms and their bond energy values are relatively low. The bond energy of the C—C bond is 80 kcal and that of C—H bond is 98 kcal because there is a slight difference in electronegativity between carbon and hydrogen. The bonds with smaller bond energies are easier to break. Thus these bonds are weaker and less stable.

Bond Length

In molecular geometry, bond length is the average distance between nuclei of two covalently bonded atoms in a molecule. Bond lengths are determined in molecules by X-ray diffraction of solids, by electron diffraction, and by spectroscopic methods. The bond lengths range from the shortest of 74 pm for H—H to some 200 pm for large atoms.

For covalent bonds, bond energies and bond lengths depend on many factors, electron affinities, sizes of atoms involved in the bond, differences in their electronegativity, and the overall structure of the molecule. For a given pair of atoms, triple bonds are shorter than double bonds, which, in turn, are shorter than single bonds. For example, the length of C≡C is shorter than that of C=C. The shorter multiple bonds are also more stable than single bonds. Bond length is also inversely related to bond strength. There is a general trend in that the shorter the bond length, the higher the bond energy.

Although many factors affect bond lengths, in general bond lengths are very consistent. Bond lengths of the same bond order (see Section 3.10) for the same pair of atoms in various molecules are remarkably consistent. Since bond lengths are consistent, bond energies of similar bonds are also consistent. Table 3.2 shows some selected bond lengths in picometers and bond energies in kJ/mol.

Table 3.2 Some selected bond length and bond energy

Bond	Length/pm	Energy /(kJ/mol)	Bond	Length/pm	Energy/(kJ/mol)
H—H	74	436	H—C	109	413
C—C	154	348	H—N	101	391
N—N	145	170	H—O	96	366
O—O	148	145	H—F	92	568
F—F	142	158	H—Cl	127	432
Cl—Cl	199	243	C—O	143	360
C=C	134	614	O=O	121	498
C≡C	120	839	N≡N	110	945

Bond Angle

The angle between two bonds sharing a common atom is known as the bond angle, that is, the angle formed by three atoms bonded together. The geometry and size of a molecule can be characterized by bond angles and lengths, for example, the shape of H_2O molecule (see Figure 3.6). And therefore they affect many physical properties of a substance. The angle between two bonds sharing a common atom through a third bond is known as the torsional or dihedral angle. As shown in Figure 3.7, the torsional angle for H_2O_2 molecule is the angle between the plane formed by the first three atoms and the plane formed by the last three atoms. Bond angles depend on the identity of the atoms bonded together and the type of chemical bonding involved. They vary considerably from molecule to molecule.

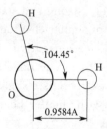

Figure 3.6 The shape of H_2O molecule

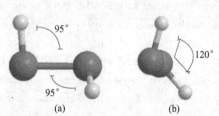

Figure 3.7 Bond angle versus torsional angle for H_2O_2
(a) Two O—O—H bond angles in hydrogen peroxide;
(b) The same molecule when viewed along the
O—O bond, showing the torsional
angle (the angle between the two O—H bonds)

3.7 Molecular Geometry

The geometry of a molecule or polyatomic ion is one of its most important characteristics. Molecular geometry is the three-dimensional arrangement of the atoms that constitute a molecule, inferred from the spectroscopic studies of the compound. It determines several properties of a substance including its reactivity, polarity, phase of matter, color, magnetism, and biological activity. In the solid state the molecular geometry can be measured by X-ray crystallography.

Molecular geometry can be described by the positions of atoms in space, determined by bond lengths of two joined atoms, bond angles of three connected atoms, and torsional angles of three consecutive bonds. Table 3.3 lists six basic shape types for molecules.

Table 3.3 Six basic geometry types for molecules

No.	Molecule Geometries	Number of Bonds on Central Atom	Bond Angles	Examples
1	Linear shape	2 (two bonding pairs of electrons or two double bond pairs)	180°	BeH_2, CO_2, $Ag(NH_3)_2^+$
2	Trigonal planar shape	3	120°	BH_3, BF_3
3	Tetrahedral shape	4	109.5°	CH_4, NH_4^+
4	Octahedral shape	6	90°	SF_6, $[PCl_6]^-$
5	Trigonal pyramid shape	three bond pairs and one lone pair	107.3°	NH_3
6	Non-linear/bent/angular shape	two bond pairs and two lone pairs	approximately 109° (104.5° for H_2O)	H_2S, FO_2, NH_2^-, H_2O

Molecular geometry is determined by the quantum mechanical behaviour of the electrons. As we will see later, molecular geometry can be understood by the valence bond approximation (see Section 3.8, the VSEPR model), the orbital hybridization process (see Section 3.9), and molecular orbital theory (see Section 3.10).

3.8 The VSEPR Model

While Lewis dot structures can tell us how the atoms in molecules are bonded to each other, they don't tell us the geometry of the molecule. The geometries of molecules and polyatomic ions can be observed experimentally. However, it is convenient to predict geometry qualitatively through the valence shell electron pair repulsion model, which is often abbreviated by the first letters, VSEPR. It is a very simple way of predicting the shapes of species that have main group elements as central atoms. According to the VSEPR model, pairs of electrons repel each other because of their negative charges. Therefore, electron pairs will be as far apart from each other in three-dimensional space as possible. So the basic principles of the VSPER model is that the molecule will assume the shape that most minimizes electron pair repulsions. In attempting to achieve that, two types of electron sets must be considered, electrons can exist in bond pairs, or lone pairs.

We define the total coordination number of an atom as the total number of atoms and sets of lone pairs around a central atom. As far as the geometry predicted by VSEPR is concerned, multiple bonds (double bonds or triple bonds) are similar to single bonds in their effect on geometry. This seems reasonable because the electrons of a multiple bond must be located between the atoms joined by the bond. So the total coordination number of the central atom is also equal to the total number of single bonds, multiple bonds, and lone pairs. According to the VSEPR model, the electron-domain geometry can be deduced, as shown in Table 3.4. The electron-domain geometry represents a single atom with all of the electrons that would be associated with it as a result of the bonds it forms with other atoms plus its lone pairs. However, since atoms in a molecule can never be considered alone, the geometry of the actual molecule might be different from what you'd predict based on its electron-domain geometry.

Using VSEPR Model to Predict Geometry

Use the electron-domain geometry to determine the molecular geometry by following steps.

(i) According to the Lewis structure of the molecule, find the total coordination number by counting the atoms and lone pairs that are attached to the central atom. For the VSEPR model multiple bonds count as one effective electron pair.

(ii) Determine the electron-domain geometry for the molecule by arranging the electron pairs so that they are as far from each other as possible in three-dimensional space and the repulsions are minimized (based on Table 3.4).

(iii) Remembering that lone pairs occupy more space than shared pairs, place the atoms around the central atom.

(iv) The positions of the atoms determine the geometry of the molecule or polyatomic ion.

Table 3.4 below summarizes all of the commonly occurring molecular geometries predicted by VSEPR. The geometry of the molecule is determined by the positions of the atoms that are bonded to the central atom. It can be seen that the positions of the atoms that are bonded to the central atom depend both on total coordination number and on the number of lone pairs on the central atom.

Table 3.4 All of the commonly occurring molecular geometries predicted by VSEPR

Total Coordination Number of the Central Atom	Electron-Domain Geometry	Outer Atoms	Bonding pairs	Lone pairs	Molecular Geometry	Ideal Bond Angle	Example
2	Linear 180°	2	2	0	Linear 180°	180°	$BeCl_2$, CO_2
3	Trigonal Planar 120°	3	3	0	Trigonal Planar 120°	120°	BF_3
		2	2	1	Bent 120°	120°	SO_2, NO_2^-

Table 3.4 (Continued)

Total Coordination Number of the Central Atom	Electron-Domain Geometry	Outer Atoms	Bonding pairs	Lone pairs	Molecular Geometry	Ideal Bond Angle	Example
4	Tetrahedral (109.5°)	4	4	0	Tetrahedral (109.5°)	109.5°	CH_4
		3	3	1	Trigonal Pyramidal (107.5°)	107.5°	NH_3
		2	2	2	Bent (145°)	104.5°	H_2O
5	Trigonal Bipyramidal (90°, 120°)	5	5	0	Trigonal Bipyramidal (90°, 120°)	90°, 120°	PCl_5
		4	4	1	Seesaw (90°, 120°)	90°, 120°	SF_4
		3	3	2	T-Shaped (90°)	90°	ClF_3
		2	2	3	Linear (180°)	180°	XeF_2

Table 3.4 (Continued)

Total Coordination Number of the Central Atom	Electron-Domain Geometry	Outer Atoms	Bonding pairs	Lone pairs	Molecular Geometry	Ideal Bond Angle	Example
6	Octahedral	6	6	0	Octahedral	90°	SF_6
		5	5	1	Square Pyramidal	90°	BrF_5
		4	4	2	Square Planar	90°	XeF_4

From the above table, it can be seen that if no lone pairs are attached to the central atom, the molecules or ions have the geometry as the electron-domain geometry. For instance, methane, CH_4, with four bonding pairs and have no lone pairs, is tetrahedral. If some of the electron pairs are lone pairs instead of bonding pairs, what will happen? In this case, the molecular geometry is determined by the position of atoms because when we experimentally look at molecules, we see how the atoms are arranged but don't see the lone electron pairs. For example, an ammonia molecule has one lone pair and its molecular geometry is trigonal pyramidal.

Experiments show that an ammonia molecule is trigonal pyramidal and the angle between the H—N bonds in ammonia is 107.3°. A water molecule is bent or angular with an angle between the H—O bonds of 104.5°. How does valence shell electron pair repulsion account for the geometry of the ammonia and water molecules?

According to the VSEPR model, it is assumed that lone pairs have more repulsive force than do bonding airs, and they repel bonding pairs more than bonding pairs repel each other. Thus lone pairs force the bonding pairs to squeeze more closely together. So the angle between N—H bonds in NH_3 is smaller than the angles between C—H bonds in CH_4. The angle between H—O bonds in water (104.5°) is even smaller than the angles between H—N bonds in ammonia because repulsion between lone pairs is greater than repulsion between a bond pair and a lone pair. Two lone pairs push the remaining bonding pairs even closer to each other than one lone pair does. So based on the VSEPR model, lone pairs are always placed where they will have the most space in the trigonal plane. For example, in XeF_4, there are two lone pairs and four bonding pairs surrounding Xe. If the lone pairs are 180° apart, which minimizes their repulsion, the molecule will be more stable. So in XeF_4, 4 bonding pairs surround Xe and the molecular geometry is square planar.

Sample 3.9 Deduce the geometry of SF_4 and NF_3.

Solution:

Molecules	Total Coordination Number of the Central Atom	Outer Atoms	Bonding Pairs	Lone Pairs	Molecular Geometry
SF_4	5	4	4	1	Seesaw
NF_3	4	3	3	1	Trigonal Pyramidal

Although the VSEPR model is very simple, its predictions are usually quite accurate. Naturally, there exist exceptions, and VSEPR occasionally predicts the wrong geometry for a molecule. We should use quantum mechanics to understand chemical bonding and electronic structure. Next, two theories about molecule structure will be discussed.

3.9 Hybrid Orbitals

The hybridization theory was promoted by chemist Linus Pauling in order to explain the structure of molecules such as methane (CH_4). Historically, this concept was developed for such simple chemical systems but the approach was later applied more widely. Today it is considered as an effective heuristic for rationalizing the structures of organic compounds.

Consider the CH_4 molecule, the central carbon atom has the ground-state electron configuration $1s^2\ 2s^2\ 2p_x^1\ 2p_y^1$. Because the carbon atom has two unpaired electrons, the valence bond theory would predict that C forms two covalent bonds, i.e. CH_2 (methylene). However, methylene is a very reactive molecule and cannot exist outside of a molecular system. And in fact, carbon almost always forms four bonds in compounds such as methane (CH_4) and tetrachloromethane (CCl_4) with tetrahedral shape. Therefore, this theory alone cannot explain the existence of these molecules. Furthermore, exciting a 2s electron into a 2p orbital would theoretically allow for four bonds. But this would indicate that the bonds in CH_4 would have differing energies due to differing levels of orbital overlap. This has been experimentally disproved, the four bonds in CH_4 and CCl_4 are identical. So we need a theory to fit the facts. This theory is called hybridization theory.

In chemistry, hybridization or hybridisation is the concept of mixing atomic orbitals to form new hybrid orbitals suitable for the qualitative description of atomic bonding properties. Hybrid orbitals are models used to account for molecular shapes that are observed by experiment. Hybrid orbitals are very useful in the explanation of geometry for molecules made up from carbon, nitrogen and oxygen (and to a lesser extent, sulfur and phosphorus). It is an integral part of valence bond theory. Its explanation starts with the way bonding is organized in methane.

sp^3 Hybrid Orbitals

Take methane (CH_4) molecule as example. In this case, carbon attempts to bond with four hydrogens, four identical orbitals are required. The hybridization process is illustrated in Figure 3.8(a). First promote an electron from the 2s orbital to the 2p orbital. Then imagine that the 2s orbital combines with the three 2p orbitals to form new orbitals. This procedure is called hybridization. These new orbitals are called hybrid orbitals pictured as resulting from combinations of atomic orbitals on the same atom. Because these hybrid orbitals resulted from the combination of an s orbital and three p orbitals, they are called sp^3

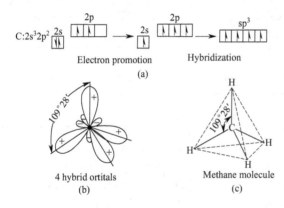

Figure 3.8　The formation of CH_4 molecule with sp^3 hybridization

hybrid orbitals or simply sp^3 orbitals. The number of hybrid orbitals formed must be the same as the number of atomic orbitals combined. So there are four sp^3 orbitals because four orbitals were combined. Each of the four hybrid orbitals is equivalent to the others. Therefore, each sp^3 orbital has one-quarter s character and three-quarters p character. And the energies of the sp^3 orbitals are lower than the energies of the p orbitals but slightly higher than the energy of the s orbital. Notice that no change occurred with the 1s orbital.

Each sp^3 orbital has the shape shown in Figure 3.9. Like a p orbital, it has two lobes, a large lobe and a small lobe. But unlike a p orbital, one lobe is much larger than the other and the electron will spend almost all of its time in the larger portion of the orbital. The large lobes are the ones that are important for forming bonds because they are more directed in space than unhybridized p or s orbitals and overlap better with other orbitals, forming stronger bonds. In the simplified diagram usually used for showing the formation of molecules, the small lobe is omitted, only the larger lobes are shown.

After hybridization, the four electrons of the carbon atom are placed in the four degenerate sp^3 orbitals according to Hund's rule. Thus, the sp^3 hybridized carbon atom has four identical unpaired electrons to use for bond formation. Figure 3.8(b) shows the orbitals which point out from the center of the atom in the same direction as the corners of a tetrahedron and the large lobes of the four sp^3 orbitals are pictured as being at 109.5 angles to each other. Four sp^3

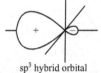

Figure 3.9　The shape of sp^3 orbital

hybridized orbitals are overlapped by hydrogen's 1s orbital [Figure 3.8(c)], yielding four s bonds with the same length and strength. So the angles between the C—H bonds in CH_4 are 109.5°.

sp^2 Hybrid Orbitals

In the boron trifluoride (BF_3) molecule with planar geometry, only three electron pairs are arranged around the central boron atom. So we only need three hybrid orbitals. In this case, the 2s orbital is combined with only two of the 2p orbitals forming three hybrid orbitals. These hybridized orbitals are called sp^2 orbitals because they are a combination of one s and two p orbitals. Each sp^2 orbital looks about like an sp^3 orbital. To minimize the repulsion of these electrons, the large lobes of the three sp^2 orbitals lie in a trigonal plane with bond angles of 120°. The other p-orbital remains unhybridized and is at right angles to the trigonal planar arrangement of the hybrid orbitals.

Figure 3.10 shows the formation of BF_3 molecule by sp^2 hybridization. The valence electrons arrangement of B is $2s^2 2p^1$. First, we promote a 2s electron to an empty 2p orbital. Mixing the 2s orbital with the two 2p orbitals generates three sp^2 hybrid orbitals that lie in the same plane. The angle between any two of them is 120°. Each of the B-F bonds is formed by the overlap of a boron sp^2 hybrid orbital and a fluorine 2p orbital. The BF_3 molecule is planar with all the F-B-F angles equal to 120°. This result conforms to experimental findings.

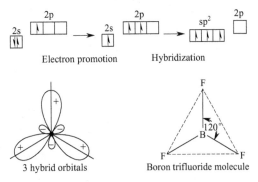

Figure 3.10 The formation of BF₃ molecule by sp² hybridization

The concept of hybridization is also useful for molecules with double and triple bonds. Ethylene, $H_2C\!\!=\!\!CH_2$, is the simplest compound with a carbon-carbon double bond. C_2H_4 has planar geometry. Both the geometry and the bonding can be understood if we assume that each carbon atom is sp² hybridized.

First promote a 2s electron to an empty 2p orbital. Then combining one s and two p orbitals giving three hybrid orbitals. The unchanged p orbital is perpendicular to the plane of the hybrid orbitals. Each of the three hybrid orbitals and the remaining unhybridized p orbital contains one unpaired electron.

The large lobe of one sp² orbital from one carbon atom overlaps the large lobe of one sp²

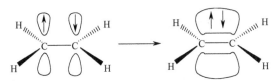

Figure 3.11 The formation of C_2H_4 molecular

orbital from the other carbon atom as shown in Figure 3.11 and form a regular sigma bond. The leftover p orbitals will undergo parallel overlap and form one pi bond perpendicular to the molecular plane. The carbon atoms joined by a double bond with a less bond length of 134 pm than the normal C-C single bond distance of 154 pm. Each carbon atom uses the other two sp² hybrid orbitals to form two sigma bonds with the two hydrogen 1s orbitals. The atoms joined by the double bond and the four atoms attached to them all lie in a plane.

Here are some important structural consequences of a double bond.

(ⅰ) The double bonded carbon atoms and the atoms bonded to them all lie in a flat plane.

(ⅱ) The pi bond sticks out above and below that plane.

(ⅲ) The pi part of a double bond does not allow for rotation around a double bond. You could not twist and turn those two carbon atoms without breaking the pi bond.

sp Hybrid Orbitals

The beryllium chloride ($BeCl_2$) molecule is linear with two identical Be-Cl bonds in every respect. The valence electrons in Be is $2s^2$. Since only two groups are attached to beryllium, we only will have two hybrid orbitals. Figure 3.12 shows the formation of $BeCl_2$ molecule with sp hybridization. First promote a 2s electron to a 2p orbital. Then the 2s and 2p orbitals are hybridized to form two equivalent sp hybrid orbitals. An sp orbital looks very much like an sp³ orbital. The two hybrid orbitals will be arranged as far apart as possible from each other and lie on the same line, so that the angle between them is 180. Each of the Be-Cl bonds is then formed by the overlap of a Be sp hybrid orbital and a Cl 3p orbital, and the resulting $BeCl_2$ molecule has a linear geometry. The two unhybridized p-orbitals stay in their respective positions (at right angles to each other) and perpendicular to the linear molecule.

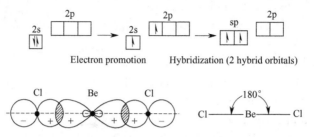

Figure 3.12　The formation of BeCl₂ molecule with sp hybrodization

sp hybridization can explain the shapes observed for compounds with triple bonds. The carbon-carbon triple bond in acetylene, HC≡CH, is an example. Because the acetylene molecule is linear, we can explain its geometry and bonding by assuming that each C atom is sp hybridized. Figure 3.13 shows the formation of acetylene molecule with sp hybridization. Both carbon atoms will be sp hybridized and the two sp orbitals lie on a straight line. Two uncombined p orbitals are left and have one electron in each of them. The two sp orbitals are perpendicular to the plane of the two leftover p orbitals.

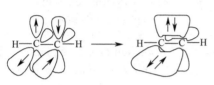

Figure 3.13　The formation of acetylene molecule with sp hybridization

The large lobe of one sp orbital from one C atom overlaps head-on the large lobe of one sp orbital from the other C atom forming one sigma bond. At the same time, the unhybridized p orbitals will undergo parallel overlap to form two pi bonds at right angles to each other. Thus the C≡C bond is made up of one sigma bond and two pi bonds in different planes. The other sp hybrid orbitals of each C atom form one sigma bond with a hydrogen 1s orbital forming a linear molecule.

The following rule helps us predict hybridization in molecules containing multiple bonds. If the central atom forms a double bond, it is sp^2 hybridized. if it forms two double bonds or a triple bond, it is sp hybridized. Note that this rule applies only to atoms of the second-period elements. Atoms of third-period elements and beyond that form multiple bonds present a more complicated picture.

Hybridization in NH₃ and H₂O Molecule

Hybrid orbitals are often used to account for the observed shapes of water and ammonia. Both water and ammonia have a total coordination number of 4, and the VSEPR model predicts that the four electron pairs of water and ammonia will be tetrahedrally arranged around the central atom. So for NH₃ and H₂O molecule the number of the hybrid orbitals must be equal to the numbers of atoms and lone electron pairs surrounding the central atom.

Both the nitrogen in ammonia and the oxygen in water can be pictured as sp^3 hybridized. In the case of ammonia, the three 2p orbitals of the nitrogen atom are combined with the 2s orbital to form four sp^3 hybrid orbitals. The lone pair will occupy a hybrid orbital. The other three hybrid orbital are overlapped by hydrogen's 1s orbital, yielding three σ bonds. Figure 3.14 shows the electron-domain geometry and the molecular geometry of NH₃ molecule. Ammonia exists as a distorted tetrahedron (trigonal pyramidal) rather than a trigonal plane.

In the case of water, the three 2p orbitals of the oxygen atom are combined with the 2s orbital to form four sp³ hybrid orbitals. The two lone pairs occupy two hybrid orbitals. The other two hybrid orbital are overlapped by hydrogen's 1s orbital, yielding two σ bonds. Figure 3.15 shows the electron-domain geometry and the molecular geometry of H₂O molecule. Water exists as a distorted tetrahedron (bent) rather than a linear molecule.

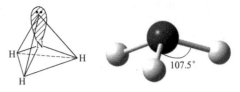

Figure 3.14 The electron-domain geometry and the molecular geometry of NH₃

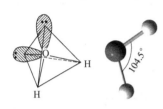

Figure 3.15 The electron-domain geometry and the molecular geometry of H₂O

According to this picture, the bond angles in H₂O and NH₃ molecule are less than the ideal tetrahedral angle of 109.5 because lone pairs occupy more space than bonding pairs and push the hydrogen atoms closer together compared to the angles found in methane. The bond angle in water is even smaller than the bond angle in ammonia because water has two lone pairs. This is improved by experimental evidence that the H—N—H bond angles in ammonia (NH₃) are 107° and the H—O—H bond angles in water are 105°.

Hybridization of s, p, and d Orbitals

As we discussed earlier, some elements in the third period and beyond can accommodate more than eight electrons around the central atom such as PF₅ and SF₆. These atoms will also be hybridized in which five or six hybrid orbitals are needed. To do this, d orbitals must be used because the total number of s and p orbitals is four. After hybridization they will have very specific arrangements of the attached groups (including bonded atoms and lone pairs) in space such as trigonal bipyramidal and octahedral geometries.

To form five hybrid orbitals, one d orbital must be used, a hybrid orbital formed by combination of one s, three p, and one d orbitals is called an sp³d orbital. The groups will be arranged in a trigonal bipyramidal arrangement with sp³d hybridization. Bond angles will be 120° in the plane with two groups arranged vertically above and below this plane. To form six hybrid orbitals, two d orbitals must be used. A hybrid orbital formed by combination of one s, three p, and two d orbitals is called an sp³d² orbital. There will be an octahedral arrangement with sp³d² hybridization. All bond angles are at 90°. Individual sp³d and sp³d² orbitals look about like the sp³ orbital. Figure 3.16 shows models of five sp³d and six sp³d² orbitals.

Figure 3.16 The models of five sp³d and six sp³d² orbitals

Consider the SF₆ molecule. The octahedral shape observed for SF₆ requires that the six sp³d² orbitals be arranged around the central atom in an octahedron. The ground-state electron configuration of S is [Ne] 3s²3p⁴. Because the 3d level is quite close in energy to the 3s and 3p levels, we can promote 3s and 3p electrons to two of the 3d orbitals. Mixing the 3s, three 3p, and two 3d orbitals generates six sp³d² hybrid orbitals.

The six S-F bonds are formed by the overlap of the hybrid orbitals of the S atom and the 2p orbitals of the F atoms. Because there are 12 electrons around the S atom, the octet rule is violated. Second-period elements do not have 2d energy levels, so they can never expand their valence shells. Hence for elements in the second period of the periodic table, eight is the maximum number of electrons that an atom of any of these elements can accommodate in the valence shell. It is for this reason that the octet rule is usually obeyed by the second-period elements.

Procedure for Hybridizing Atomic Orbitals

In essence, hybridization simply extends Lweis theory and the VSEPR model. To assign a suitable state of hybridization to the central atom in a molecule, we use the following steps.

(ⅰ) Get some idea about the geometry of the molecule through VSEPR model or experiment.

(ⅱ) Predict the overall arrangement of the electron pairs (including both bonding pairs and lone pairs) using the VSEPR model.

(ⅲ) Deduce the hybridization of the central atom by matching the arrangement of the electron pairs with those of the hybrid orbitals.

Summarize Hybridization

(ⅰ) The concept of hybridization is not applied to isolated atoms. It is a theoretical model used only to explain covalent bonding.

(ⅱ) Hybridization is the mixing of at least two nonequivalent atomic orbitals, for example, s and p orbitals. Therefore, a hybrid orbital is not a pure atomic orbital. Hybrid orbitals and pure atomic orbitals have very different shapes.

(ⅲ) The number of hybrid orbitals generated is equal to the number of pure atomic orbitals that participate in the hybridization process.

(ⅳ) Hybridization requires an input of energy. however, the system recovers this energy during bond formation.

Information about the five common types of hybrid orbitals is summarized in Table 3.5. This holds if there are no lone electron pairs on the central atom. If there are, the molecular geometry will only consider the arrangement of atoms, and bond angles become smaller due to increased repulsion. For example, as discussed above, in water (H_2O), the oxygen atom has two bonds with H and two lone pairs. Then the shape for water is a bent structure, with a bond angle of 104.5 degrees (the two lone pairs are not visible).

Table 3.5 Hybridization and molecular geometry

Number of Atoms Attached to a Central Atom	Molecular Geometry	Hybridization of the Central Atom	Number of Hybrid Orbitals	Bond Angles	Examples
2	Linear	sp	2	180°	BeF_2, C_2H_2
3	Trigonal planar	sp^2	3	120°	BF_3, C_2H_4
4	Tetrahedral	sp^3	4	109.5°	CH_4
5	Trigonal bipyramidal	sp^3d	5	120°(3); 90°(2)	PF_5
6	Octahedral	sp^3d^2	6	90°	SF_6

3.10 Molecular Orbitals

Valence bond theory is one of the two quantum mechanical approaches that explain bonding in molecules. In this theory, a single electron pair bond between two atoms is resulted from the overlap of atomic orbitals which are centered on the nuclei joined by the bond. It accounts for the stability of the covalent bond and hybridization theory can explain molecular geometries predicted by the VSEPR model. However, in some cases, valence bond theory cannot satisfactorily account for observed properties of molecules. Consider the oxygen molecule, whose Lewis structure is

$$:\ddot{O}::\ddot{O}:$$

It can be seen that all the electrons in O_2 are paired and the molecule should therefore be diamagnetic. But experiments have shown that the oxygen molecule is paramagnetic, with two unpaired electrons. So we need an alternative bonding approach to account for the properties of molecules that do not match the predictions of valence bond theory.

This quantum mechanical theory of chemical bonding distinct from those employed in valence bond theory is called molecular orbital (MO) theory, first introduced by Friedrich Hund and Robert S. Mulliken in 1927 and 1928. Molecular orbital theory describes covalent bonds in terms of molecular orbital (or MO), a region in which an electron may be found in a molecule, which results from interaction of the atomic orbitals of the bonding atoms. The difference between a molecular orbital and an atomic orbital is that an atomic orbital is associated with only one atom, while molecular orbitals are associated with the entire molecule. Molecular orbital must describe the motion of an electron in the field of more than one nucleus, as well as in the average field of the other electrons. Some properties are as follows.

(ⅰ) In molecules the electrons occupy similar molecular orbitals which surround the molecule. A molecular orbital will in general encompass all the nuclei in the molecule, rather than being centered on a single nucleus as in the atomic case.

(ⅱ) Most commonly an MO is represented as a linear combination of atomic orbitals. This Linear Combination of Atomic Orbitals (LCAO) approach for molecular orbitals approximation was introduced in 1929 by Sir John Lennard-Jones.

(ⅲ) The number of molecular orbitals is equal to the number the atomic orbitals included in the linear combination.

(ⅳ) Only atomic orbitals of about the same energy interact to a significant degree forming molecular orbitals.

(ⅴ) When two atomic orbitals combine, they interact in two extreme ways to form two molecular orbitals, a bonding molecular orbital and an antibonding molecular orbital. The more stable the bonding molecular orbital, the less stable the corresponding antibonding molecular orbital.

(ⅵ) Within a particular representation, the symmetry adapted atomic orbitals mix more if their atomic energy level are closer.

(ⅶ) Once the forms and properties of the molecular orbitals are known, the electronic configuration and properties of the molecule are again determined by assigning electrons to the molecular orbitals in the order of increasing energy.

Formation of Molecular Orbitals

The simplest molecule is hydrogen, which is made up of two separate protons and electrons. According to MO theory, the combination of the 1s orbitals of two hydrogen atoms leads to the formation of two molecular orbitals, a bonding molecular orbital and an antibonding molecular orbital.

We have known that electrons in orbitals have wave characteristics. In bonding molecular orbital, wave functions of two atom orbitals are enhanced, which is similar to the enhancement of two in-phase light waves. This enhancement leads to an increase in the intensity of the negative charge between the nuclei and an increase in the attraction between the electron charge and the nuclei of the atoms in the bond. The greater attraction results lower potential energy than that of separate atomic orbitals. So the molecular orbital is more stable than two separated atomic orbitals and keeps the atoms together in the molecule. This molecular orbital is symmetrical with respect to rotation around the molecular axis of the bond (no change) and is called sigma, σ, molecular orbitals. The symbol σ_{1s} is used to describe the bonding molecular orbital formed from two 1s atomic orbitals. Placing an electron in this orbital yields a stable covalent bond, therefore stabilizes the H_2 molecule, so we call this orbital a bonding molecular orbital. Two electrons in a sigma bonding MO form a sigma bond.

In antibonding molecular orbital, wave function is counteracted, which leads to a decrease in negative charge of the nuclei and a decrease in the attraction of the electron charge and the nuclei of the atoms in the bond. The lesser attraction leads to higher potential energy. Electrons placed in this orbital spend most of their time away from the region between the two nuclei. Placing an electron in this orbital makes the molecule less stable and this molecular orbital is also symmetrical about the axis of the bond. So this orbital is therefore an antibonding, or σ^*, molecular orbital. The asterisk indicates an antibonding molecular orbital. The symbol σ_{1s}^* is used to describe the antibonding molecular orbital formed from two 1s atomic orbitals.

Figure 3.17 shows the molecular orbital energy level diagram of the bonding and antibonding molecular orbitals formed from the interaction of two 1s atomic orbitals. Each line in the molecular orbital diagram represents a atomic/molecular orbital. The two atomic orbitals are depicted on the left and on the right. The vertical axis always represents the orbital energies. A bonding molecular orbital (or bonding MO) has lower energy and greater stability than the atomic orbitals which it was formed, while an antibonding molecular orbital (or antibonding MO) has higher energy and lower stability than the atomic orbitals from which it was formed. The electrons in the bond MO are called bonding electrons and those in the antibonding orbital are called antibonding electrons. The reduction in energy of these electrons is the driving force for chemical bond formation.

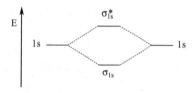

Figure 3.17 The molecular orbital energy level diagram of σ_{1s} and σ_{1s}^* molecular orbitals

The MO theory can be equally applicable to other molecules. When two larger atoms combine to form a diatomic molecule (like O_2, F_2), more atomic orbitals interact. For these molecules, the orbital energies are different enough so only orbitals of the same energy interact to a significant degree.

For all s orbitals, the process is the same as for 1s orbitals. Thus the 2s orbitals on one atom combine with the 2s orbitals on another to form a σ_{2s} bonding MO and a σ_{2s}^* antibonding MO, just like the σ_{1s} and σ_{1s}^* orbitals formed from the 1s atomic orbitals. Both σ_{2s} and σ_{2s}^* molecular orbitals are higher energy and larger than the σ_{1s} and σ_{1s}^* molecular orbitals.

The three p orbitals have equal energy and are oriented mutually perpendicular. The p atomic orbitals of the two atoms can interact with each other in two different ways, endwise or sidewise. The resulting molecular orbitals are different for each type of interaction. As shown in Figure 3.18(a), if we arbitrarily define the X axis as the axis along which the bond forms, the $2p_x$ orbitals on the adjacent atoms will meet head-on to form a σ_{2p_x} bonding and a

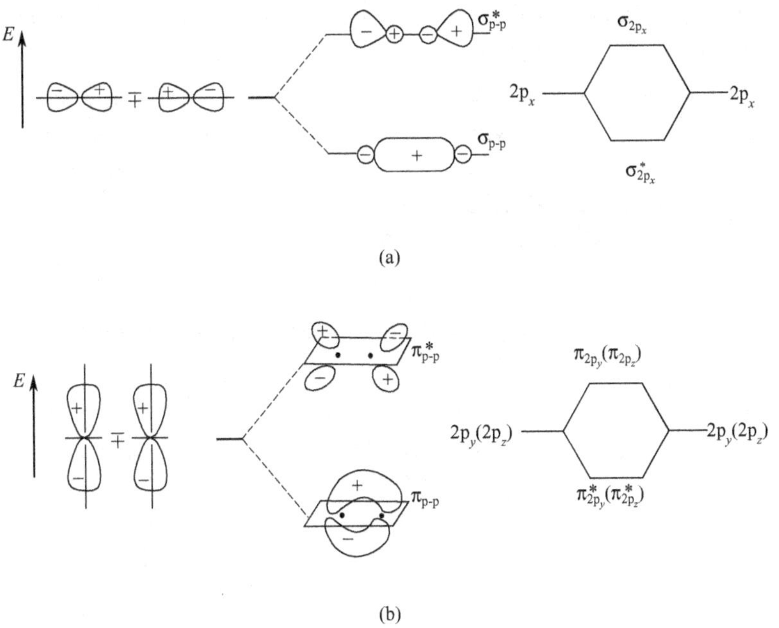

Figure 3.18 The formation of p orbitals forming sigma MO and pi MO

$\sigma_{2p_x}^*$ antibonding molecular orbital, which are symmetrical about the axis of the bond.

The $2p_y$ and $2p_z$ orbitals are perpendicular to the X-axis, and they will overlap sideway to give two pi molecular orbitals. The $2p_z$ orbital on one atom interact with the $2p_z$ orbital on the other atom to form a bonding and an antibonding pi molecular orbital (π_{2p_z} and $\pi_{2p_z}^*$), as shown in Figure 3.18(b). The $2p_y$-$2p_y$ overlap is similar to the $2p_z$-$2p_z$ overlap and generates another pair of molecular orbitals (π_{2p_y} and $\pi_{2p_y}^*$). Because the space arrangement of $2p_y$ and $2p_z$, the π_{2p_z} and π_{2p_y} are perpendicular to each other. This is the same for the $\pi_{2p_z}^*$ and $\pi_{2p_y}^*$. Because there is no difference between the energies of the $2p_x$ and $2p_y$ atomic orbitals, there is no difference between the energies of the π_{2p_z} and π_{2p_y} or the $\pi_{2p_z}^*$ and $\pi_{2p_y}^*$ molecular orbitals. Two electrons in a pi molecular orbital form a pi bond.

There is less overlap for the parallel atomic orbitals forming pi orbitals. This leads to less electron charge between the nuclei for the pi bonding MO than for the sigma bonding MO. As a result the pi bonding MO is less stable, and has higher potential energy compared to the sigma bonding MO.

To summary, if define the X axis as the bond axis, the interaction of four atomic orbitals on one atom (2s, $2p_x$, $2p_y$ and $2p_z$) with a set of four atomic orbitals on another atom leads to the formation of a total of eight molecular orbitals, σ_{2s}, σ_{2s}^*, σ_{2p_x}, $\sigma_{2p_x}^*$, π_{2p_x}, π_{2p_y}, $\pi_{2p_z}^*$, $\pi_{2p_y}^*$.

Molecular Orbitals of Homonuclear Diatomic Molecules of the Second Energy Level

Before write the electron configuration of molecule, we should sort out the relative energies of the eight molecular orbitals. We have known that only atomic orbitals of about the same energy can combine to form molecular orbitals. Both σ_{2s} and σ_{2s}^* molecular orbitals are higher energy than the σ_{1s} and σ_{1s}^* molecular orbitals. Since the energy of 2s orbital is lower than that of 2p orbitals, the σ_{2s} and σ_{2s}^* orbitals both lie at lower energies than the σ_{2p_x},

$\sigma_{2p_x}^*$, π_{2p_z}, π_{2p_y}, $\pi_{2p_z}^*$ and $\pi_{2p_y}^*$ orbitals. Next we would consider the interaction between 2s and 2p orbitals. There are two different cases.

Oxygen and fluorine have very high effective nuclear charges. the difference in energy between the 2s and 2p orbitals in oxygen and fluorine atoms is large, for example, 2500 kJ/mol for F. So for oxygen and fluorine, there is no interaction between the 2s and 2p orbitals. Furthermore, as discussed above, overlap of the two p orbitals is normally greater in a σ MO than in a π MO. As a result, the σ_{2p_x} orbital lies at a lower energy than the π_{2p_z} and π_{2p_y} orbitals, and the $\sigma_{2p_x}^*$ orbital lies at higher energy than the $\pi_{2p_z}^*$ and $\pi_{2p_y}^*$ orbitals, as shown in Figure 3.19(a). For F_2, O_2 and Ne_2, the energies of molecular orbitals actually increase as follows.

For F_2, O_2 and Ne_2: $\sigma_{1s} < \sigma_{1s}^* < \sigma_{2s} < \sigma_{2s}^* < \sigma_{2p_x} < \pi_{2p_y} = \pi_{2p_z} < \pi_{2p_y}^* = \pi_{2p_z}^* < \sigma_{2p_x}^*$

For Li_2, Be_2, B_2, C_2 and N_2: $\sigma_{1s} < \sigma_{1s}^* < \sigma_{2s} < \sigma_{2s}^* < \pi_{2p_y} = \pi_{2p_z} < \sigma_{2p_x} < \pi_{2p_y}^* = \pi_{2p_z}^* < \sigma_{2p_x}^*$

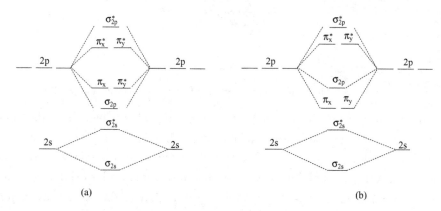

Figure 3.19　The molecular orbital energy level diagram for homonuclear diatomic molecules of the second energy level (ignore the σ_{1s} and σ_{1s}^* orbitals)
(a) For F_2, O_2 and Ne_2; (b) For Li_2, Be_2, B_2, C_2 and N_2

MO treatment for the lighter molecular Li_2, Be_2, B_2, C_2 and N_2 is different from that of O_2 and F_2. Compared with oxygen and fluorine, the elements Li through N have low effective nuclear charges and there is small difference in energy between the 2s and 2p orbitals, for example, only 200 kJ/mol for lithium. As a result, for the elements Li through N, interaction between the 2s and 2p orbitals is significant. This interaction makes the energy of the σ_{2p_x} orbital higher than the energy of the π_{2p_y} and π_{2p_z} orbitals shown in Figure 3.19(b). So for Li_2, Be_2, B_2, C_2 and N_2, the energies of molecular orbitals actually increase as above.

Molecular Electron Configurations

To understand properties of molecules, we must know how electrons are distributed among molecular orbitals, that is, the molecular electron configurations. We can use them to predict whether the molecule is paramagnetic or diamagnetic. If all the electrons are paired, the molecule is diamagnetic. If one or more electrons are unpaired, the molecule is paramagnetic.

The stability of the molecule is described with bond order, which is equal to the number of bonding electrons minus the number of antibonding electrons, divided by 2.

bond order = $\frac{1}{2}$(number of electrons in bonding MOs − number of electrons in antibonding MOs)

Then

(i) A bond order of greater than zero suggests stable covalent bonds and a stable molecule.

(ii) Bond order can be a fraction.

(iii) If the bond order for a molecule is equal to zero (or a negative value), the bond has no stability and the molecule is unstable and cannot exist.

(iv) The higher the bond order is, the more stable the bond is.

1. Rules Governing Molecular Electron Configuration and Stability

Only the ground-state electron configurations are concerned in these cases. The procedure for determining the electron configuration of a molecule is analogous to the one we use to determine the electron configurations of atoms. We use the following procedures when writing the electron orbital configuration of a molecule.

(i) Determine the total number of electrons in the molecule, not just valence electrons. The number of electrons per atom can be obtained from their atomic number on the periodic table.

(ii) Arrange the molecular orbitals in order of increasing energy according to the principle discussed earlier.

(iii) The filling of molecular orbitals with electrons proceeds from low to high energies. As a result, in a stable molecule, the number of electrons in bonding molecular orbitals is always greater than that in antibonding molecular orbitals.

(iv) Like an atomic orbital, each molecular orbital can accommodate a maximum of two electrons with opposite spins in accordance with the Pauli exclusion principle.

(v) When electrons are added to molecular orbitals of the same energy, Hund's rule states that the most stable arrangement is that the electrons fill one MO at a time with parallel spins. So orbitals of equal energy are half filled with parallel spin before they begin to pair up.

(vi) Check the configuration and make sure that the number of electrons in the molecular orbitals is equal to the sum of all the electrons on the bonding atoms.

2. Molecular Electron Configuration for Hydrogen and Helium Molecules

(1) Dihydrogen, H_2

As we discussed earlier, when the 1s atomic orbitals of two hydrogen atoms are combined, two molecular orbitals are formed. The two electrons associated with a pair of hydrogen atoms are placed in the lowest energy, σ_{1s} molecular orbital, as shown in the Figure 3.20. This diagram suggests that both of the two electrons go in the bonding orbital and the energy of an H_2 molecule is lower than that of a pair of isolated atoms. As a result, the H_2 molecule is more stable than a pair of isolated atoms.

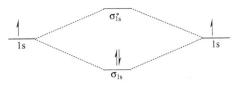

The molecular electron configuration for, H_2, is $(\sigma_{1s})^2$. The bond order for dihydrogen is $(2-0)/2=1$, and there is a single covalent bond between the two hydrogen atoms. Because there are no unpaired electrons, H_2 is diamagnetic.

Figure 3.20 The electron configuration diagram for H_2 molecule

(2) Dihelium, He_2

He_2 is a hypothetical species and this molecular orbital model can be used to explain why He_2 molecules don't exist. Similar to the molecule H_2, in He_2 molecule two molecular orbitals (one bonding MO and one antibonding MO) are formed from the combination of two 1s atomic orbitals. In its neutral ground state, each Helium atom contains two electrons in its 1s orbital, combining for a total of four electrons. The electron configuration of He_2 would be $(\sigma_{1s})^2 (\sigma_{1s}^*)^2$. Two electrons fill the bonding MO, while the remaining two fill the antibonding MO. The total energy of an He_2 molecule would be essentially the same as the energy of a pair of isolated helium atoms, and there would be nothing to hold the helium atoms together to form a molecule. The bond order for He_2 is $(2-2)/2=0$ and no bond exists. He_2 is unstable.

Sample 3.10 Predict the relative stabilities of the species, H_2^+, H_2, He_2^+ and He_2.
Solution:

Molecule	Total Electrons Number in the Molecule	Molecular Electron Configuration	Bond Order	Stability of Molecule
H_2^+	1	$(\sigma_{1s})^1$	$\frac{1}{2}$	Stable
H_2	2	$(\sigma_{1s})^2$	$\frac{1}{2}(2-0)=1$	Stable
He_2^+	3	$(\sigma_{1s})^2(\sigma_{1s}^*)^1$	$\frac{1}{2}(2-1)=\frac{1}{2}$	Stable
He_2	4	$(\sigma_{1s})^2(\sigma_{1s}^*)^2$	$\frac{1}{2}(2-2)=0$	Unstable

To summarize, we can arrange our example in order of decreasing stability, $H_2 > H_2^+ > He_2^+ > He_2$. H_2^+ is somewhat more stable than He_2^+, because there is only one electron in H_2^+ and therefore it has no electron-electron repulsion. Furthermore, H_2^+ also has less nuclear repulsion than He_2^+.

3. Molecular Electron Configuration of Homonuclear Diatomic Molecules of Second-Period Elements

We will discuss the configuration, stability and magnetic properties of homonuclear diatomic molecules, diatomic molecules containing atoms of the same elements.

(1) Dilithium, Li_2

MO theory correctly predicts that dilithium is a stable molecule. Using the order of increasing energies of the molecular orbitals for elements Li through N, the ground-state molecular configuration of Li_2 is $(\sigma_{1s})^2(\sigma_{1s}^*)^2(\sigma_{2s})^2$. The 1s MO's are completely filled and do not participate in bonding. The bond order of Li_2 is $(4-2)/2=1$. So it is stable. Dilithium is a gas-phase molecule with a much lower bond strength than dihydrogen, which can be explained by the configuration because the 2s electrons are further removed from the nucleus.

(2) Diberyllium, Be_2

The electron configuration of beryllium is $1s^2 2s^2$. So there are eight electrons in Be_2 with ground-state molecular configuration of $(\sigma_{1s})^2(\sigma_{1s}^*)^2(\sigma_{2s})^2(\sigma_{2s}^*)^2$. Like He_2 molecule, the electrons in the bonding MO and the antibonding MO are all paired. The bond order is $1/2(4-4)=0$. No bond would form and Be_2 cannot exist.

(3) Diboron, B_2

The ground-state electron configuration of B is $1s^2 2s^2 2p^1$, thus there are 10 electrons in B_2. The σ_{1s}, σ_{1s}^*, σ_{2s} and σ_{2s}^*, orbitals can only accommodate 8 electrons, which requires the introduction of an atomic orbital overlap model for p orbitals. The ground-state electron configuration of B_2 is

$$(\sigma_{1s})^2(\sigma_{1s}^*)^2(\sigma_{2s})^2(\sigma_{2s}^*)^2(\pi_{2p_y})^1(\pi_{2p_z})^1$$

Ignoring the σ_{1s}, σ_{1s}^*, σ_{2s}, and σ_{2s}^* orbitals (because their net effects on bonding are zero), we calculate the bond order of B_2 using Equation 3.1.

$$\text{bond order} = \frac{1}{2} \times 2 = 1$$

Diboron is stable. In diboron only the single electrons in π_{2p_z} and π_{2p_y} participate in bonding. Because the electrons have equal energy diboron is a diradical. Since the spins are parallel the compound is paramagnetic.

(4) Dicarbon, C_2

Dicarbon is a reactive gas-phase molecule. The ground-state electron configuration of C is $1s^2 2s^2 2p^2$; thus there are 12 electrons in C_2. The ground-state electron configuration of C_2 is

$$(\sigma_{1s})^2(\sigma_{1s}^*)^2(\sigma_{2s})^2(\sigma_{2s}^*)^2(\pi_{2p_y})^2(\pi_{2p_z})^2$$

Like B_2, we calculate the bond order of C_2 using Equation 3.1.

$$\text{bond order} = \frac{1}{2} \times 4 = 2$$

Therefore, C_2 molecule is stable. There is no unpaired electron, so C_2 molecule is diamagnetic.

(5) Dinitrogen, N_2

The ground-state electron configuration of N is $1s^2 2s^2 2p^3$, thus there are 14 electrons in N_2. The ground-state electron configuration of N_2 is

$$(\sigma_{1s})^2(\sigma_{1s}^*)^2(\sigma_{2s})^2(\sigma_{2s}^*)^2(\pi_{2p_y})^2(\pi_{2p_z})^2(\sigma_{2p_x})^2$$

Calculate the bond order of N_2 using Equation 3.1.

$$\text{bond order} = \frac{1}{2} \times 6 = 3$$

Therefore, three covalent bonds (one sigma bond and two pi bond) form and N_2 is very stable. The electrons are all paired, so N_2 molecule is diamagnetic.

(6) Dioxygen, O_2

The ground-state electron configuration of O is $1s^2 2s^2 2p^4$, thus there are 16 electrons in O_2. Using the order of increasing energies of the molecular orbitals for F_2 and O_2, we write the ground-state electron configuration of O_2 as

$$(\sigma_{1s})^2(\sigma_{1s}^*)^2(\sigma_{2s})^2(\sigma_{2s}^*)^2(\sigma_{2p_x})^2(\pi_{2p_y})^2(\pi_{2p_z})^2(\pi_{2p_y}^*)^1(\pi_{2p_z}^*)^1$$

We calculate the bond order of O_2.

$$\text{bond order} = \frac{1}{2} \times (6-2) = 2$$

Therefore, the O_2 molecular is stable. The oxygen atoms are joined by a double bond including one sigma bond and one pi bond. According to Hund's rule, the last two electrons enter the $\pi_{2p_y}^*$ and $\pi_{2p_z}^*$ orbitals with parallel spins. The two unpaired electrons show that O_2 is paramagnetic, a prediction that corresponds to experimental observations.

(7) Difluorine, F_2

The molecular electron configuration for a diatomic fluorine molecule, F_2, is

$$(\sigma_{1s})^2(\sigma_{1s}^*)^2(\sigma_{2s})^2(\sigma_{2s}^*)^2(\sigma_{2p_x}^*)^2(\pi_{2p_y})^2(\pi_{2p_z})^2(\pi_{2p_y}^*)^2(\pi_{2p_z}^*)^2$$

The bond order of F_2 is $(10-8)/2=1$. So the fluorine molecule is stable. Because all of the electrons are paired, F_2 is diamagnetic.

(8) Dineon, Ne_2

The molecular configuration for a diatomic neon molecule, Ne_2, is

$$(\sigma_{1s})^2(\sigma_{1s}^*)^2(\sigma_{2s})^2(\sigma_{2s}^*)^2(\sigma_{2p_x})^2(\pi_{2p_y})^2(\pi_{2p_z})^2(\pi_{2p_y}^*)^2(\pi_{2p_z}^*)^2(\sigma_{2p_x}^*)^2$$

The bond order of Ne₂ is (10−10)/2=0. Like dihelium, in Ne₂ the number of bonding electrons equals the number of antibonding electrons and this compound does not exist.

From the above discussion, it is indicated that the core orbitals on an atom make no contribution to the stability of the molecules that contain this atom. The only orbitals that are important are those formed when valence-shell orbitals are combined.

Valence Bond Model vs. Molecular Orbital Theory

The Valence Bond Theory introduces a completely new bond that is different from ionic bond and metallic bond. But the Valence Bond Theory can't adequately explain the fact that some molecules contain two equivalent bonds with a bond order between that of a single bond and a double bond. Its extension, Hybridization Theory, remain widespread in synthetic organic chemistry. But one specific problem with hybridization is that it incorrectly predicts the photoelectron spectra of many molecules, including fundamental species such as methane and water. Hybridization approach tends to over-emphasize the localisation of bonding electrons and does not effectively embrace molecular symmetry as does MO Theory. Molecular Orbital Theory is more powerful than Valence Bond Theory because the orbitals reflect the geometry of the molecule to which they are applied. And it has been widely used in other branches of chemistry.

3.11 Intermolecular Force

Molecular Polarity

We have learned that, in chemical bonds, polarity refers to an unequal distribution of electron pairs between the two bonded atoms, one of the atoms is slightly more electronegative than the other. However, molecules can be polar too. And polarity is an important property of molecules because both physical properties, such as melting point, boiling point, solubility, and chemical properties depend on molecular polarity.

Molecular polarity results from the uneven partial charge distribution between various atoms in a compound. The polarity of a molecule can be measured quantitatively by dipole moment, μ. Dipole moments are vectors, that is, they have direction as well as magnitude as shown in Figure 3.21. The magnitude of μ is equal to the product of q and d.

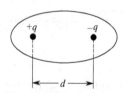

Figure 3.21 The dipole moment of a polar molecule

Molecular polarity is dependent on bond polarity in a compound and the asymmetry of the compound's structure. A molecule may be polar either as a result of polar bonds due to differences in electronegativity (for example, HCl molecule) or as a result of an asymmetric arrangement of nonpolar covalent bonds and lone pairs of electrons. For example, we have known that water (H_2O) molecule is bent. The two lone pairs on the oxygen atom establish a negative pole, while the bound hydrogen atoms constitute a positive pole. As a result H_2O is thought to be polar because of the unequal sharing of its electrons.

In a similar manner, a nonpolar molecule occurs either because there is no polarity in the bonds or because of the symmetrical arrangement of polar bonds. In the oxygen, O_2, molecule, O—O bond does not have polarity because of equal electronegativity, hence there is no polarity in the molecule.

Molecules can also contain polar bonds and not be polar. If a molecule contains more than one polar bond, however, whether the molecule as a whole is polar depends on the shape of the molecule. Carbon dioxide is a perfect example. Because oxygen and carbon have different electronegativity, both of the C—O bonds are polar with a partial positive charge on carbon and a partial negative charge on-oxygen. However, carbon dioxide is linear, so the dipole moments of the two C—O bonds cancel each other. So the carbon dioxide molecule is nonpolar despite the fact that it has two polar bonds. Compare these two triatomic molecules, CO_2 and H_2O. CO_2 is linear, whereas H_2O is bent. Both carbon-oxygen and hydrogen-oxygen bonds are polar. But the dipole moments of the two H—O bonds in water do not cancel because water is bent. So water molecules are polar and have net dipole moment.

Take another example, the methane (CH_4) molecule. Each C—H bond has polarity and the four C—H bonds are arranged tetrahedrally around the carbon atom. Since the bonds are arranged symmetrically there is no overall dipole in the molecule. So methane is a nonpolar molecule.

Most nonpolar molecules are water insoluble (hydrophobic) at room temperature because water is a polar solvent. However many nonpolar organic solvents, such as turpentine, are able to dissolve nonpolar substances. Polar and nonpolar compounds also show different physical properties. Polar molecules have been called as dipole that consists of a positive end and a negative end. So the dipole-dipole interaction between polar molecules leads to their higher boiling point when comparing a polar and nonpolar molecule with similar molar masses.

Sample 3.11 Predict the polarity of following molecules: (a) CO; (b) BF_3; (c) NH_3
Solution

(a) Polar. Because the carbon-oxygen bond is polar.

(b) Nonpolar. BF_3 molecule has a trigonal planar arrangement of three polar bonds at 120°. This results in no overall dipole in the molecule. So BF_3 is nonpolar.

(c) Polar. In the ammonia, NH_3, molecule the three N—H bonds have only a slight polarity. However, the molecule has one lone pairs, that points towards the fourth apex of the approximate tetrahedron. This orbital is electron rich which results in a powerful dipole across the whole ammonia molecule.

Intermolecular Forces

We have learned that chemical bonds hold atoms together in a molecule and stabilize individual molecules. So chemical bonds can be thought as intramolecular forces, whereas intermolecular forces are attractive force between molecules that hold two or more molecules together other than those due to covalent bonds or to the electrostatic interaction of ions with one another or with neutral molecules. Intermolecular forces are primarily responsible for the non-ideal behaviour of gasses, the existence of the condensed states of matter, liquids and solids, and the bulk properties of matter (for example, melting point and boiling point).

Intermolecular forces must exist between all molecules because all gases condense to liquids if the temperature is low enough and the pressure high enough. They are not nearly as strong as intramolecular forces. The energies required to break covalent bonds ordinarily range from 150 to 800 kJ/mol, whereas the energies required to overcome intermolecular attractions are normally of the order of 1~40 kJ/mol. Meanwhile, the strength of intermolecular attractions decreases rapidly with increasing distance between molecules. Therefore, intermolecular attractions are not very important in gases, where the molecules are far apart, but they are very important in liquids and solids, where the molecules are close together.

Many properties of liquids, including their boiling points, reflect the strengths of the intermolecular forces. The stronger the attractive forces, the higher is the temperature at which the liquid boils. Similarly, the melting points of solids increase with the strengths of the intermolecular forces.

To understand the properties of condensed matter, we must first get some ideas with the different types of intermolecular forces: dipole-dipole, dipole-induced dipole, and dispersion forces (London forces). They make up what chemists commonly refer to as van der Waals forces named after Johannes van der Waals, who developed the equation for predicting the deviation of gases from ideal behaviour. He also tried to develop a model that would explain the behavior of liquids by including terms that reflected the size of the atoms or molecules in the liquid and the strength of the bonds between these atoms or molecules. Hydrogen bond is a particularly strong type of dipole-dipole interaction. Because only a few elements can participate in hydrogen bond formation, it is treated as a separate category. As we consider these forces, notice that each is electrostatic in nature, involving attractions between positive and negative species.

Dipole-Dipole Forces

Dipole-dipole forces, also called Keesom interactions or Keesom forces, are named after Willem Hendrik Keesom, who produced the first mathematical description in 1921. They are the dipole-dipole interactions that act between polar molecules with permanent dipoles. Consider hydrogen chloride as an example.

The electro negativity of chlorine (3.16) is greater than the electronegativity of hydrogen (2.18). therefore, the H—Cl bond is polar and the hydrogen chloride molecule is also a polar molecule with a dipole moment. Because of the force of attraction between oppositely charged particles, there is a small dipole-dipole force of attraction between adjacent HCl molecules. A point dipole consists of two equal charges of opposite sign δ^+ and δ^- (Figure 3.22).

$$\overset{\delta^+}{H}—\overset{\delta^-}{Cl}\cdots\overset{\delta^+}{H}—\overset{\delta^-}{Cl}$$

Figure 3.22 The small dipole-dipole force of attraction between adjacent HCl molecules

Because the positive and negative charges on hydrogen and chlorine are only partial charges, the dipole-dipole attraction between hydrogen chloride molecules is not nearly as strong as the attraction between sodium ions and chloride ions, which have whole charge. The dipole-dipole interaction in HCl is only 3.3 kJ/mol, so small that hydrogen chloride boils at −85.0 ℃. As a result, hydrogen chloride is a gas under ordinary conditions, whereas sodium chloride is a high melting solid.

The origin of dipole-dipole forces is electrostactic force. The larger the dipole moments, the greater the force. In the solid state, molecules are touching. The positive ends of dipoles are attracted to the negative ends of dipoles in other molecules as shown in Figure 3.23. In liquids the molecules are not held as rigidly as in a solid, but they tend to align in such a way that, on the average, the attractive interaction is at a maximum.

Note that almost always the dipole-dipole interaction between two atoms is zero, be-

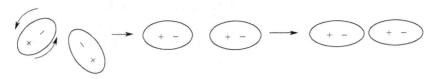

Figure 3.23 The arrangement of the dipoles in solid state

cause atoms rarely carry a permanent dipole.

Dipole-Induced Dipole Forces

If we place an ion or a polar molecule near an atom (or a nonpolar molecule), the electron distribution of the atom (or molecule) is distorted by the force exerted by the ion or the polar molecule. The resulting dipole in the atom (or molecule) is said to be an induced dipole because the separation of positive and negative charges in the atom (or a nonpolar molecule) is due to the proximity of an ion or a polar molecule. The attractive interaction between an ion and the induced dipole is called ion-induced dipole forces which are not van der Waals forces, and the attractive interaction between a polar molecule and the induced dipole is called dipole-induced dipole forces [Figure 3.24(a)]. For example, argon has no dipole moment because the electrons on an argon atom are distributed homogeneously around the nucleus of the atom. But these electrons are in constant motion. The electron cloud is diffuse and spread over an appreciable volume. When an argon atom comes close to a polar HCl molecule, the electrons can shift to one side of the nucleus to produce a very small dipole moment that lasts for only an instant. So by distorting the distribution of electrons around the argon atom, the polar HCl molecule induces a small dipole moment on this atom, which creates a weak dipole-induced dipole force of attraction between the HCl molecule and the Ar atom [Figure 3.24(b)]. This force is very weak, with a energy of about 1 kJ/mol.

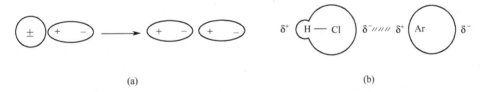

(a)   (b)

Figure 3.24 The dipole-induced dipole force between polar molecule and induced dipole

Dipole-induced dipole forces can attribute to the solubility of some substances. An example is chlorine dissolving in water.

(Permanent Dipole) H—O—H----Cl—Cl (Induced Dipole)
δ^+ δ^- δ^+ δ^- δ^+

Sketched is an interaction between the permanent dipole on water and an induced dipole on chlorine.

The likelihood of a dipole moment being induced depends not only on the charge on the ion or the strength of the dipole but also on the polarizability of the atom or molecule. Polarizability is the ease with which the electron distribution in the atom (or molecule) can be distorted. Generally, the larger the number of electrons and the more diffuse the electron cloud in the atom or molecule, the greater its polarizability.

Dispersion Forces (London Forces)

Even substances with nonpolar molecules such as H_2, Cl_2, CH_4, and He condense to liquids if the temperature is low enough and the pressure is high enough. All of they are nonpolar molecules and neither dipole-dipole nor dipole-induced forces can explain the fact. What is the nature of the attractive forces between nonpolar molecules? Polarizability of atoms and molecules.

Polarizability enables gases containing atoms or nonpolar molecules (for example, He and N_2) to condense. The helium atom is the simplest example. By itself, a helium atom is perfectly symmetrical. On average, the charge density distribution of a helium atom is spherically symmetrical. But at any instant the electrons are moving at some distance from the nucleus. In these cases the charge density distribution can be unsymmetrical.

So at any instant it is likely that the atom has a dipole moment created by the movement and the specific positions of the electrons. This dipole moment is called an instantaneous dipole because it lasts for just a tiny fraction of a second. In the next instant the electrons are in different locations and the atom has a new instantaneous dipole and so on. An instantaneous dipole of one He atom can induce a dipole in its nearest He atom. At the next moment, a different instantaneous dipole can create temporary dipoles in the surrounding He atoms and so on. Movement of the electrons around the nuclei of a pair of neighboring helium atoms can become synchronized so that each atom simultaneously obtains an induced dipole moment. This kind of interaction between instantaneous dipoles results in an attraction between the He atoms [Figure 3.25(a)]. At very low temperatures (and reduced atomic speeds), this attraction is enough to hold the atoms together, causing the helium gas to condense.

The attractive forces that exist between atoms and nonpolar molecules as a result of instantaneous dipoles induced in the atoms or molecules are called dispersion forces, London forces or induced dipole-dipole forces [Figure 3.25(b)]. London forces are named after Fritz London, a German physicist, who made a quantum mechanical interpretation of instantaneous dipoles in 1930. He showed that the magnitude of this attractive interaction is directly proportional to the polarizability of the atom or molecule. Dispersion forces, may be quite weak. This is certainly true of helium, only 0.076 kJ/mol, which has a boiling point of only 4.3 K, or −269 ℃. Note that helium has only two electrons, which are tightly held in the 1s orbital. Therefore, the helium atom has a small polarizability.

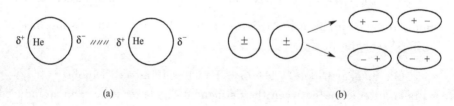

(a)   (b)

Figure 3.25 Formation of dispersion forces

But atoms or molecules become more polarizable as they become larger. The electron clouds are more dispersed and there are more electrons to be polarized. So dispersion forces usually increase with molar mass because molecules with larger molar mass tend to have more electrons. This trend is exemplified by the halogens (F_2, Cl_2, Br_2, I_2). Fluorine and chlorine are gases at room temperature, bromine is a liquid, and iodine is a solid. The London forces also become stronger with larger amounts of surface contact. Greater surface area means closer interaction between different molecules.

Keep in mind that dispersion forces exist among species of all types, whether they are neutral or bear a net charge, whether they are atoms or molecules, and whether they are polar or nonpolar. It is the only reason for rare-gas atoms to condense at low temperature because dispersion force is the only attractive intermolecular force at large distances present between neutral atoms.

In many cases, dispersion forces are comparable to or even greater than the dipole-dipole forces between polar molecules. Compare the boiling points of CH_3F (-78.4 ℃) and CCl_4 (76.5 ℃). Although CH_3F has a dipole moment, it boils at a much lower temperature than CCl_4, a nonpolar molecule because CCl_4 contains more electrons. As a result, the dispersion forces between CCl_4 molecules are stronger than the dispersion forces plus the dipole-dipole forces between CH_3F molecules.

Sample 3.12 Which of the following molecules have a permanent dipole moment: H_2O; CO_2; CH_4; N_2; CO; NH_3

Solution:

H_2O; CO; NH_3

CO_2, CH_4, and N_2 are symmetric molecules, hence they are nonpolar and have no permanent dipole moments. Only molecules with polar bonds unsymmetrically arranged will possess a permanent dipole.

Van der Waals forces are relatively weak compared to normal chemical bonds, but play a fundamental role in fields as diverse as supramolecular chemistry, structural biology, polymer science, nanotechnology, surface science, and condensed matter physics. Van der Waals forces define the chemical character of many organic compounds. They also define the solubility of organic substances in polar and nonpolar media.

Hydrogen Bond

Early evidence of hydrogen bond came from the study of the boiling points of compounds. Normally, the boiling points of a series of similar compounds containing elements in the same group increase with increasing molar mass because dispersion forces become greater. For example, if you plot the boiling points of the hydrides of the group ⅣA elements, you find that the boiling points increase as you go down the group. But as Figure 3.26 shows, the hydrogen compounds of group ⅤA, ⅥA, and ⅦA elements do not follow this trend. In each of these series, the boiling point of the hydride of the first element (NH_3, H_2O, HF) in each group is abnormally high. For example, hydrogen sulphide (H_2S), hydrogen selenide (H_2Se), and hydrogen telluride (H_2Te) are all gases under ordinary conditions.

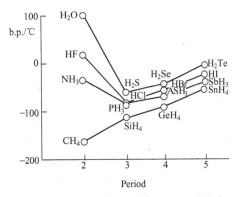

Figure 3.26 The boiling points of the hydrides of the group ⅣA elements

Water (H_2O) is, of course, a liquid. Its boiling point is higher than any of them. So in the cases of NH_3, H_2O and HF there must be some additional intermolecular forces of attraction, requiring significantly more heat energy to break. This fact is explained by hydrogen bond, a relatively powerful intermolecular force.

Notice that in each of these molecules with abnormal high boiling point, the hydrogen is attached directly to one of the most electronegative elements (fluorine, oxygen, and nitrogen), causing the hydrogen to acquire a significant amount of positive charge. So the covalent bonds between these elements and hydrogen are more polar than bonds between hydrogen and any other elements. In addition, hydrogen is the smallest atom. Atoms of fluorine, oxygen, and nitrogen are smaller than atoms of any other elements except hydrogen. As a result of the large partial charges and the small sizes of these atoms, dipole-dipole attractions between molecules with hydrogen-fluorine, hydrogen-oxygen, and hydrogen-nitrogen bonds are unusually strong. This force of attraction is called a hydrogen bond, resulting from a dipole-dipole force between an electronegative atom and a hydrogen atom bonded to nitrogen, oxygen or fluorine. Molecules that have hydrogen bonded to nitrogen, oxygen, or fluorine have unusually strong hydrogen bonds. For example, in liquid HF the molecules are hydrogen bonded to one another and the molecules are more difficult to break apart, so liquid HF has an unusually high boiling point.

Hydrogen bonds are commonly indicated by a dashed line as shown in Figure 3.27. The three atoms joined by a hydrogen bond usually lie on a straight line. The hydrogen bond is directed toward one of the lone pairs of electrons on fluorine.

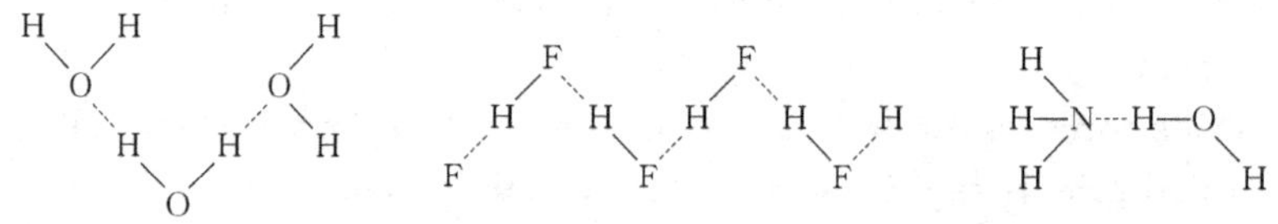

Figure 3.27 Formation of hydrogen bonds

Many properties of water stem from hydrogen bonding, high melting point, high boiling point, high heat of vaporization, low density of ice compared to water, high specific heat, high ionic conductance of hydronium and hydroxide ions and many others.

High water solubility of many compounds such as ammonia can also be explained by hydrogen bonding with water molecules. For example, the hydrogens in formaldehyde are attached to carbon (not F, O, or N) so that formaldehyde molecules can't hydrogen bond to other formaldehyde molecules. When dissolved in water, the oxygen atom in formaldehyde can accept a hydrogen bond from a water molecule. This type of hydrogen bonding is important in the formation of aqueous solutions.

Sample 3.13 Ethanol (C_2H_5OH, molar mass 46) boils at 351 K, but water (H_2O, molar mass 18) boils at higher temperature, 373 K. Explain it.
Solution:

Water has stronger hydrogen bonds than ethanol because the strength of a hydrogen bond depends upon the electronegativities and sizes of the two atoms.

Sample 3.14 Ethanol C_2H_5OH and methyl ether CH_3OCH_3 have the same molar mass. Which has a higher boiling point?
Solution:

Ethanol.

R-OH group can form hydrogen bonding with each other. Methyl groups have very weak hydrogen bonding, if any.

Hydrogen bonds can vary in strength from very weak (1~2 kJ/mol) to extremely strong (>155 kJ/mol). But the average energy of a hydrogen bond is quite large (up to 40 kJ/mol). Typical values are listed in Table 3.6. Thus, hydrogen bonds are a powerful force in determining the structures and properties of many compounds. Hydrogen bond is significant in almost all biologically important molecules. for example, the shapes of large biological molecules such as proteins are determined largely by hydrogen bond. Hydrogen bonds can occur between molecules (intermolecularly), or within different parts of a single molecule (intramolecularly). The hydrogen atoms and nitrogen atoms of adjacent DNA base pairs generate intramolecular forces that improve binding between the strands of the molecule. The double helical structure of DNA, is due largely to hydrogen bonds between the base pairs, which link one complementary strand to the other and enable replication. Intramolecular hydrogen bond is also partly responsible for the secondary, tertiary, and quaternary structures of proteins and nucleic acids.

Table 3.6 The typical values for the bond strength of hydrogen bonds

Hydrogen Bonds	Values for the Bond Strength/(kJ/mol)
F—H⋯F	155
O—H⋯N	29
O—H⋯O	21
N—H⋯N	13
N—H⋯O	8
HO—H⋯OH$_3^+$	18

Key Words

chemical bond ['kemikəl bɔnd] 化学键
ionic bond [ai'ɔnik bɔnd] 离子键
covalent bond [kəu'veilənt bɔnd] 共价键
metallic bond [mi'tælik bɔnd] 金属键
geometry [dʒi'ɔmitri] 几何形状
intermolecular force [ˌintə(:)mə'lekjulə fɔ:s] 分子间力
octet rule [ɔk'tet ru:l] 八隅律
resonance ['rezənəns] 共振体
nonpolar [nɔn'pəulə] 非极性的
polar ['pəulə] 极性的
dipole ['daipəul] 偶极
dipole moment ['daipəul 'məumənt] 偶极矩
torsional angle ['tɔ:ʃənəl æŋgl] 扭角
dihedral angle [dai'hedrəl, dai'hi:drəl æŋgl] 二面角
bond pair [bɔnd pɛə] 成键电子对
lone pair [ləun pɛə] 孤对电子
hybridization [ˌhaibridai'zeiʃən] 杂化
hybrid orbital ['haibrid 'ɔ:bitl] 杂化轨道

molecular orbital ['mɔlikjul 'ɔ:bitl] 分子轨道
bonding molecular orbital ['bɔndiŋ 'mɔlikjul 'ɔ:bitl] 成键分子轨道
antibonding molecular orbital ['ænti,bɔndiŋ 'mɔlikjul 'ɔ:bitl] 反键分子轨道
diamagnetic ['daiəmægnetik] 逆，反，抗磁性
paramagnetic [ˌpærəmægnetik] 顺磁性
bond order [bɔnd 'ɔ:də] 键级
homonuclear [ˌhɔmə'nju:kliə (r)] 同核的
diatomic molecule ['mɔlikju:l] 双原子分子
polarity [pəu'læriti] 极性
dipole-dipole force ['daipəul 'daipəul fɔ:s] 取向力
dipole-induced dipole force ['daipəul in'dju:sd 'daipəul fɔ:s] 诱导力
dispersion force [dis'pə:ʃen fɔ:s] 色散力
hydrogen bond ['haidrəudʒən bɔnd] 氢键
induced dipole [in'dju:sd 'daipəul] 诱导偶极
polarizability ['pəulə,raizə'biləti] 极化

Exercises

1. In which of the following compounds is the bonding essentially ionic, in which is the bonding essentially covalent, and in which are both types of bonding represented? (a) PCl$_3$

(b) $(NH_4)_2S$ (c) MgO (d) $Ba(CN)_2$ (e) CuS (f) CH_3CH_2OH

Answer: (a) Covalent (b) both (c) ionic (d) both (e) ionic (f) covalent

2. Use electron configurations to describe the formation of the following compounds. (a) compound of barium and oxygen (b) compound of barium and chlorine (c) compound of calcium and hydrogen.

3. Write the Lewis structure for the following molecules and decide which do not obey the octet rule. Explain why. (a) BCl_3 (b) PH_3 (c) PCl_3 (d) PCl_5 (e) SO_2 (f) SO_3 (g) SCl_6 (h) F_2 (i) CO (j) CCl_4

4. Why is rotation about pi bonds not free?

5. For each molecule, tell the total number of covalent bonds, the number of sigma bonds, and the number of pi bonds: (a) CO, (b) CO_2, (c) HCN, (d) CH_3CHO, (e) $H_2C=CHC\equiv CH$

Answer:

Molecules	The total number of covalent bonds	The number of sigma bonds	The number of pi bonds
CO	2	1	1
CO_2	4	2	2
HCN	4	2	2
CH_3CHO	7	6	1
$H_2C=CHC\equiv CH$	10	7	3

6. Arrange the bonds in each of the following sets in order of increasing polarity: (a) C—F, O—F, Be—F; (b) N—Br, P—Br, O—Br; (c) N—O, B—O, S—O

Answer: (a) Be—F, C—F, O—F (b) O—Br, P—Br, N—Br (c) B—O, S—O, N—O

7. What is the difference between the electron-domain geometry and the molecular geometry of a molecule? Use the ammonia molecule as an example in your discussion.

8. Write Lewis structures for the following substance. For each species, what is the total coordination number? What is the electron-domain geometry? How many lone pairs are around the central atom? How many bonding pairs? Use VSEPR to predict the geometry of them. (a) H_3O^+ (b) ClO^- (c) NH_4^+ (d) SO_3 (e) SO_3^{2-} (f) NF_3 (g) CH_2O (h) XeF_4 (i) ICl_3 (j) BrF_5 (k) PCl_3 (l) CO_2 (m) XeF_2 (n) SF_6

Answer: (a) Pyramidal (b) Bent (c) Terahedral (d) Trigonal planar (e) Pyramidal (f) Pyramidal (g) Trigonal planar (h) Square planar (i) T-shaped (j) Square pyramid (k) Triangular pyramid (l) Linear (m) Linear (n) Octahedral

9. The observed H-C-H angle in the molecule $H_2C=O$ is 111°, rather than 120°. Explain this observation.

10. Which of the following molecules has two lengths of bonds? (a) CO_2 (b) SO_3 (c) SF_4 (d) XeF_4

Answer: (c) SF_4 Seesaw

11. What is a hybrid orbital? In your answer, use a hybrid orbital of carbon as an example. Why is the idea of hybrid orbital used?

12. Indicate the type of hybrid orbitals of the central atom and the molecular geometry. (a) $\underline{Be}F_2$ (b) $\underline{C}Cl_4$ (c) $\underline{Be}H_2$ (d) $\underline{S}F_6$ (e) $H\underline{C}N$ (f) $\underline{C}O_2$ (g) $\underline{P}Cl_3$ (h) $\underline{S}O_2$ (i) $\underline{Si}H_4$ (j) $\underline{C}S_2$ (k) $\underline{Xe}F_2$ (l) $\underline{Xe}F_4$

Answer: (a) sp, linear (b) sp^3, tetrahedral (c) sp, linear (d) sp^3d^2, octahedral (e) sp, linear (f) sp, linear (g) sp^3, triangular pyramid (h) sp^2, bent (i) sp³, tetrahedral　(j) sp, linear　(k) sp³d, linear　(l) sp³d², square planar

13. What are the hybridization states of each carbon atom in the following molecules. (a) $CH_3CH_2CH_2CH_3$　(b) $CH_2{=}CH{-}CH{=}CH_2$　(c) $CH_3CH{=}CHCH_3$　(d) $H{-}C{\equiv}C{-}H$　(e) CH_3CHO　(f) $H_2C{=}CHC{\equiv}CH$

Answer: (a) All sp³　(b) all sp²　(c) sp³ sp² sp² sp³　(d) both sp　(e) sp³ sp²　(f) sp² sp² sp sp

14. According to molecular-orbital theory, would Be_2^+, Ne_2^+ be expected to exist? Explain.

Answer: The bond order of Be_2^+ and Ne_2^+ are both 1/2. So both of them would exist.

15. Discuss the magnetism of the following ions (paramagnetic or diamagnetic). (a) O_2^+　(b) N_2^{2-}　(c) Li_2^+　(d) O_2^{2-}　(e) Be_2^{2-}　(f) C_2^+?

Answer: (a) paramagnetic　(b) paramagnetic　(c) paramagnetic　(d) diamagnetic　(e) diamagnetic　(f) paramagnetic

16. Draw the molecular orbital energy level diagrams for the Li_2 molecule, the Be_2^+ ion, and the Be_2 molecule in their ground states. Compare the stabilities of these three species.

Answer: Bond order: Li_2: 1, Be_2^+: 1/2, Be_2: 0　Stability: $Li_2 > Be_2^+ > Be_2$

17. Compare the structure of O_2 predicted by the valence bond and molecular orbital theories. Which structure agrees with the properties observed by experiment? Explain it.

18. Predict whether the following molecules are polar or nonpolar: CCl_4　$SiCl_4$　NH_3　H_2S　IF　SO_2　SO_3　SF_6　IF_5　CH_3Cl　CH_2Cl_2

Answer: Polar molecules: NH_3　H_2S　IF　SO_2　CH_3Cl　CH_2Cl_2
Nonpolar molecules: CCl_4　$SiCl_4$　SO_3　SF_6　IF_5

19. True or false: Molecules that contain a polar bond are not necessarily polar compounds. Because if polar bonds in a molecule are symmetrically arranged, then their polarities will cancel and they will be nonpolar.

Answer: T

20. The dipole moments of SO_2 and CO_2 are 5.37×10^{-30} C·m and zero, respectively. What can be said about the shapes of the two molecules?

21. Distinguish between a polar bond and a polar molecule. To which does them dipole refer?

22. Which member of the following pairs has the highest boiling point? (a) H_2O or H_2S　(b) CH_4, CCl_4, CH_3Cl or CH_3Br　(c) CH_3CH_2Cl, or $CH_3CH_2CH_2Cl$　(d) CH_3CH_2OH OR CH_3OCH_3　(e) PH_3, NH_3 or $(CH_3)_3N$　(f) H_2, He, Ne, Xe or CH_4　(g) N_2 OR NO　(h) CH_2COOH or $HCOOCH_3$

Answer: (a) H_2O　(b) CH_3Br　(c) $CH_3CH_2CH_2Cl$　(d) CH_3CH_2OH　(e) NH_3　(f) Xe　(g) NO　(h) CH_2COOH

23. Select the specie that is best described by the statement to the right.
(a) CO_2　NH_3　CO　which has a zero dipole moment?
(b) CH_4　NH_3　HF　which has the highest boiling point?
(c) Cl_2　Br_2　I_2　which has the lowest boiling point?
(d) Br_2　ICl　Cl_2　which has the higher boiling point?

Answer: (a) CO_2　(b) HF　(c) Cl_2　(d) ICl

Chapter 4 Thermochemistry

The production of energy is one of the most important aspects of chemistry. The study of energy and its transformation is known as thermodynamics and the aspect of thermodynamics is called thermochemistry. In this chapter, we will study the relationships between chemical reactions and energy changes.

4.1 First Law of Thermodynamics

System and Surroundings

When we use thermodynamics to study energy changes, some terms must be defined precisely. Two fundamental notions are system and surroundings. The system is a well-defined and limited part of universe that we pay attention on. That is to say, the portion we are interested in is called the system, everything else is called surroundings.

There are three types of systems, open system, closed system and isolated system. An open system can exchange both mass and energy with its surroundings. A closed system allows the transfer of energy but not mass. An isolated system allows the transfer of neither mass nor energy.

State Functions

Properties such as pressure, volume and temperature that depend only on the initial and final states of the system and not on how the system gets from one state to another are called state functions.

The term state has a very precise meaning and a system is said to be in a certain state when its properties have certain values in thermodynamics.

Heat and Work

There are two common ways to express energy changes in everyday lives. The energy which is transferred to cause the motion of an object against a force is called work, which is represented with the symbol W. The units of W are J or kJ. When work is done on the system by the surroundings, it has a positive value. Conversely, when the work is done on the surroundings by the system, it has negative values.

As shoun in Figure 4.1, the work W can be defined as force F multiplied by distance d.

$$W = Fd = -p\Delta V \tag{4.1}$$

Where ΔV means the change in volume; $\Delta V = V_{final} - V_{initial} = Al$; p means pressure.

The energy which is transferred from a hotter object to a colder one is called heat, which is represented with the symbol Q. When heat is transferred from the surroundings to the system, it has a positive value. Conversely, when the heat is transferred from the system to the surroundings, it has negative values.

According to the transfer of heat to and from the system, it includes endothermic and exothermic processes. When a process occurs in a system in which heat is absorbed,

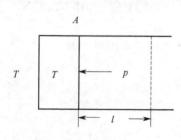

Figure 4.1 The description of work W

we call that this process is endothermic. In an endothermic process, the heat flows from the surroundings to the system. While in an exothermic process, the heat flows from the system to the surroundings. The exothermic process results in the evolution of the heat.

Internal Energy

In order to use the first law of thermodynamics, we must first define the internal energy of the system. The internal energy means the sum of all the kinetic and potential energy of all the components of the system. We represent the internal energy with the symbol U. The units of U are J or kJ.

When we begin with a system with an initial internal energy, $U_{initial}$, then, the system undergoes a change, which may involve heat being transferred or work being done. The final internal energy of the system is U_{final} after the change. So we define the change in internal energy with the symbol ΔU. The actual values of $U_{initial}$ or U_{final} for any system cannot be determined. But the value of ΔU can be determined through the law.

$$\Delta U = U_{final} - U_{initial} \tag{4.2}$$

First Law of Thermodynamics

We also see that energy can be transferred from one part of the universe to another, or it can be transferred back and forth in the forms of work and heat between a system and its surroundings. The first law of thermodynamics is well known as the most important observation that energy can be neither created nor destroyed. Any energy which is lost by the surroundings must be gained by the system, and vice versa. The first law of thermodynamics is based on the law of conservation of energy, relating the internal energy change of a system to the heat change and the work done.

The expression of the first law of thermodynamics is as follows.

$$\Delta U = Q + W \tag{4.3}$$

When a system undergoes any physical or chemical change, its internal energy, ΔU, is given by the heat added to or liberated from the system, Q, plus the work done on or by the system, W.

Sample 4.1 When a gas is compressed in a cylinder, the work done was 460J in this process. There is a heat transfer of 120J from the gas to the surroundings. Calculate the energy change for this process.

Solution:

According to the first law of thermodynamics.

$$\Delta U = Q + W = -120 \text{ J} + 460 \text{ J} = 340 \text{ J}$$

As the result of work done and heat transferred, the energy of the gas increases by 340 J.

4.2 Enthalpy and Enthalpy Change

Usually, in our living systems, most of physical and chemical changes take place under the essentially constant pressure of Earth's atmosphere. The next step is to see how the first law of thermodynamics can be applied to processes carried out under different conditions. There are two main situations in which the volume of the system is kept constant or in which the pressure applied on the system is kept constant. When the volume of the system is kept constant, $\Delta V = 0$, so no work will result from this change. According to the first law of thermodynamics.

$$\Delta U = Q + W = Q_V \qquad (4.4)$$

The subscript V means that this is a constant-volume condition, and Q_V means the heat in a constant-volume process. So the heat change can have only a specific value, which is equal to ΔU. Constant-volume conditions are often impossible to achieve.

Enthalpy

Most reactions occur under constant-pressure conditions. There are three conditions. If a reaction results in a net increase in the number of moles of a gas, then the system does work on the surroundings. Conversely, if a reaction results in a net decrease in the number of moles of a gas, then the surroundings does work on the system. Finally, if there is no net change in the number of moles of gases from reactants to products, no work is done.

According to the first law of thermodynamics.
$$\Delta U = Q + W$$
For the constant-pressure process.
$$\Delta U = Q + W = Q_p - p\Delta V$$
$$Q_p = \Delta U + p\Delta V \qquad (4.5)$$
Where the subscript p means constant-pressure condition.

Now we introduce a new function of the system, which is called enthalpy. We represent the internal energy with the symbol H. The units of H are J or kJ. The enthalpy is defined by the equation.
$$H = U + pV \qquad (4.6)$$
For any process, the change in enthalpy for the constant-pressure process is given by
$$\Delta H = \Delta(U + pV) = \Delta U + \Delta(pV) = \Delta U + p\Delta V \qquad (4.7)$$
So we see that $Q_p = \Delta H$ for a constant-pressure process, that is to say, the heat change at constant pressure is equal to ΔH.

Enthalpy of Reactions

Most reactions are in constant pressure processes, so we can equate the heat change in these cases to change in enthalpy.

For any reaction,
$$\text{reactants} \rightarrow \text{products}$$
The change in enthalpy is called the enthalpy of reactions, with the symbol ΔH.
$$\Delta H = H(\text{products}) - H(\text{reactants}) \qquad (4.8)$$
The enthalpy of reaction can be positive or negative, depending on the endothermic or exothermic processes. ΔH is positive for the endothermic and ΔH is negative for the exothermic.

4.3 Hess's Law

Thermochemical Reaction Equation

For many reactions, the thermal energy releases is the chief reason for carrying out the reaction. The equation that shows the enthalpy change as well as the mass relationships is called thermochemical equations.

For example, when methane is burned with oxygen to product carbon dioxide and water, thermal energy is given off.
$$CH_4(g) + 2O_2(g) \longrightarrow CO_2(g) + 2H_2O(l) \quad \Delta_r H_m^{\ominus} = -890.32 \text{kJ/mol}$$
Where $\Delta_r H_m^{\ominus}$ means the standard enthalpy change of reaction.

These guidelines are helpful in writing and interpreting thermochemical equations.

(1) Thermochemical equations are interpreted in terms of moles. For example,
$$1/2CH_4(g) + O_2(g) \longrightarrow 1/2CO_2(g) + H_2O(l) \quad \Delta_r H_m^\ominus = -445.16 kJ/mol$$

(2) When we reverse an equation, we change the roles of reactants and products. Consequently, the magnitude of ΔH for the equation remains the same, but its sign changes. For example,
$$CO_2(g) + 2H_2O(l) \longrightarrow CH_4(g) + 2O_2(g) \quad \Delta_r H_m^\ominus = +890.32 kJ/mol$$

(3) When writing thermochemical equations, we must always specify the physical states of all reactants and products, because the actual enthalpy changes are determined by the physical states. For example,
$$CH_4(g) + 2O_2(g) \longrightarrow CO_2(g) + 2H_2O(g) \quad \Delta_r H_m^\ominus = -802.32 kJ/mol$$

The enthalpy change is $-802.32 kJ/mol$ rather than $-890.32 kJ/mol$, because 88kJ/mol are needed to convert 2 moles of H_2O (l) to 2 moles of H_2O (g)

Standard Enthalpy of Formation and Reaction

A formation reaction produces 1 mol of a chemical substance from the elements in their most stable forms at $T = 298K$ and $p = 1$ atm❶. There are two examples, the formation reactions for H_2O and CH_4.
$$2H_2(g) + O_2(g) \longrightarrow 2H_2O(l)$$
$$C(s) + 2H_2(g) \longrightarrow CH_4(g)$$

There are some features in formation reactions.

(1) There is a single product with a stoichiometric coefficient of one.

(2) All the starting materials are elements which are in the form that are in their most stable forms at $T = 298K$ and $p = 1$ atm.

(3) Fractional stoichiometric coefficients are common in formation reactions which must generate 1 mol of product.

The elements are said to be in the standard state at 1 atm. Although the standard state does not specify a temperature, we always mean standard state at $T = 298K$. The change of enthalpy that accompanies the formation of 1 mol of a chemical substance from pure elements in their most stable forms at standard state is called the standard enthalpy of formation, with a symbol $\Delta_f H_m^\ominus$. The units of $\Delta_f H_m^\ominus$ are J/mol or kJ/mol. The standard enthalpy of formation of any element in its most stable form is zero. For example, we write $\Delta_f H_m^\ominus(O_2) = 0$, but $\Delta_f H_m^\ominus(O_3) \neq 0$.

Standard enthalpy of formation can be used to calculate the enthalpy for any reaction that occurs under standard state. For example,
$$2NO_2(g) \longrightarrow N_2O_4(g)$$

The enthalpy change of above reaction is the sum of the formation of enthalpies of the products minus the sum of the formation of enthalpies of the reactants.
$$\Delta_r H_m^\ominus = \Delta_f H_m^\ominus(N_2O_4) - 2\Delta_f H_m^\ominus(NO_2) \tag{4.9}$$

Furthermore, the reaction enthalpy for any reaction can be calculated by summing the standard enthalpies of formation for all the products and then subtracting the sum of the standard enthalpies of reactants. The equation is
$$\Delta_r H_m^\ominus = \sum_i v_i \Delta_f H_m^\ominus(\text{products}) - \sum_i v_i \Delta_f H_m^\ominus(\text{reactants}) \tag{4.10}$$

---

❶ 1atm=101325Pa。

Where v_i is stoichiometric coefficient.

Sample 4.2 Please calculate the $\Delta_r H_m^\ominus$ of the following reaction
$$C_6H_{12}O_6(s) + 6O_2(g) \longrightarrow 6CO_2(g) + 6H_2O(l)$$

Solution:
According to the data
$$\Delta_f H_m^\ominus [C_6H_{12}O_6(s)] = -1274.5 \text{ kJ/mol}$$
$$\Delta_f H_m^\ominus [CO_2(g)] = -393.51 \text{ kJ/mol}$$
$$\Delta_f H_m^\ominus [H_2O(l)] = -285.83 \text{ kJ/mol}$$
$$\Delta_r H_m^\ominus = 6\Delta_f H_m^\ominus [CO_2(g)] + 6\Delta_f H_m^\ominus [H_2O(l)] - \Delta_f H_m^\ominus [C_6H_{12}O_6(s)] - 6\Delta_f H_m^\ominus [O_2(g)]$$
$$= -2801.6 \text{ kJ/mol}$$

Hess's Law

Many compounds cannot be directly synthesized from their elements and in these cases, $\Delta_f H_m^\ominus$ can be determined by an indirect approach, which is based on Hess's law. Hess's law means that when reactants are converted to products, the change in enthalpy is the same whether the reaction takes place in one step or in a series of steps. That is to say, we can calculate $\Delta_r H_m^\ominus$ for the overall reactions by breaking down the reaction of interest into a series of reactions for which $\Delta_r H_m^\ominus$ can be measured.

Let's calculate the standard enthalpy of the formation of CO
$$C(\text{graphite}) + 1/2 O_2 = CO \quad \Delta_r H_m^\ominus$$

However we cannot measure the enthalpy change of CO directly. We must use indirect route based on to calculate it.
$$C(\text{graphite}) + O_2 = CO_2 \qquad \Delta_r H_1^\ominus = -393.5 \text{ kJ/mol}$$
$$CO_2 = CO + 1/2 O_2 \qquad \Delta_r H_2^\ominus = 283.0 \text{ kJ/mol}$$

The equation for the one-step reaction is the sum of the equations for the two steps. So the enthalpy change for the one-step reaction is the sum of the $\Delta_r H_m^\ominus$ for the two steps
$$\Delta_r H_m^\ominus = \Delta_r H_1^\ominus + \Delta_r H_2^\ominus = (-393.5 + 283.0) \text{ kJ} = -110.5 \text{ kJ}$$

Sample 4.3 Please calculate the $\Delta_r H_m^\ominus$ of the reaction
(1) C+D=A+B   (2) 2C+2D=2A+2B   (3) A+B=E
Suppose A+B=C+D $\Delta_r H_1^\ominus = -40.0 \text{ kJ/mol}$
         C+D=E $\Delta_r H_2^\ominus = +60.0 \text{ kJ/mol}$

Solution:
According to the Hess's Law
(1) $\Delta_r H_m^\ominus = -\Delta_r H_1^\ominus = +40.0 \text{ kJ/mol}$
(2) $\Delta_r H_m^\ominus = 2(-\Delta_r H_1^\ominus) = +80.0 \text{ kJ/mol}$
(3) $\Delta_r H_m^\ominus = \Delta_r H_1^\ominus + \Delta_r H_2^\ominus = (-40.0) + (+60.0) = +20.0 \text{ kJ/mol}$

Sample 4.4 Please calculate the $\Delta_r H_m^\ominus$ of the reaction
$$FeO(s) + CO(g) = Fe(s) + CO_2(g) \qquad *$$
(1) $Fe_2O_3(s) + 3CO(g) = 2Fe(s) + 3CO_2(g)$ $\Delta_r H_1^\ominus = -27.6 \text{ kJ/mol}$
(2) $3Fe_2O_3(s) + CO(g) = 2Fe_3O_4(s) + CO_2(g)$ $\Delta_r H_2^\ominus = -58.58 \text{ kJ/mol}$
(3) $Fe_3O_4(s) + CO(g) = 3FeO(s) + CO_2(g)$ $\Delta_r H_3^\ominus = +38.07 \text{ kJ/mol}$

Solution:
According to the Hess's Law
$$* = [(1) \times 3 - (2) - (3) \times 2]/6$$

$$\Delta_r H_m^\ominus = [3\Delta_r H_1^\ominus - \Delta_r H_2^\ominus - 2\Delta_r H_3^\ominus]/6$$
$$= [3\times(-27.6)-(-58.58)-2\times(+38.07)]/6$$
$$= -16.73 \text{ kJ/mol}$$

4.4 Entropy and the Second Law of Thermodynamics

Entropy

One of the main objectives in studying thermodynamic is to predict if a reaction will occur under a special set of conditions. A reaction that can occur under the given set of conditions (for example, at a certain pressure, temperature and concentration) is called a spontaneous reaction.

The spontaneous reaction includes many of the examples.

For example, water freezes spontaneously below 0 ℃ and ice melts spontaneously above 0 ℃. A waterfall runs downhill spontaneously, but never up. Heat transfers from a hotter object to a colder one spontaneously.

A spontaneous reaction is the process that occurs spontaneously in one direction but cannot take place spontaneously in the opposite direction under the same conditions.

In order to predict the spontaneity of a reaction, we need to know two things about the system, the change in enthalpy and entropy. The entropy is a measure of the disorder or randomness of a system with a symbol S. The units of S are J/K or kJ/K. The more ordered a system, the smaller its entropy. On the contrary, the greater disorder of a system, the greater its entropy.

The standard entropy, with a symbol $S_m^\ominus$, means the entropy of a substance in a standard state. The units of $S_m^\ominus$ are J/K · mol or kJ/K · mol. Elements and compounds have nonzero entropies and the standard entropies of all the substances are positive.

These guidelines are helpful in charging standard entropies.

(1) As long as substances are in the same physical state, entropy increases down groups in the periodic chart.

(2) If species having similar structures are compared, the more atoms that are joined together, the greater the entropy.

Sample 4.5 Predict the sign of the entropy change for each of the following change and explain you answer.

(1) $H_2O(s) \longrightarrow H_2O(l)$
(2) $CH_4(g) + 2O_2(g) \longrightarrow CO_2(g) + 2H_2O(l)$
(3) $N_2(g) + 3H_2(g) \longrightarrow 2NH_3(g)$

Solution

(1) $\Delta S > 0$ The state of H_2O change from solid to liquid, so $\Delta S > 0$
(2) $\Delta S < 0$ When $CH_4(g)$ and $O_2(g)$ react to $CO_2(g)$ and $2H_2O(l)$, the volume decreases and $\Delta S < 0$
(3) $\Delta S < 0$ When $N_2(g)$ and $H_2(g)$ react to $NH_3(g)$, the volume decreases and $\Delta S < 0$

Entropy is a state function, a standard entropy change is the change in entropy that accompanies the conversion of reactions in their standard states to products in their standard states.

For any reaction

$$\text{Products} \rightarrow \text{Reactants}$$

We can calculate the standard entropy change of the reaction

$$\Delta_r S_m^\ominus = \sum_i v_i S_m^\ominus (\text{products}) - \sum_i v_i S_m^\ominus (\text{reactants}) \qquad (4.11)$$

Where v_i is stoichiometric coefficient.

The Second Law of Thermodynamics

If we assume that spontaneous processes occur so as to decrease the energy of a system, we can explain why a large number of exothermic reactions are spontaneous. For example,

$$CH_4(g) + 2O_2(g) = CO_2(g) + 2H_2O(l) \quad \Delta_r H_m^\ominus = -890.4 \text{ kJ/mol}$$

The combustion of methane, which is an exothermic reaction, is spontaneous. But consider this reaction

$$H_2O(s) = H_2O(l) \quad \Delta_r H_m^\ominus = 6.01 \text{ kJ/mol}$$

In this case, ice melts spontaneously above 0 ℃ even if the reaction is endothermic.

From the study of many examples, we come to the conclusion. It is possible for an endothermic reaction to be spontaneous and also possible for an exothermic reaction to be spontaneous. In other words, we cannot decide if a reaction will occur spontaneously solely according to the energy changes in the system.

The second law of thermodynamics expresses the connection between entropy and the spontaneity of a reaction. The entropy of the universe increases in a spontaneous process and remains unchanged in an equilibrium process.

The second law of thermodynamics can be expressed as follows.

$$\Delta S_{univ} = \Delta S_{sys} + \Delta S_{surr} > 0 \quad \text{for a spontaneous process}$$
$$\Delta S_{univ} = \Delta S_{sys} + \Delta S_{surr} = 0 \quad \text{for an equilibrium process}$$

Where ΔS_{univ}, ΔS_{sys}, and ΔS_{surr} represent the changes in entropy that occur in the universe, in the system, and in the surroundings respectively.

The second law of thermodynamics tells us that ΔS_{univ} must be greater than zero for a spontaneous process, but it does not set a restriction on either ΔS_{sys} or ΔS_{surr}.

4.5 Spontaneous Processes and Gibbs Free Energy

The second law of thermodynamics says that ΔS_{surr} is greater than zero in a spontaneous process. Thus, we must calculate both ΔS_{sys} and ΔS_{surr} to determine the sign of ΔS_{surr}. However, We are usually concerned only with what happens in a particular system or reaction, and the calculation of ΔS_{surr} is quite difficult. Therefore, we need another thermodynamic function to help us determine whether a process or a reaction will occur spontaneously if we consider only the system itself.

From above, we know that for a spontaneous process it has

$$\Delta S_{univ} = \Delta S_{sys} + \Delta S_{surr} > 0$$
$$\Delta S_{surr} = -\Delta H_{sys}/T$$

So that

$$\Delta S_{univ} = \Delta S_{sys} - \Delta H_{sys}/T > 0$$

For convenience, we change the style of the equation

$$-T\Delta S_{univ} = \Delta H_{sys} - T\Delta S_{sys} < 0$$

The equation says that for a process carried out at constant pressure and temperature, if the change in ethalpy and entropy of the system are such that $\Delta H_{sys} - T\Delta S_{sys}$ is less than zero, the process must be spontaneous.

We can use another thermodynamic function which is called Gibbs free energy (G), or simply free energy to express the spontaneity of a process more directly. The units of G are J/mol or kJ/mol.

$$G = H - TS \tag{4.12}$$

Where H is enthalpy, T is temperature and S is entropy. The G has units of both H and TS. Like H and S, G is a state function. The change in free energy for any system to other thermodynamic changes.

$$\Delta G = \Delta H - \Delta(TS)$$

For a constant-temperature process, the change in G of a system is

$$\Delta G = \Delta H - T\Delta S \tag{4.13}$$

In this context free energy is the energy cavailable to do work. Thus, if a particular reaction is accompanied by a release of usable energy (that is, if ΔG is negative), this fact alone guarantees that it is spontaneous, and there is no need to worry about what happens to the rest of the universe.

We can now summarize that if ΔG of a process is negative ($\Delta G < 0$), it guarantees that the process is spontaneous. If ΔG of a process is positive ($\Delta G > 0$), this process is nonspontaneous and the process is spontaneous in the opposite direction. The system is at equilibrium when ΔG of a process is zero ($\Delta G = 0$).

If all substances are in their standard states, then

$$\Delta_r G_m^\ominus = \Delta_r H_m^\ominus - T\Delta_r S_m^\ominus \tag{4.14}$$

4.6 Temperature and Direction of Spontaneous Change

Temperature and Gibbs Free Energy

We are usually interested in examining reaction at different temperatures. But how is the change of free energy affected by the change in temperature?

From the equation

$$\Delta G = \Delta H - T\Delta S$$

The expression for ΔG is the sum of two contributions, ΔH and $-T\Delta S$. The sign of ΔG will depend on the signs and magnitudes of ΔH and $-T\Delta S$. When both ΔH and $-T\Delta S$ are positive, ΔG will always be positive, the process is nonspontaneous. When both ΔH and $-T\Delta S$ are negative, ΔG will always be positive, the process is spontaneous. Generally, ΔH and ΔS change very little with temperature, so when ΔH and $-T\Delta S$ have opposite signs, the sign of ΔG will depend on the magnitudes of these two terms, in which instances temperature is an important consideration. The value of T directly affects the magnitude of $-T\Delta S$. As the temperature increases, the magnitude of $-T\Delta S$ increases and it will become relatively more important in determining the sign and magnitude of ΔG.

At equilibrium, $\Delta G = 0$. This is useful to solve problems. For example, in any equilibrium for a normal boiling point both the liquid and the vapor are in their standard states. As a result, $\Delta G = \Delta G^\ominus = 0$. So that we can use the equation below to calculate the normal melting points and normal sublimation points of any solids and the normal boiling point of any other pure liquid.

$$\Delta G^\ominus = \Delta H^\ominus - T\Delta S^\ominus = 0 \tag{4.15}$$

Solving the equation for T, we get $T = \Delta H^\ominus / \Delta S^\ominus$.

Temperature and Direction of Spontaneous Change

We can often predict how ΔG will change in temperature. If we assume that the values of ΔH and ΔS do not change with temperature, we can estimate the value of ΔG at temperatures other than 298K.

Sample 4.6 Please calculate the highest temperature of spontaneous reaction under standard state.
$$N_2(g)+3H_2(g)=2NH_3(g)$$

Materials	$\Delta_f H_m^\ominus/(kJ/mol)$	$S_m^\ominus/[J/(K \cdot mol)]$
$N_2(g)$	—	191.6
$H_2(g)$	—	130.7
$NH_3(g)$	−45.9	192.8

Solution:
$$\Delta_r H_m^\ominus = (-45.9) \times 2 = -91.8 \text{ kJ/mol}$$
$$\Delta_r S_m^\ominus = 192.8 \times 2 - (191.6 + 130.7 \times 3) = -189.9 \text{ J/(K} \cdot \text{mol)}$$

According to the equation (4.14) $\Delta_r G_m^\ominus = \Delta_r H_m^\ominus - T\Delta_r S_m^\ominus$
If the process is spontaneous, $\Delta_r G_m^\ominus \leqslant 0$
or
$$T_{max} \leqslant \Delta_r H_m^\ominus / \Delta_r S_m^\ominus = \frac{-91.8 \times 1000}{-189.9} = 483 \text{ K}$$

Sample 4.7 According to the data of $\Delta_f H_m^\ominus$ and $S_m^\ominus$ (298K), please calculate $\Delta_r H_m^\ominus$ (298K), $\Delta_r S_m^\ominus$ (298K) and $\Delta_r G_m^\ominus$ (850K) of the reaction below,
$$MgCO_3(s) = MgO(s) + CO_2(g)$$

Materials	$\Delta_f H_m^\ominus/(kJ/mol)$	$S_m^\ominus/[J/(K \cdot mol)]$
$MgCO_3(s)$	−1096	65.7
$CO_2(g)$	−393	214
$MgO(s)$	−601.6	27

Solution:

According to the equation,
$$\Delta_r H_m^\ominus = \sum_i v_i \Delta_f H_m^\ominus(\text{products}) - \sum_i v_i \Delta_f H_m^\ominus(\text{reactants})$$
$$= (-601.6) + (-393) - (-1096) = 101.4 \text{ kJ/mol}$$
$$\Delta_r S_m^\ominus = \sum_i v_i S_m^\ominus(\text{products}) - \sum_i v_i S_m^\ominus(\text{reactants})$$
$$= 27 + 214 - 65.7 = 175.3 \text{ J/(K} \cdot \text{mol)}$$
$$\Delta_r G_m^\ominus = \Delta_r H_m^\ominus - T\Delta_r S_m^\ominus$$
$T = 850K$, $\Delta_r G_m^\ominus = 101.4 - 850 \times 175.3 \times 10^{-3} = -47.6 \text{ kJ/mol}$

4.7 Standard Free Energy of Formation and Standard Free Energy of Reaction

The standard free energy of formation of a substance is the free energy change that results when one mole of the substance is formed from its elements with all substances in their standard states. Its symbol is $\Delta_f G_m^\ominus$ and its unit is kJ/mol. Just as the standard enthalpies of formation, the standard free energies of elements in their most stable form are zero.

The standard free energy of reaction ($\Delta_r G_m^\ominus$) is the free-energy change for a reaction when it occurs under standard state.

Now, we can calculate $\Delta_r G_m^\ominus$ of the reaction according to $\Delta_f G_m^\ominus$.
$$\text{Reactants} \rightarrow \text{Products}$$

$$\Delta_r G_m^\ominus = \sum_i v_i \Delta_f G_m^\ominus(\text{products}) - \sum_i v_i \Delta_f G_m^\ominus(\text{reactants}) \qquad (4.16)$$

Where v_i is stoichiometric coefficient. The term $\Delta_f G_m^\ominus$ is the standard free energy of formation of a compound.

Sample 4.8 According to the data of $\Delta_f G_m^\ominus$ (298K), please calculate $\Delta_r G_m^\ominus$ of the reactions below.

Materials	HAc(aq)	Ac⁻(aq)	AgCl(s)	Ag⁺(aq)	Cl⁻(aq)	H⁺(aq)
$\Delta_f G_m^\ominus$/(kJ/mol)	−399.61	−372.46	−109.72	77.11	−131.17	0

(1) $HAc(aq) = H^+(aq) + Ac^-(aq)$ (2) $AgCl(s) = Ag^+(aq) + Cl^-(aq)$

Solution:

According to the equation 4.15,

$$\Delta_r G_m^\ominus = \sum_i v_i \Delta_f G_m^\ominus(\text{products}) - \sum_i v_i \Delta_f G_m^\ominus(\text{reactants})$$

(1) $\Delta_r G_m^\ominus = (-372.46) - (-399.61) = 27.15$ kJ/mol

(2) $\Delta_r G_m^\ominus = (-131.17) + 77.11 - (-109.72) = 55.66$ kJ/mol

Sample 4.9 According to the data of $\Delta_f H_m^\ominus$、$\Delta_f G_m^\ominus$ and $S_m^\ominus$ (298K):

Materials	$\Delta_f H_m^\ominus$/(kJ/mol)	$\Delta_f G_m^\ominus$/(kJ/mol)	$S_m^\ominus$/[J/(K·mol)]
CO(g)	−110.5	−137.2	197.7
CO₂(g)	−393.5	−394.4	213.8
H₂O(g)	−241.8	−228.6	—
H₂(g)	—	—	130.7

(1) Please calculate $\Delta_r H_m^\ominus$、$\Delta_r G_m^\ominus$ and $\Delta_r S_m^\ominus$ of the reaction, $H_2O(g) + CO(g) =\!\!=\!\!= H_2(g) + CO_2(g)$.

(2) Please calculate $\Delta_r S_m^\ominus$ of $H_2O(g)$.

Solution:

According to the equations,

(1) $\Delta_r H_m^\ominus = (-393.5) - [(-110.5) + (-241.8)] = -41.2$ kJ/mol

$\Delta_r G_m^\ominus = (-394.4) - [(-137.2) + (-228.6)] = -28.5$ kJ/mol

$$\Delta_r G_m^\ominus = \Delta_r H_m^\ominus - T\Delta_r S_m^\ominus$$

$\Delta_r S_m^\ominus = (\Delta_r H_m^\ominus - \Delta_r G_m^\ominus)/T = \dfrac{[-41.2 - (-28.5)] \times 10^3}{298} = -42.6$ J/(K·mol)

(2) According to the equation:

$$\Delta_r S_m^\ominus = \sum_i v_i S_m^\ominus(\text{products}) - \sum_i v_i S_m^\ominus(\text{reactants})$$

$S_m^\ominus(H_2O) = (130.7 + 213.8) - 197.7 - (-42.6) = 188.4$ J/(K·mol)

Key Words

thermodynamics ['θə:məudai,næmiks]　热力学

thermochemistry ['θə:məu'kemistri]　热化学

internal energy [in'tə:nl'enədʒi]　内能

state [steit]　状态

function ['fʌŋkʃən]　函数

enthalpy ['enθælpi, en'θælpi]　焓

enthalpy change ['enθælpi, en'θælpi tʃeindʒ]　焓变

enthalpy of reactions　反应焓

entropy ['entrəpi]　熵

spontaneous [spɔn'teinjəs, -niəs]　自发的

Exercises

1. Calculate the energy change for this process.

(1) There is a heat transfer of 60J from the surroundings to the system and the work done was 40J in this process.

(2) There is a heat transfer of 70J from the system to the surroundings and the work done was 50J in this process.

Answer: (1)+100 kJ, (2)+20 kJ

2. Please calculate the $\Delta_r H_m^{\ominus}$ of the following reaction,
$$2Na_2O_2(s) = 2Na_2O(s) + O_2(s)$$

Answer: 177.4 kJ/mol

3. Please calculate the $\Delta_r H_m^{\ominus}$ of the following reaction,
$$CaCO_3 = CaO(s) + CO_2(g)$$

Answer: 179.2 kJ/mol

4. Please calculate the $\Delta_f H_m^{\ominus}$ of $MnO_2(s)$,

(1) $MnO_2(s) = MnO(s) + \frac{1}{2}O_2(g)$ $\Delta_r H_1^{\ominus} = +134.8$ kJ/mol;

(2) $Mn(s) + MnO_2(s) = 2MnO(s)$ $\Delta_r H_2^{\ominus} = -250.1$ kJ/mol

Answer: −519.7 kJ/mol

5. Please calculate the $\Delta_r H_3^{\ominus}$ of the reaction, $C(s) + \frac{1}{2}O_2(g) = CO(g)$

(1) $C(s) + O_2(g) = CO_2(g)$; $\Delta_r H_1^{\ominus} = -393.5$ kJ/mol

(2) $CO(g) + \frac{1}{2}O_2(g) = CO_2(g)$; $\Delta_r H_2^{\ominus} = -283.0$ kJ/mol

Answer: −110.5 kJ/mol

6. Please calculate the $\Delta_f H_m^{\ominus}$ of CuO (s),

(1) $Cu_2O(s) + \frac{1}{2}O_2(g) = 2CuO(s)$ $\Delta_r H_1^{\ominus} = -143.7$ kJ/mol;

(2) $CuO(s) + Cu(s) = Cu_2O(s)$ $\Delta_r H_2^{\ominus} = -11.5$ kJ/mol

Answer: −155.2 kJ/mol

7. Predict the sign of the entropy change for each of the following change.

(1) $Br_2(l) \rightarrow Br_2(g)$ (2) $Ar(0.1kPa) \rightarrow Ar(0.01kPa)$

(3) $HF(g) \rightarrow HCl(g)$ (4) $CH_4(g) \rightarrow C_2H_6(g)$

(5) $NH_4Cl(s) \rightarrow NH_4I(s)$ (6) $HCl(g, 298K) \rightarrow HCl(g, 1000K)$

Answer: (1) $\Delta S > 0$ (2) $\Delta S > 0$ (3) $\Delta S < 0$ (4) $\Delta S < 0$ (5) $\Delta S < 0$ (6) $\Delta S > 0$

8. Please calculate the lowest temperature of spontaneous reaction under standard state. $CaCO_3(s) = CaO(s) + CO_2(g)$

$\Delta_r H_m^{\ominus}$/(kJ/mol)	$\Delta_r S_m^{\ominus}$/[J/(K·mol)]
178.3	160.4

Answer: 1112 K

9. Please calculate the $\Delta_r H_m^{\ominus}$, $\Delta_r G_m^{\ominus}$ and $\Delta_r S_m^{\ominus}$ of the reaction.
$$H_2O(g) + CO(g) = H_2(g) + CO_2(g)$$

Materials	$\Delta_f H_m^{\ominus}$/(kJ/mol)	$\Delta_f G_m^{\ominus}$/(kJ/mol)	$S_m^{\ominus}$/[J/(K·mol)]
CO(g)	−110.5	−137.3	197.9
CO_2(g)	−393.5	−394.4	213.6
H_2O(g)	−241.8	−228.6	188.8
H_2(g)	—	—	130.6

Answer: $\Delta_r H_m^{\ominus} = -41.2$ kJ/mol $\Delta_r S_m^{\ominus} = -42.5$ J/K·mol $\Delta_r G_m^{\ominus} = -28.5$ kJ/mol

Chapter 5 Chemical Kinetics

Chemical thermodynamics is the study of the energy changes in the chemical reaction processes, the possibility of the reaction, and the extent to which the reaction occurs. But it can not tell us how fast do reactions go, because it does not concern with the time that required reaching the chemical equilibrium. Although many reactions should go to completion according to thermodynamics, they may take place too slowly to be useful. For example, at room temperature, the equilibrium for the formation of water from hydrogen and oxygen is very thermodynamically favorable, $\Delta_r G_m^{\ominus} = -237.19$ kJ/mol, but the rate of the reaction is so slow that a mixture of hydrogen and oxygen can exist indefinitely without any water forming.

Chemical kinetics is the study of the rates and mechanisms of chemical reactions. In this chapter we are primarily concerning with the rate of chemical reaction and the factors that influence it. In addition, the study of reaction rates yields further research into reaction mechanism, which is the sequence of steps by which reactants are converted to products. An understanding of reaction mechanisms and relative relates of reaction has both practical and theoretical applications.

5.1 Reaction Rates

We are all familiar with the concept of rate or speed, which is often used to describe the change in a quantity that occurs per unit of time. For example, the rate at which an object travels through space is the change in distance per unit of time, such as kilometers per hour. In chemical kinetics, the reaction rate is defined as the rate at which the concentrations of reactants and products change, which is usually measured in terms of the increase in concentration of any of the products per unit of time, or the decrease in concentration of any of the reactants per unit of time. The usual unit of concentrations is moles per liter (mol/L), and the usual unit of time is seconds.

For example, consider the reaction of the decomposition of dinitrogen pentoxide, N_2O_5, to nitrogen dioxide and oxygen.

$$2N_2O_5(g) = 4NO_2(g) + O_2(g)$$

We can express the reaction average rate in terms of the decrease in the concentration of N_2O_5 in a certain time interval,

$$\bar{v}_{N_2O_5} = -\frac{\Delta c(N_2O_5)}{\Delta t} = -\frac{c_2(N_2O_5) - c_1(N_2O_5)}{t_2 - t_1} \qquad (5.1)$$

Where $c_1(N_2O_5)$ is the concentration of N_2O_5 at time t_1, and $c_2(N_2O_5)$ is the concentration of N_2O_5 at time t_2. In the given time interval $\Delta t = t_2 - t_1$, the concentration of N_2O_5 has decrease by the amount $\Delta c(N_2O_5) = c_2(N_2O_5) - c_1(N_2O_5)$. Note the negative sign, it always occurs in a rate expression for a reactant in order to give a positive quantity for the rate. Because the concentration of N_2O_5 decreases as reaction takes place, thus $\Delta c(N_2O_5)$ is negative and $-\Delta c(N_2O_5)/\Delta t$ is positive. Table 5.1 shows the data of the concentrations of reactants and products at different times.

For example, over the interval from $t=200$s to $t=300$s, the average rate is

$$\bar{v}_{N_2O_5} = -\frac{\Delta c(N_2O_5)}{\Delta t} = -\frac{0.0240-0.0284}{300-200} = 4.40\times 10^{-5}\,\text{mol/(L}\cdot\text{s)}$$

As the reaction proceeds, the rate keeps changing as reactants are used up, as is shown in Table 5.1.

Table 5.1 Concentrations of reactants and products for the decomposition of N_2O_5 at 55℃

t/s	Concentration/(mol/L)			Δt/s	$-\Delta c(N_2O_5)$ /(mol/L)	Decomposition Rate of N_2O_5 $-\frac{\Delta c(N_2O_5)}{\Delta t}$
	N_2O_5	NO_2	O_2			
0	0.0400	0	0	—	—	—
100	0.0338	0.0126	0.0032	100	0.0062	6.2×10^{-5}
200	0.0284	0.0230	0.0058	100	0.0054	5.4×10^{-5}
300	0.0240	0.0320	0.0080	100	0.0044	4.4×10^{-5}
400	0.0202	0.0394	0.0098	100	0.0038	3.8×10^{-5}
500	0.0172	0.0458	0.0114	100	0.0030	3.0×10^{-5}
600	0.0144	0.0512	0.0128	100	0.0028	2.8×10^{-5}

We also can express the average rate in terms of the concentrations of products.

$$\bar{v}_{NO_2} = \frac{\Delta c(NO_2)}{\Delta t} \tag{5.2}$$

$$\bar{v}_{O_2} = \Delta c(O_2) \tag{5.3}$$

Note that the sign is plus for products, because the concentration of the products (NO_2 and O_2) increase with time, thus the values of $\Delta c(NO_2)$ and $\Delta c(O_2)$ are positive. Then, over the interval from $t=200$ s to $t=300$ s, the average rate of formation of NO_2 is

$$\bar{v}_{NO_2} = \frac{\Delta c(NO_2)}{\Delta t} = \frac{0.0320-0.0230}{300-200} = 9.00\times 10^{-5}\,\text{mol/(L}\cdot\text{s)}$$

The result shows that the rates of the reaction expressed by different substances are different, but they are easily related to each other by the stoichiometry of the reaction. We can see from the above equation that two N_2O_5 molecules decompose for four NO_2 molecules and each O_2 molecule produced, so the rate of formation of O_2 is only one-half the rate of decomposition of N_2O_5 and the rate of formation of NO_2 is twice of the rate of decomposition of N_2O_5. Thus, we have

$$-\frac{1}{2}\frac{\Delta c(N_2O_5)}{\Delta t} = \frac{1}{4}\frac{\Delta c(NO_2)}{\Delta t} = \frac{\Delta c(O_2)}{\Delta t} \tag{5.4}$$

Note that the fractions here are related the coefficients in the balanced chemical equation.

For any reaction

$$a\text{A}+b\text{B}=g\text{G}+h\text{H}$$

The rates are related by

$$-\frac{1}{a}\frac{\Delta c(\text{A})}{\Delta t} = -\frac{1}{b}\frac{\Delta c(\text{B})}{\Delta t} = \frac{1}{g}\frac{\Delta c(\text{G})}{\Delta t} = \frac{1}{h}\frac{\Delta c(\text{H})}{\Delta t} \tag{5.5}$$

The above equations give the average rate over the time interval Δt. If the time interval Δt is very short, the average rate becomes nearly the instantaneous rate of reaction, that is, the rate at one specific instant. The value of the instantaneous rate can be obtained by drawing a tangent to the concentration-time graph at the desired point. The instantaneous rate of reaction is equal to the negative of the slope of the tangent. Thus the instantaneous rate for the above reaction can be defined as

$$v_{N_2O_5} = \lim_{\Delta t \to 0} \left[-\frac{\Delta c(N_2O_5)}{\Delta t} \right] \quad (5.6)$$

Such a tangent line has been drawn at the 200 seconds point in Figure 5.1.

$$v_{N_2O_5} = -slope = \frac{AC}{BC} \quad (5.7)$$

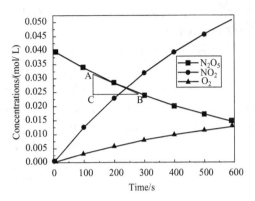

Figure 5.1 The concentrations of N_2O_5, NO_2 and O_2 vs. time

The initial rate of reaction, which means the rate of reaction at time zero when the reactants are mixed and reaction starts, is the most important instantaneous rate, which is often used in studying the relationship between reaction rate and concentrations.

5.2 Theories of Reaction Rates

The theory of reaction rates explains how chemical reactions occur and why reaction rates differ for different reactions. The individual factors which affect the rate of a reaction, such as temperature, concentration, catalyst, are discussed on separate pages.

Collision Theory

The collision theory of reaction rates, proposed by William Lewis in 1918, is based on the kinetic-molecular theory of gases. It assumes that, in order for reaction to occur, reactant molecules must collide.

Consider the reaction of gases A with B to form C. The calculation based on the kinetic molecular theory shows that, at ordinary pressures and temperatures, the collision frequency, Z, between A and B molecules is about 10^{27} collisions/(L·s). If every collision resulted in reaction, then the reaction would be over almost instantaneously, yet observation shows that most reactions occur at a much lower rate. This means that, in many cases, most collisions do not result in reaction. Therefore, besides collision frequency, there are other factors we should consider for the reaction rate.

In consequence, it is assumed further that for collision to result in reaction, molecules must collide with a certain minimum energy to break the bonds that exist in them. If molecules collide with a kinetic energy less than the minimum energy, they will merely bounce off one another without reacting. The minimum collision energy required for reaction is called the activation energy, E_a. The value of E_a depends on the particular reaction. In order to react, colliding molecules must possess a total kinetic energy equal to or greater than the activation energy. The molecules with sufficient energy for reaction are called activated molecules. The fraction of the activated molecules and the fraction of effective collisions for reaction are both related to the activation energy, it can be shown using the Maxwell-Boltzmann equations.

$$f = e^{-\frac{E_a}{RT}} \quad (5.8)$$

Where e is the base of the natural logarithm system; R is the ideal-gas constant which equals 8.31 J/(mol·K); and T is Kelvin temperature. For a particular reaction, the value of f is fixed at a given temperature.

However, not every collision with sufficient energy results in reaction. It is further assumed that no reaction will occur unless the reactant molecules collide with the proper orientation. The fraction of the collisions in which the molecules have an orientation favorable for reaction is called orientation factor, P.

For example, the reaction
$$NO_2(g) + CO(g) = NO(g) + CO_2(g)$$
Figure 5.2 shows two possible molecular orientations in the collisions of NO_2 and CO.

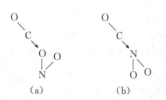

Figure 5.2 Molecular orientations in the collisions of NO_2 and CO

The reaction can only happen if the CO molecule approaches with its C atom toward the O atom of NO_2 molecule, as shown in Fig 5.2(a), because this orientation allows electrons rearrange themselves so as to favor the new CO_2 molecule. Any other collision between the two molecules will be ineffective and the molecules merely bounce off each other.

The general conclusion according to collision theory, then, is that the reaction rate is proportional to the following three factors.

(1) The collision frequency, Z.

(2) The fraction of collisions with the required energy, f.

(3) The fraction of collisions in which the reactant molecules have the proper orientations, P.

In other words
$$\bar{v} \propto ZPe^{-\frac{E_a}{RT}} \tag{5.9}$$

From the equation, we can see that the reactions with large activation energy have small reaction rates. Because the higher the activation energy, the smaller the fraction of the molecules having the energy requisite for reaction upon collision, as a result of a smaller rate law.

Transition State Theory

Collision theory is intuitionistic and easily understandable, but it is limited in that it does not explain the role of activation energy. The transition-state theory assumes that a collision with sufficient energy to react first results in combining of the colliding molecules to form a more complex structure of transitory existence, which is called an activated complex. This short-lived intermediate is not a stable molecule. it may either revert to the original reactants or yield the products. For example, in the reaction between nitrogen dioxide and carbon monoxide, an oxygen atom is transferred from NO_2 to CO.
$$NO_2(g) + CO(g) = NO(g) + CO_2(g)$$
When the two molecules approach each other, their electron clouds repel each other. Therefore, the molecules must have enough kinetic energies to overcome the repulsion. Then the nuclei can get close enough for the electron clouds to rearrange and form new bonds. When the molecules with sufficient energies come together with proper orientation, the activated complex will be formed. We can represent the formation of the activated complex this way.

$$O—N—O + C—O \leftrightarrow [O—N\cdots O\cdots C—O] \rightarrow O—N + O—C—O$$
reactants　　　　　activated complex　　　　products

Figure 5.3 shows the change in potential energy that occurs during the progress of the reaction. At first, when the rapidly moving NO_2 and CO molecules collide, the potential energy increases to a maximum corresponding to the formation of [O—N⋯O⋯C—O], the activated complex, then the potential energy decreases to that of the products, $NO+CO_2$.

The difference in energy between the activated complex and the reactant molecules is the activation energy for forward reaction. We can think of activation energy as a barrier that prevents less energetic molecules from reacting. Only if the reactant molecules have a total kinetic energy which is equal to or greater than the activation energy, it is possible for the reaction to occur.

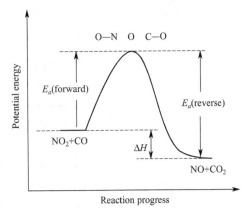

Figure 5.3 Potential energy profiles for the reaction $NO_2+CO=NO+CO_2$

The difference in energy between the reactants and products equals the enthalpy change for the reaction, ΔH, as shown in Figure 5.3. The potential energy of the products is less than that of the reactants, so the heat energy is released when the reaction goes in the forward direction, that is, $\Delta H < 0$, and the overall reaction is exothermic. On the other hand, if the potential energy of the products is higher than that of the reactants, then heat will be absorbed by the reacting mixture from the surroundings, so $\Delta H > 0$, and the overall reaction is endothermic.

5.3 Effect of Concentration on Rate of Reaction

A lot of experimental results have proved that, for many reactions, increasing the concentration of the reactants increases the rate of reaction. These cases can be explained by the collision theory. Because the fraction of molecules with sufficient energy is fixed at a constant temperature, when the concentration of the reactants increases. Then the number of molecules with sufficient energy per unit volume increases, thus the chances of collisions are greater, so that a higher reaction rate results. Before we discuss the relationship between concentration and reaction rate, we need to know the reaction mechanism firstly.

Reaction Mechanism

In many cases, a balanced chemical equation is merely a description of the overall result of a chemical reaction. However, it provides no information about how the reaction actually takes place at the molecular level. The reaction may involve a series of simple steps. The process by which a reaction occurs is called the reaction mechanism.

Elementary reactions

The elementary reactions are those that occur in a single step. For example,

$$SO_2Cl_2 = SO_2 + Cl_2 \qquad ①$$
$$NO_2 + CO = NO + CO_2 \qquad ②$$
$$2NO_2 = 2NO + O_2 \qquad ③$$

The number of molecules reacting in an elementary step determines the molecularity of a reaction. If a single molecule is involved in the reaction, the reaction is said to be unimolecular reaction. One example of a unimolecular reaction is the decomposition of SO_2Cl_2 (Equation ①). Elementary steps that involve the collision of two reactant molecules are said to be bimolecular. For example, the above reactions ② and ③ is bimolecular reactions. Elementary steps involving the participation of three molecules are called termolecular reactions, for example, $H_2 + 2I \longrightarrow HI$. Termolecular steps are rare because the probability of three molecules colliding at the same time is very small. No elementary processes with molecularity greater than three are known.

Multistep Mechanisms

In fact, most reactions are complex which have a mechanism consisting of a sequential elementary steps, that is, they have multistep mechanisms. It is often found that one of the steps is much slower than any of the others. The overall rate of a reaction is controlled by the rate of the slowest elementary step. The slowest step in a multi-step reaction is known as the rate-determining step (or rate-limiting step).

For example, the following reaction
$$H_2(g) + I_2(g) \longrightarrow 2HI(g)$$
The mechanism of this reaction has been found by experiment to involved two elementary steps as follows,

Step 1: $\quad\quad\quad\quad I_2 \longrightarrow 2I \quad\quad$ (fast step)
Step 2: $\quad\quad\quad\quad H_2 + 2I \longrightarrow 2HI \quad\quad$ (slow step)

In this reaction, the first step is found to be much faster than the second step. Therefore the step 1 is the rate-determining step. I is called intermediate, a species which is formed in step 1 but consumed in the next step. The overall equation for the reaction is found by adding the two elementary steps.

Rate Laws

The algebraic expression of the relationship between the concentrations of reactants and the reaction rate is called the rate law for the reaction. Many rate laws have the form
$$v = kc^m(\text{reactant}_1)c^n(\text{reactant}_2)\cdots \quad\quad (5.10)$$
The rate law for any reaction must be determined from experimental measurements of reaction rates. However, for elementary reaction, the rate law can be written directly according to the law of mass action. That is, the rate of elementary reaction is proportional to the product of the concentrations of reactants which are raised to some powers, and the power of each reactant is the same as the stoichiometric coefficient in the reaction equation. For example, for a general elementary process
$$aA + bB = cC + dD$$
The rate law is
$$v = kc_A^a c_B^b \quad\quad (5.11)$$
Where, k is the proportionality between the reaction rate and the concentrations of the reactants, which called the rate constant. It is affected by temperature, but it is independent of reactant concentrations.

Note that the law of mass action is only applicable for elementary reactions. But for multi-step reactions, there is no necessary simple relationship between the rate law and the reaction equation, and the rate law must be determined by experiment. The initial rate method is the simplest way to determine rate laws. To see how this method works, we consider the following reaction,

$$2H_2 + 2NO = 2H_2O + N_2$$

The initial concentrations and corresponding initial rates obtained from three experiments are given in the table 5.2.

Table 5.2 Initial concentrations and corresponding initial rates for the reaction $2H_2 + 2NO \longrightarrow 2H_2O + N_2$

Experiment	Initial Concentration of H_2/(mol/L)	Initial Concentration of NO/(mol/L)	Initial Rate /[(mol/L·s)]
1	1.00×10^{-3}	1.00×10^{-3}	8.86×10^{-5}
2	2.00×10^{-3}	1.00×10^{-3}	1.77×10^{-4}
3	1.00×10^{-3}	2.00×10^{-3}	3.54×10^{-4}

The general form of the rate law is

$$v = kc^m(H_2)c^n(NO) \tag{5.12}$$

Experiment 1 and 2 indicate that doubling the concentration of H_2 will double the rate when the concentration of NO is fixed. Using the rate law and the experimental data give

$$\frac{v_2}{v_1} = \frac{k(2.00 \times 10^{-3})^m (1.00 \times 10^{-3})^n}{k(1.00 \times 10^{-3})^m (1.00 \times 10^{-3})^n} = 2^m = 2$$

So the value of m is 1.

Compared Experiment 1 with 3, we can find that when the concentration of NO is doubled, the rate goes up 4 times. Then, we have

$$\frac{v_3}{v_1} = \frac{k(1.00 \times 10^{-3})^m (2.00 \times 10^{-3})^n}{k(1.00 \times 10^{-3})^m (1.00 \times 10^{-3})^n} = 2^n = 4$$

We can get the value of n is 2.

Therefore, the rate law is

$$v = kc(H_2)c^2(NO) \tag{5.13}$$

We can determine the value of the rate constant k by using the rate law and the data from any one of the experiments. Using values from experiment 1, we have

$$k = \frac{v}{c(H_2)c(NO)} = \frac{8.86 \times 10^{-5}}{(1.00 \times 10^{-3})(1.00 \times 10^{-3})^2} = 8.86 \times 10^4 \ L^2/(mol^2 \cdot s)$$

Note that, even though the experimentally determined rate law is the same as that wrote by the law of mass action, we can not get the conclusion that the reaction is elementary reaction. For example, the reaction we have mentioned above

$$H_2 + I_2 = 2HI$$

The experimentally determined rate law is

$$v = kc(H_2)c(I_2) \tag{5.14}$$

This equation is consistent with that wrote by the law of mass action, but the reaction is actually a two-step reaction.

Reaction Order

The reaction order with respect to a given reactants equals the exponents of the concentration of that species in the rate law. The value of the reaction order can be used to show how the concentrations affect the rate of the reaction. The overall order of a reaction is found by adding up the individual orders.

For example, the elementary reaction

$$NO_2 + CO = NO + CO_2$$

The rate law is

$$v = kc(H_2)c(NO) \tag{5.15}$$

Thus the reaction is first order in NO, first order in CO, and second order overall. Obviously, for the elementary reaction, the overall reaction order is equal to the molecularity of the reaction, and is also equal to the sum of coefficients in the chemical equations.

However, for multi-step reactions, you can't deduce anything about the reaction order just by looking at the equation for the reaction. The values of the reaction orders must be determined by experiment. Some examples are giving in the following.

For example, the reaction

$$2Na + 2H_2O = 2NaOH + H_2$$

The experimentally determined rate law is as follows.

$$v = k \quad (5.16)$$

This reaction is zero order with respect to both Na and H_2O, because the concentrations of Na and H_2O don't affect the rate of the reaction. The reaction is also zero order overall.

And the reaction

$$CHCl_3 + Cl_2 \rightleftharpoons CCl_4 + HCl$$

The experimentally determined rate law is as follows.

$$v = k c_{CHCl_3} c_{Cl_2}^{\frac{1}{2}} \quad (5.17)$$

In this case, the order of reaction with respect to $CHCl_3$ is 1, the order of reaction with respect to Cl_2 is 1/2, and the overall order of this reaction is 1.5.

Although reaction orders frequently have positive integer values (particularly 1 or 2), they can be fractional and zero, or even be negative. We can classify a reaction by its orders.

Pay attention to the difference between the sometimes confusing terms "order of reaction" and "molecularity of reaction". Each elementary step has its own molecularity, but molecularities of reactions should not be applied to multi-step reactions. Note, also, that molecularities, unlike reaction orders, must be integers.

The unit of k is dependent on the overall reaction order. In a reaction that is n order overall, the unit of the rate constant must satisfy

Unit of rate = (unit of rate constant) (unit of concentration)^n

The usual unit of rate is given as $mol/(L \cdot s)$, then the unit of rate constant is

$$\frac{\text{units of rate}}{(\text{units of concentration})^n} = \frac{mol/(L \cdot s)}{(mol/L)^n} = mol^{1-n} \cdot L^{n-1} \cdot s^{-1}$$

For example, for zero order reactions ($n=0$), the unit of rate constant is $mol/(L \cdot s)$, and for first order reactions ($n=1$), the units of rate constant are s^{-1}, and so on. Similarly, we can also deduce the overall reaction orders from the units of the rate constant.

5.4 Rate and Temperature

As a rough approximation, for many reactions happening at around room temperature, the rate of reaction doubles for every 10℃ rise in temperature.

According to the kinetic molecular theory of gases, an increase in temperature makes the molecules move faster and so collide more frequently, which will speed up the rate of reaction. However, the calculation results show that a 10℃ rise in temperature produces an increase of about 2% in collision frequency. But experiments found that the rate of reaction will probably have doubled for that increase in temperature, that is to say the effect of increasing collision frequency on the rate of the reaction is very minor. The more important effect is that, as the temperature increases, the fraction of molecules with sufficient energy increases dramatically resulting in more effective collisions.

In 1889, the Swedish chemist Svante Arrhenius summarized the relationship between temperature and rate constant in the form

$$k = A e^{-\frac{E_a}{RT}} \tag{5.18}$$

This is called the Arrhenius equation. In the equation, k is the rate constant, and A is called the frequency factor or the pre-exponential factor, which includes factors like the frequency of collisions and their orientation. It varies slightly with temperature and is often treated as a constant for a given system across small temperature ranges. And e is the base of the natural logarithms, which has a value of $2.71828\cdots$, E_a is the activation energy of the reaction, it has to be expressed in J/mol to fit into the equation, R is the gas constant, T is absolute temperature. The expression, $e^{-\frac{E_a}{RT}}$, counts the fraction of molecular collisions with energies equal to or in excess of activation energy at a particular temperature T.

According to the equation, the rate constant depends exponentially on temperature and activation energy. Furthermore, because of the minus sign associated with the exponent E_a/RT, the rate constant increases with increasing temperature and decreases with increasing activation energy.

This equation can be expressed in a more useful form, which is given by taking the natural logarithm of both sides of the Arrhenius equation.

$$\ln k = -\frac{E_a}{RT} + \ln A \tag{5.19}$$

or, expressed in terms of logarithms to the base 10,

$$\lg k = -\frac{E_a}{2.303RT} + \lg A \tag{5.20}$$

Thus, a plot of $\lg k$ versus $1/T$ gives a straight line. The slop of this line is equal to $-E_a/2.303R$, and the intercept with the ordinate is equal to $\lg A$.

The activation energy can be calculated using the Arrhenius equation if you have values for the rate constant at different temperatures. For example, suppose that at two different temperatures T_1 and T_2, the rate constants are k_1 and k_2 respectively. Then, we have

$$\lg k_1 = -\frac{E_a}{2.303RT_1} + \lg A \tag{5.21}$$

$$\lg k_2 = -\frac{E_a}{2.303RT_2} + \lg A \tag{5.22}$$

Thus the A term is eliminated by subtracting theses equations.

$$\lg \frac{k_2}{k_1} = \frac{E_a}{2.303R}\left(\frac{T_2 - T_1}{T_1 T_2}\right) \tag{5.23}$$

If the activation energy is known, we can also use the equation to find the rate constant at another temperature.

Sample 5.1 The rate constant of a reaction is 1.48×10^{-9} s^{-1} at 470 K and 1.22×10^{-4} s^{-1} at 570 K, calculate the activation energy E_a. What is the value of the rate constant at 650 K?

Solution:

Substituting the data given in the problem into the equation

$$\lg \frac{k_2}{k_1} = \frac{E_a}{2.303R}\left(\frac{T_2 - T_1}{T_1 T_2}\right)$$

$$\lg \frac{1.22\times 10^{-4} s^{-1}}{1.48\times 10^{-9} s^{-1}} = \frac{E_a}{2.303\times 8.314 \text{ J/(mol·K)}} \times \left(\frac{570 \text{ K}-470 \text{ K}}{470 \text{ K}\times 570 \text{ K}}\right)$$

Hence,
$$E_a = 2.52\times 10^5 \text{ J/mol}$$

To find the rate constant at 650 K, we substitute $E_a = 2.52\times 10^5$ J/mol and the data $k_1 = 1.48\times 10^{-9}$ s^{-1}, $T_1 = 470$ K, $k_2 = ?$, $T_2 = 650$ K into the equation

$$\lg \frac{k_2}{k_1} = \frac{E_a}{2.303R}\left(\frac{T_2-T_1}{T_1 T_2}\right)$$

We get

$$\lg \frac{k_2}{1.48\times 10^{-9} s^{-1}} = \frac{2.52\times 10^5 \text{ J/mol}}{2.303\times 8.314 \text{ J/(mol·K)}} \times \left(\frac{650 \text{ K}-470 \text{ K}}{470 \text{ K}\times 650 \text{ K}}\right) = 7.75$$

Hence,
$$k_2 = 0.084 \text{ s}^{-1}$$

5.5 Catalysis

An increase in temperature increases the rate of the reaction, but the addition of heat to a reaction adds to production costs. Furthermore, high-temperature reactions introduce safety concerns, and many chemical reactions can not process at high temperature, such as chemical processes in living organisms. Thus another way for increasing the rate of chemical reaction as the result of addition of a catalyst has been developed, this process is called catalysis. A catalyst is a substance that changes the rate of a chemical reaction without undergoing a permanent chemical change itself as a result of the reaction. Catalysts that speed the reaction are called positive catalysts. Catalysts that slow the reaction are called negative catalysts. Frequently only a trace of catalyst is needed. The production of most experimentally and industrially important chemicals involves catalysis. For example, in the laboratory the preparation of oxygen by heating potassium chlorate, $KClO_3$, is a reaction that is strongly affected by catalysts.

$$2KClO_3(s) = 2KCl(s) + 3O_2(g)$$

In the absence of a catalyst, this thermal decomposition is very slow, even on strong heating. Upon the addition of a small amount of manganese dioxide, MnO_2, a black powdery substance, the rate of decomposition can be increased dramatically. All the MnO_2 can be recovered at the end of the reaction. Thus, MnO_2 acts as a catalyst for this reaction.

A catalyst speeds up a reaction by providing a new mechanism of reaction with a lower activation energy than the corresponding uncatalyzed reaction, as shown in Figure 5.4. From the Arrhenius equation we know that the smaller the activation energy E_a, the greater the rate constant k, and thus a faster reaction rate. Notice that the total energies of either the reactants or the products are not affected by the catalyst. It decreases the kinetic

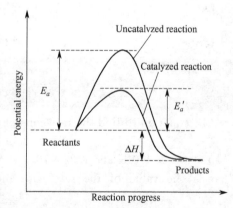

Figure 5.4 Energy profiles for a given reaction when catalyzed and uncatalyzed

barrier by decreasing the difference in energy between reactants and transition state. Therefore, ΔH is not changed. Also notice that because the activation energies for both the forward and the reverse reactions are lowered, a catalyst increases the rate of the reverse reaction to the same extent as it does the rate of the forward reaction. It do not alter the chemical equilibrium of a reaction, it just make the chemical equilibrium reach faster. Furthermore, a catalyst can only catalyze the thermodynamically favorable reaction. There is no catalyst that can cause the thermodynamically unfavorable reaction to occur.

Catalysts are highly selective, that is, the catalyst which catalyzes one reaction well may have no effect on another. Sometimes the same reactants may have several parallel reactions that produce different products. In such cases, we can choose the proper catalyst and conditions to make one of these reactions to proceed at a much more rapid rate than the others, and therefore lead to higher yields of desired products. For instance, the dehydrogenation of ethanol mainly forms acetaldehyde and hydrogen in the presence of copper powders, whereas the principal product is a mixture of ethene and water in the presence of Al_2O_3.

$$C_2H_5OH \xrightarrow{473 \sim 523K,\ Cu} CH_3CHO + H_2$$

$$C_2H_5OH \xrightarrow{623 \sim 633K,\ Al_2O_3} C_2H_4 + H_2O$$

However, the activity of a catalyst is destroyed when some substances mixed with the catalyst. Such substances are known as catalytic poisons, and the inhibition process is called poisoning.

Similarly, most biochemically significant processes are catalyzed. Catalysts in biology are called enzymes which are very large, complex protein molecules. Like all catalysts, enzymes increase the rate of the reaction by lowering the activation energy for a reaction. However, enzymes have the following characteristics which differ from most other catalysts.

(1) Highly specificity

Enzymes are much more specific than the other catalysts. An enzyme only catalyzes a certain reaction while leaving the rest of the system unaffected. For example, α-amylase just catalyzes the hydrolysis of the main chain of starch into maltose and β-amylase only hydrolyzes the branched chain of starch into dextrin.

(2) Highly catalytic activity

Enzymes are so amazingly efficient catalysts that most enzyme reaction rates are millions of times faster than those of comparable un-catalyzed reactions. Enzymes are much more efficient than the ordinary nonbiochemical catalysts. For example, all plants and animals can synthesize proteins at ordinary temperature, but the synthesis of nylon, which has similar structure as proteins, requires a much more higher temperature of 280~300℃.

(3) Dependent on conditions

The activity of an enzyme is affected by temperature and chemical environment (e.g., pH). Many drugs and poisons are enzyme inhibitors that can decrease enzyme activity. Conditions that denature the protein markedly decrease or even abolish enzyme activity, such as high temperatures or extremes of pH.

Key Words

chemical kinetics ['kemikəl kai'netiks] 化学动力学	reaction rate [ri(:)'ækʃən reit] 反应速率
	average rate ['ævəridʒ reit] 平均反应速率

instantaneous rate ['instən'teinjəs reit] 瞬时反应速率
collision theory [kə'liʒən 'θiəri] 碰撞理论
transition state theory [træn'ziʒən steit 'θiəri] 过渡状态理论
activated complex ['æktiveitid 'kɔmpleks] 活化配合物
intermediate [intə'mi:djət] 中间体
rate-determining step [reit di'tə:miniŋ step] 限速步骤
frequency factor ['fri:kwənsi 'fæktə] 频率因子
pre-exponential factor [pri:,ekspəu'nenʃəl 'fæktə] 指前因子
activated molecule ['æktiveitid 'mɔlikju:l] 活化分子
effective collision [i'fektiv kə'liʒən] 有效碰撞
collision probability, f [kə'liʒən,prɔbə'biliti] 碰撞分数
collision frequency, Z [kə'liʒən 'fri:kwənsi] 碰撞频率
orientation factor, P [,ɔ(:)rien'teiʃən 'fæktəl] 取向因子

activation energy [,ækti'veiʃən 'enədʒi] 活化能
energy barrier ['enədʒi 'bæriə] 能垒
reaction mechanism [ri(:)'ækʃən 'mekənizəm] 反应机理
unimolecular [ju:nimɔu'lekjulə] 单分子的
bimolecular [,baimə'lekjulə] 双分子的
termolecular [tə:mɔu'lekjulə] 三分子的
elementary reaction [,eli'mentəri ri(:)'ækʃən] 基元反应
complex reaction ['kɔmpleks ri(:)'ækʃən] 复杂反应
rate law [reit lɔ:] 速率方程
rate constant, k [reit 'kɔnstənt] 速率常数
law of mass action [lɔ:v mæs 'ækʃən] 质量作用定律
molecularity [mɔu,lekju'læriti] 分子数
reaction order [ri(:)'ækʃən 'ɔ:də] 反应级数
Arrhenius equation [ə'ri:niəs, i'kweiʃən] 阿伦尼乌斯方程
catalysis [kə'tælisis] 催化
catalyst ['kætəlist] 催化剂
enzyme ['enzaim] [生化] 酶

Exercises

1. What are the units for the rate constants of zero-order, first-order, and second-order reaction?

2. Compare the difference between the terms "order of reaction" and "molecularity of reaction".

3. How does a catalyst speed up a reaction?

4. For the following elementary reactions,

① $SO_2Cl_2 \longrightarrow SO_2 + Cl_2$

② $H_2 + 2I \longrightarrow 2HI$

③ $2N_2O \longrightarrow 2N_2 + O_2$

(1) What is the molecularity of each of these reactions?

(2) Write the rate law for each.

Answer: ①1, $v = kc(SO_2Cl_2)$; ②3, $v = kc(H_2)c^2(I)$; ③2, $v = kc^2(N_2O)$

5. The reaction $N_2(g) + 3H_2(g) = 2NH_3(g)$, the initial concentrations of N_2 and H_2 are 1.5 mol/L and 4.5 mol/L respectively, after 2 min, the concentration of N_2 is 0.8 mol/L. Calculate the average rate of the reaction in terms of the concentrations of N_2, H_2 and NH_3 respectively, and indicate the relationship of these reaction rates.

Answer: $\bar{v}_{N_2} = 0.35$ mol/(L·min), $\bar{v}_{H_2} = 1.05$ mol/(L·min), $\bar{v}_{NH_3} = 0.7$ mol/(L·min); $\bar{v}_{N_2} : \bar{v}_{H_2} : \bar{v}_{NH_3} = 1:3:2$.

6. For the reaction $2NO + 2H_2 = N_2 + 2H_2O$, the observed rate law is $v = kc^2(NO)c$

(H_2).

(1) What is the reaction order with respect to each reactant? What is the overall order?

(2) What is the unit for the rate constant?

Answer: The reaction is second order with respect to NO, and first order with respect to H_2, and third order overall; $L^2/(mol^2 \cdot s)$

7. The reaction $2A+2B=C+2D$ is first order with respect to A and second order with respect to B. When the initial concentrations are $c(A)=2.0\times10^{-3}$ mol/L and $c(B)=3.5\times10^{-2}$ mol/L, the reaction rate is 2.8×10^{-8} mol/(L $\cdot$ s).

(1) Write the rate law for the reaction.

(2) Calculate the rate constant of the reaction.

Answer: $v=kc(A)c^2(B)$; $1.1\times10^{-2} L^2/(mol^2 \cdot s)$

8. The initial rates of a reaction, $NO(g)+O_2(g)=2NO_2(g)$, were obtained from three experiments using different initial concentrations. The results are given in the table.

Experiment	Initial concentrations/(mol/L)		Initial Rate/[mol/(L $\cdot$ s)]
	$c(NO)$	$c(O_2)$	
1	0.0115	0.0252	0.0282
2	0.0230	0.0252	0.1128
3	0.0115	0.0504	0.0564

(1) Write the rate law for the reaction. (2) What is the reaction order with respect to each reactant? (3) What is the value of rate constant? (4) What will be the initial rate in an experiment using initial concentrations $c(NO)=0.0300$ and $c(O_2)=0.0450$?

Answer: $v=kc^2(NO)c(O_2)$; The reaction is second order with respect to NO, and first order with respect to O_2; $8461 L^2/(mol^2 \cdot s)$; $0.3427 mol/(L \cdot s)$

9. The activation energy of a particular reaction is 120.4 kJ/mol, how many times faster will the reaction occur at 300 K than at 450 K?

Answer: 9.7×10^6

10. The rate constant of a reaction is 9.53×10^{-6} L/(mol $\cdot$ s) at 350 K and 3.45×10^{-3} L/(mol $\cdot$ s) at 550 K. Calculate the activation energy E_a for the reaction. What is the value of the rate constant at 450K?

Answer: 4.72×10^4 J/mol; 3.50×10^{-4} L/(mol $\cdot$ s)

11. For the reaction $2A(g)+2B(g)=C(g)+D(g)$, the rate law is $v=kc^2(A)c(B)$, what effect will each of the following changes have upon their rate of reaction?

(1) The concentration of A is doubled.

(2) Adding catalyst.

(3) Lowering temperature.

(4) The volume of the container is doubled.

Answer: (1) 4 times; (2) increase; (3) decrease; (4) 1/8

Chapter 6 Chemical Equilibrium

6.1 Concept of Equilibrium

Few chemical reactions proceed in only one direction. Under proper conditions most are, at least to some extent, reversible. Reversible reaction is a chemical reaction that results in an equilibrium mixture of reactants and products. For example, consider the reaction between H_2 and I_2 in the gas phase.

$$H_2(g) + I_2(g) \rightleftharpoons 2HI(g)$$

At first, the reaction proceeds toward the formation of products HI. As soon as some product HI molecules are formed, the reverse process, that is, the formation of reactant molecules H_2 and I_2 from product molecules HI, begins to decompose. At the start of the reversible process, both the concentrations of reactants H_2, I_2 and the concentrations of products HI change with time. However some times later, when the rates of the forward and reverse reactions are equal, the concentrations of the reactants H_2, I_2, and the concentrations of products HI no longer change with time, chemical equilibrium is reached. In chemical equation of reversible reaction, arrows pointing both ways are used to indicate equilibrium.

6.2 Equilibrium Constant

Chemical equilibrium describes the state in which the rates of forward and reverse are equal and the concentrations of the reactants and products remain unchanged with time. This state of dynamic equilibrium is characterized by an equilibrium constant.

Depending on the nature of reacting species, the equilibrium constant can be expressed in terms of molarities (for solutions) or partial pressure (for gases). The equilibrium constant provides information about the net direction of a reversible reaction and the concentrations of the equilibrium mixture.

Experiments Equilibrium Constant

Experiment equilibrium constant, K, is based on the experiments. From observations of many chemical reactions, in 1864 two Norwegian chemists, Gato Maximilian Guldberg (1836~1902) and Peter Waage (1833~1900) proposed the law of mass action which expresses the relationship between the concentrations of the reactants and products at equilibrium in any reaction.

Suppose we have the general equilibrium equation

$$aA + bB \rightleftharpoons gG + hH$$

Where A, B are the reactants; G, H are the products; The letter a, b, g and h are the number of moles of each substances involved in the balanced chemical equation.

According to the law of mass action, the equilibrium condition is expressed by the equation

$$K = \frac{[G]^g [H]^h}{[A]^a [B]^b} \tag{6.1}$$

The square brackets indicate the concentrations of the chemical species at equilibrium, and K is called experiment equilibrium constant.

For chemical reaction involving gases, we often find it convenient to express the concentrations in terms of the partial pressures of the gases. That is, in this case we may write K_p to represent the equilibrium constant in which the concentrations of gases are expressed in partial pressures in atmospheres.

$$K_p = \frac{(p_G)_{eq}^g (p_H)_{eq}^h}{(p_A)_{eq}^a (p_B)_{eq}^b} \tag{6.2}$$

Similarly, we may write K_c with a subscript c to denote that the equilibrium constant is expressed by concentrations.

The Relationship between K_c and K_p

The relationship between the concentration and the pressure of a gas can be seen from the ideal gas equation

$$pV = nRT \quad \text{or} \quad p = (n/V)RT = cRT$$

For a gas system reaction

$$aA(g) + bB(g) = gG(g) + hH(g)$$

the equilibrium expression can be written in terms of concentrations

$$K_c = \frac{[G]^g [H]^h}{[A]^a [B]^b}$$

or in the terms of equilibrium partial pressures of the gases

$$K_p = \frac{(p_G)_{eq}^g (p_H)_{eq}^h}{(p_A)_{eq}^a (p_B)_{eq}^b}$$

Substituting $p = cRT$ into this equation we can get

$$K_p = \frac{[G]^g [H]^h}{[A]^a [B]^b} (RT)^{\Delta n}$$

where Δn is the sum of the coefficients of gaseous products in the chemical reaction equation minus the sum of the coefficients of gaseous reactants $[\Delta n = (g+h) - (a+b)]$.

So the relationship between K_c and K_p is

$$K_p = K_c (RT)^{\Delta n} \tag{6.3}$$

Some Notes of the Equilibrium Constant Expressions

1. The concentrations of pure solids, pure liquids (in heterogeneous equilibrium), and solvents (in homogeneous equilibrium) do not appear in the equilibrium constant expressions. An example is the decomposition reaction of calcium carbonate, $CaCO_3(s) \rightleftharpoons CaO(s) + CO_2(g)$. We would write its equilibrium expression

$$K_c = [CO_2] \quad \text{or} \quad K_p = p(CO_2)_{eq}$$

2. The equilibrium constant expression for a reverse reaction is the reciprocal of that for the forward reaction. For example, the equilibrium constant for the Haber synthesis of ammonia $N_2(g) + 3H_2(g) \rightleftharpoons 2NH_3(g)$, can be expressed in two ways

$$K_{\text{forward}} = \frac{(p_{NH_3})_{eq}^2}{(p_{N_2})_{eq} (p_{H_2})_{eq}^3} \qquad K_{\text{reverse}} = \frac{(p_{N_2})_{eq} (p_{H_2})_{eq}^3}{(p_{NH_3})_{eq}^2}$$

So, the two equilibrium constants are reciprocals of each other

$$K_{\text{forward}} = \frac{1}{K_{\text{reverse}}} \tag{6.4}$$

3. When the balanced equation for a reaction is multiplied by a factor n, the equilibrium expression for the new reaction is the original equilibrium expression raised to the nth power. For example, there is a reaction as following.

$$a\text{A} + b\text{B} \rightleftharpoons g\text{G} + h\text{H}$$

The equilibrium constant is

$$K = \frac{[\text{G}]^g [\text{H}]^h}{[\text{A}]^a [\text{B}]^b}$$

There is a new equation for the reaction which is multiplied by n

$$na\text{A} + nb\text{B} \rightleftharpoons ng\text{G} + nh\text{H}$$

The new equilibrium constant is

$$K_{\text{new}} = \left(\frac{[\text{G}]^g [\text{H}]^h}{[\text{A}]^a [\text{B}]^b}\right)^n$$

So
$$K_{\text{new}} = K^n \tag{6.5}$$

4. In a multiple equilibrium system, if a given reaction equation can be obtained by taking the sum of other reaction equations, the equilibrium constant for the given equation equals the product of the other equations.

As an application of this multiple equilibrium rule, consider the following reactions

(1) $SO_2(g) + 1/2 O_2(g) \rightleftharpoons SO_3(g)$ K_1

(2) $NO_2(g) \rightleftharpoons NO(g) + 1/2 O_2(g)$ K_2

Taking the sum of these two equations, we can get

(3) $SO_2(g) + NO_2(g) \rightleftharpoons SO_3(g) + NO(g)$ K_3

According to the previous rule, the equilibrium constant for this reaction(3), K_3, is

$$K_3 = K_1 K_2 \tag{6.6}$$

Standard Equilibrium Constant $K^\ominus$

As we discussed in Chapter 4, under non-standard state, the spontaneity of a reaction is determined by $\Delta_r G_m$ (how the free energy changes with a change in concentration or partial pressure). This is related to the standard free energy change, $\Delta_r G_m^\ominus$, by the equation

$$\Delta_r G_m = \Delta_r G_m^\ominus + RT \ln Q$$

Where Q is the reaction quotient for the system, which will be discussed later. At equilibrium, $\Delta_r G_m = 0$, $Q = K^\ominus$,

so
$$\Delta_r G_m^\ominus = -RT \ln K^\ominus \tag{6.7}$$

Where $K^\ominus$ is called standard equilibrium constant or thermodynamic equilibrium constant. This equation shows that the standard equilibrium constant $K^\ominus$ is determined by the standard Gibbs free energy changes for the reaction. If $\Delta_r G_m^\ominus < 0$, $K^\ominus > 1$, and the reaction is spontaneous under standard conditions. If $\Delta_r G_m^\ominus > 0$, $K^\ominus < 1$, and the reaction is not spontaneous under standard conditions. If $\Delta_r G_m^\ominus = 0$, $K^\ominus = 1$, and the reaction is at equilibrium under standard conditions. This equation is very useful because it permits us to determine $K^\ominus$ from either measured or calculated values of $\Delta_r G_m^\ominus$.

The equilibrium expression for the standard equilibrium constant $K^\ominus$ is different with the expression for the experiment equilibrium constant K. For a general equilibrium equation

$$a\text{A} + b\text{B} \rightleftharpoons g\text{G} + h\text{H}$$

The standard equilibrium constant $K^\ominus$ is

$$K^\ominus = \frac{([\text{G}]/c^\ominus)^g ([\text{H}]/c^\ominus)^h}{([\text{A}]/c^\ominus)^a ([\text{B}]/c^\ominus)^b} \tag{6.8}$$

$c^\ominus$ is the standard concentration, and $c^\ominus = 1$ mol/L.

If the reaction involving gases

$$a\text{A}(g) + b\text{B}(g) \rightleftharpoons g\text{G}(g) + h\text{H}(g)$$

The standard equilibrium constant $K^\ominus$ can be expressed as the following equation

$$K_p^\ominus = \frac{[(p_G)_{eq}/p^\ominus]^g [(p_H)_{eq}/p^\ominus]^h}{[(p_A)_{eq}/p^\ominus]^a [(p_B)_{eq}/p^\ominus]^b} \tag{6.9}$$

$p^\ominus$ is the standard pressure, and $p^\ominus = 1.00 \times 10^5$ Pa or $p^\ominus = 1.00 \times 10^2$ kPa.

Sample 6.1 The standard free energy of formation for $I_2(g)$ and $HI(g)$ are 19.359 kJ/mol and 1.7 kJ/mol. Please calculate the equilibrium constant $K_p^\ominus$ of the reaction.

$$H_2(g) + I_2(g) \rightleftharpoons 2HI(g)$$

Solution:

$$H_2(g) + I_2(g) \rightleftharpoons 2HI(g)$$

$\Delta_f G_m^\ominus$ 0   19.359   1.7 (kJ/mol)

$$\Delta_r G_m^\ominus = 2 \times 1.7 - 19.359 = -15.959 \text{ kJ/mol}$$

According to the equation

$$\Delta_r G_m^\ominus = -RT \ln K_p^\ominus$$

$$\therefore \ln K_p^\ominus = \frac{-(-15.959) \times 10^3}{8.314 \times 298} = 6.44 \quad \therefore K_p^\ominus = 628$$

Applications of Equilibrium Constants

1. Predicting the extent of reaction

The magnitudes of equilibrium constants vary over a tremendous range and depend on the nature of the reaction and on the temperature of the system.

Many reactions have large equilibrium constants (more than 10), the equilibrium mixture is mostly products, and the equilibrium lies to the right. For example, the reaction between H_2 and Br_2 to form HBr, has a huge equilibrium constant.

$$H_2(g) + Br_2(g) \rightleftharpoons 2HBr(g) \quad K = 5.4 \times 10^{18}$$

The large value for this equilibrium constant indicates that the reaction goes virtually to completion.

Other reactions have small (less than 0.1), the equilibrium mixture is mostly reactants, and the equilibrium lies to the left. For example, the reaction

$$F_2(g) \rightleftharpoons 2F(g) \quad K = 2.1 \times 10^{-22}$$

The extremely small value of this equilibrium constant indicates that a sample of fluorine at standard conditions consists almost entirely of F_2 molecules. Nevertheless, the equilibrium constant is not zero, so tiny fluorine atoms are also present in the gas.

When the equilibrium constants are neither large nor small (between 0.1 to 10), neither products nor reactants are all favored. The equilibrium mixture contains significant amounts of all substances in the reaction. For example, the dimerization of NO_2 has a moderate equilibrium constant at 298 K.

$$2NO_2(g) \rightleftharpoons N_2O_4(g) \quad K = 8.77$$

An equilibrium mixture of NO_2 and N_2O_4 at 298 K contains enough of each molecule for readily measured partial pressures of both to exist. Then equilibrium calculations are essential to the description of chemical composition and behavior when the equilibrium constant of a reaction falls in the range where the concentrations of reactants and products both can be measured readily.

2. Predicting the direction of reaction

When a chemical reaction is at equilibrium, the ratio of product concentrations to reactant concentrations is equal to the equilibrium constant, K.

$$aA + bB \rightleftharpoons gG + hH \qquad K = \frac{[G]^g[H]^h}{[A]^a[B]^b}$$

When a chemical reaction system is not at equilibrium, the ratio of product concentrations to reactant concentrations is called the concentration quotient (Q). The concentration quotient has the same form as the equilibrium concentration ratio, but the concentrations are not equilibrium values, consequently, Q can has any value.

$$aA + bB \rightleftharpoons gG + hH \qquad Q = \frac{(c_G)^g(c_H)^h}{(c_A)^a(c_B)^b} \qquad (6.10)$$

The NO_2/N_2O_4 system can be used to show how Q is related to K.

$$2NO_2(g) \rightleftharpoons N_2O_4(g) \qquad K = 8.77$$

At any time, to determine in which direction this system will shift to reach equilibrium, we can compare the values of Q and K. There are there possible cases.

If $Q < K$, the reaction will go to the right to make products N_2O_4.
If $Q = K$, the reaction is at equilibrium and will not shift.
If $Q > K$, the reaction will go to the left to make reactants NO_2.

The relationship between Q and K signals the direction of a chemical reaction. The Gibbs free energy change, $\Delta_r G_m$, also signals the direction of a chemical reaction. These two criteria can be compared as following.

Reaction goes right when $Q < K$, $\Delta_r G_m < 0$.
Reaction is at equilibrium when $Q = K$, $\Delta_r G_m = 0$.
Reaction goes left when $Q < K$, $\Delta_r G_m > 0$.

The similarities suggest a link among Q, $K^\ominus$, and $\Delta_r G_m$. This link can be found from the equations of thermodynamics.

$$\Delta_r G_m = \Delta_r G_m^\ominus + RT\ln Q = -RT\ln K^\ominus + RT\ln Q = RT\ln\frac{Q}{K^\ominus}$$

so
$$\Delta_r G_m = RT\ln\frac{Q}{K^\ominus} \qquad (6.11)$$

Calculation of Equilibrium Concentrations

If we know the equilibrium constant for a chemical reaction, we can calculate the concentrations in the equilibrium mixture from knowledge of the initial concentrations.

For example, A mixture of 0.500 mol H_2 and 0.500 mol I_2 was placed in a 1.00 L stainless-steel flask at 703 K. The equilibrium constant K_c for the reaction $H_2(g) + I_2(g) \rightleftharpoons 2HI(g)$ is 54.3 at this temperature. Calculate concentrations of H_2, I_2, and HI at equilibrium.

We follow the given procedure in the Sample 6.2 to calculate the equilibrium concentrations.

Sample 6.2 Calculate the equilibrium concentration of the reaction, $H_2(g) + I_2(g) \rightleftharpoons 2HI(g)$

Solution:

Step 1 The stoichiometry of the reaction is 1 mol H_2 reacting with 1 mol I_2 to yield 2 mol HI. Let x be the depletion in concentration (mol/L) of H_2 and I_2 at equilibrium. It follows that the equilibrium concentration of HI must be $2x$. We summarize the changes in concentrations as following.

	$H_2(g)$	+	$I_2(g)$	$\rightleftharpoons$	$2HI(g)$
Initial/(mol/L)	0.500		0.500		0.000

Change/(mol/L)	$-x$	$-x$	$2x$
Equilibrium/(mol/L)	$(0.500-x)$	$(0.500-x)$	$2x$

Step 2 The equilibrium constant is given by

$$K_c = \frac{[HI]^2}{[H_2][I_2]}$$

Substituting, we get

$$54.3 = \frac{(2x)^2}{(0.500-x)(0.500-x)}$$

Taking the square root of both sides, we get

$$7.37 = \frac{2x}{0.500-x}$$

$$x = 0.393 \text{ mol/L}$$

Step 3 At equilibrium, the concentrations are

$$[H_2] = 0.500 - 0.393 = 0.107 \text{ mol/L}$$
$$[I_2] = 0.500 - 0.393 = 0.107 \text{ mol/L}$$
$$[HI] = 2 \times 0.393 = 0.786 \text{ mol/L}$$

6.3 Le Chatelier's Principle

Chemical equilibrium represents a balance between forward and reverse reactions. In most cases, this balance is quite delicate. Changes in experimental conditions may disturb the balance and shift the equilibrium position so that more or less of the desired product is formed. When we say that an equilibrium position shifts to the right, for example, we mean that the net reaction is now from left to right. At our disposal are the following experimentally controllable variables, concentration, pressure, volume, and temperature. Here we will examine how each of these variables affects a reacting system at equilibrium. In addition, we will also examine the effect of a catalyst on equilibrium.

Le Châtelier's Principle

There is a general rule that helps us to predict the direction in which an equilibrium reaction will move when a change in concentration, pressure, volume, or temperature occurs. The rule, known as Le Châtelier's Principle (after the French chemist Henri Le Châtelier), states that if an external stress is applied to a system at equilibrium, the system adjusts in such a way that the stress is partially offset as it tries to reestablish equilibrium. The word "stress" here means a change in concentration, pressure, volume, or temperature that removes a system from the equilibrium state. We will use Le Châtelier's Principle to assess the effects of such changes.

Changes in Concentrations

Changes in concentrations do not change the value of the equilibrium constant. As long as temperature is constant, the value of the equilibrium constant remains the same. However, changes in concentrations do make a difference in the composition of the equilibrium mixture, that is, in the position of equilibrium.

To see how we can predict the effects of a change in concentrations on a system at equilibrium, we choice a sample as following.

Sample 6.3 At 993 K, the equilibrium constant K_c for the reaction $N_2(g) + 3H_2(g) \rightleftharpoons$

$2NH_3(g)$ is 2.37×10^{-3}. In a certain experiment, the equilibrium concentrations are $[N_2] = 0.683$ mol/L, $[H_2] = 8.80$ mol/L, $[NH_3] = 1.05$ mol/L. Suppose some of the NH_3 is added to the mixture so that its concentration is increased to 3.65 mol/L. (a) Use Le Châtelier's Principle to predict the shift in direction of the net reaction to reach a new equilibrium. (b) Confirm your prediction by calculating the reaction quotient Q and comparing its value with K_c.

Solution:

(a) The stress applied to the system is the addition of NH_3. To offset this stress, some NH_3 reacts to produce N_2 and H_2 until a new equilibrium is established. The net reaction shifts from right to left.

(b) At the instant when some of the NH_3 is added, the system is no longer at equilibrium. The reaction quotient is given by

$$Q = \frac{(c_{NH_3})^2}{(c_{N_2})(c_{H_2})^3} = \frac{(3.65)^2}{(0.683) \times (8.80)^3} = 2.86 \times 10^{-2}$$

Because this value of Q is greater than the value of K_c, 2.37×10^{-3}, the net reaction shifts from right to left until Q equals K_c.

Now, we can restate the effects of a change in concentrations on a system at equilibrium by using Le Châtelier's Principle. When the concentration of any of the reactants or products in a chemical reaction at equilibrium are changed, the composition of the equilibrium mixture changes so as to offset the change in concentrations that were made.

Changes in Pressure and Volume

Changes in pressure ordinarily do not affect the concentration of reacting species in condensed phases (say, in an aqueous solution) because liquids and solids are virtually incompressible. On the other hand, concentrations of gases are greatly affected by changes in pressure. Let us look again at the idea gas equation

$$pV = nRT \quad \text{or} \quad p = (n/V)RT = cRT$$

Thus p and V are related to each other inversely, The greater the pressure, the smaller the volume, and vice versa. Note, too, that the term (n/V) is the concentration of the gas in moles per liter, and it varies directly with pressure.

Suppose the equilibrium system $N_2O_4(g) \rightleftharpoons 2NO_2(g)$ is in a cylinder fitted with a movable piston. What happens if we increase the pressure on the gases by pushing down on the piston at constant temperature? Because the volume decreases, then the concentration (n/V) of both N_2O_4 and NO_2 increases. Because the concentration of NO_2 is squared, the increase in pressure increases the numerator more than the denominator. The system is no longer at equilibrium, so we calculate

$$Q = \frac{(c_{NO_2})^2}{c_{N_2O_4}}$$

Thus $Q > K_c$, and the net reaction will shift to the left until $Q = K_c$. Conversely, a decrease in pressure (increase in volume) would result in $Q < K_c$, the net reaction will shift to the right until $Q = K_c$.

In general, an increase in pressure (decrease in volume) favors the net reaction that decreases the total number of moles of gases (the reverse reaction, in the preceding case), and a decrease in pressure (increase in volume) favors the net reaction that increases the total number of moles of gases (here, the forward reaction in the preceding case). For reactions in which there is no change in the number of moles of gases, a pressure (or volume) change has no effect on the position of equilibrium.

It is possible to change the pressure of a system without changing its volume. Suppose the NO_2-N_2O_4 system is contained in a stainless-steel vessel whose volume is constant. We can increase the total pressure in the vessel by adding an inert gas (helium, for example) to the equilibrium system. Adding helium to the equilibrium mixture at constant volume increases the total gas pressure and decreases the mole fraction of both NO_2 and N_2O_4. But the partial pressure of each gas, which is given by the product of its mole fraction and total pressure, dose not change. Thus the presence of an inert gas in such a case does not affect the equilibrium.

We can predict the direction of the shift by comparing Q and K when the pressure of a chemical reaction system is changed. However, there is an easier way to predict the shift direction by using Le Châtelier's Principle. When the volume of the container holding a gaseous system is reduced, the system responds by reducing its pressure. This is done by decreasing the total number of gaseous molecules in the system.

Changes in Temperature

A change in concentration, pressure, or volume may change the equilibrium position, but it does not change the value of the equilibrium constant. Only a change in temperature can alter the equilibrium constant.

The studies of the effect of temperature on equilibrium reveal a consistent pattern. The equilibrium constant of an exothermic reaction decreases with increasing temperature, whereas the equilibrium constant of an endothermic reaction increases with increasing temperature. Equation 6.7 provides a thermodynamic explanation for this behavior. Recall that Gibbs free energies related to the enthalpy and entropy change through Equation 4.14.

$$\Delta_r G_m^\ominus = \Delta_r H_m^\ominus - T\Delta_r S_m^\ominus$$

This equality can be substituted into Equation 6.7 to show how $\ln K$ depends on $\Delta_r H_m^\ominus$, $\Delta_r S_m^\ominus$ and T

$$-RT\ln K = \Delta_r H_m^\ominus - T\Delta_r S_m^\ominus$$

which rearranges to give

$$\ln K = -\frac{\Delta_r H_m^\ominus}{RT} + \frac{\Delta_r S_m^\ominus}{R} \tag{6.12}$$

An exothermic reaction has a negative $\Delta_r H_m^\ominus$, making the first term on the right of Equation 6.12 positive. As T increases, this term decreases, causing K to decrease.

An endothermic reaction, in contrast, has a positive $\Delta_r H_m^\ominus$, making the first term on the right of Equation 6.12 negative. As T increases, this term becomes less negative, causing K to increase. These variations in K with temperature, which can be substantial, can be estimated using Equation 6.12 and standard thermodynamic functions.

However, even if we do not know the values of the equilibrium constant at different temperatures, Le Châtelier's Principle can tell us to predict the direction in which an equilibrium position will be shifted by a temperature change. And to make this prediction we need only know whether the reaction is exothermic or endothermic. The conclusion from Le Châtelier's Principle regarding temperature effects on an equilibrium can be summarized as follows, For an exothermic reaction, the amounts of products are increased at equilibrium by decrease in temperature, and for an endothermic reaction, the amounts of products are increased at equilibrium by an increase in temperature.

Effect of a Catalyst

For a reversible reaction, a catalyst affects the rate in the forward and reverse direction to the same extent. Therefore, the presence of a catalyst neither does not change the equilibrium constant, nor does it can shift the position of an equilibrium reaction system. Adding a catalyst to a reaction mixture that is not at equilibrium will speed up both the forward and reverse rates to achieve an equilibrium mixture much faster.

So far, we have discussed four ways to affect a reacting system at equilibrium. It is important to note that, of the four factors, only a change in temperature changes the value of the equilibrium constant. Changes in pressure, concentration, and volume can only alter the equilibrium concentrations of the reacting mixture, but they cannot change the value of the equilibrium constant as long as the temperature keeps constant. A catalyst can make the reacting system establish equilibrium faster, but it has no effect on the equilibrium constant or on the equilibrium concentrations of reactants and/or products.

Key Words

reversible [ri'və:səbl] 可逆的
forward reaction ['fɔ:wəd ri(:)'ækʃən] 正反应
reverse reaction [ri'və:s ri(:)'ækʃən] 逆反应
chemical equilibrium ['kemikəl ˌi:kwi'libriəm] 化学平衡
equilibrium constant [ˌi:kwi'libriəm 'kɔnstənt] 平衡常数
reaction quotient [ri(:)'ækʃən 'kwəuʃənt] 反应商
Le Châtelier's Principle 勒沙特列原理
reversibility [riˌvə:sə'biliti] 可逆性
shift [ʃift] 移动，轮班，移位，变化

Exercises

1. Write K_c and K_p for this reaction $2N_2O_5(g) \rightleftharpoons 4NO_2(g) + O_2(g)$

Answer: $K_c = \dfrac{[NO_2]^4[O_2]}{[N_2O_5]^2}$, $K_p = \dfrac{p_{NO_2}^4 \, p_{O_2}}{p_{N_2O_5}^5}$

2. The equilibrium constant K_p for the decomposition of phosphorus pentachloride to phosphorus trichloride and molecular chlorine $PCl_5(g) \rightleftharpoons PCl_3(g) + Cl_2(g)$ is found to be 1.05 at 523 K. If the equilibrium partial pressure of PCl_5 and PCl_3 are 0.875 atm and 0.463 atm, respectively, what is the equilibrium partial pressure of Cl_2 at 523 K?

Answer: 1.98 atm

3. For the reaction $N_2(g) + 3H_2(g) \rightleftharpoons 2NH_3(g)$ K_p is 4.3×10^{-4} at 648 K. Calculate K_c for the reaction.

Answer: 1.2

4. Consider the following equilibrium at 295 K $NH_4HS(s) \rightleftharpoons NH_3(g) + H_2S(g)$, the partial pressure of each gas is 0.265 atm. Calculate K_c and K_p for the reaction.

Answer: 1.20×10^{-4}, 0.0702

5. A mixture of 0.623 mol/L H_2, 0.414 mol/L I_2 and 2.24 mol/L HI was placed in a stainless-steel flask at 703 K. The equilibrium constant K_c for the reaction $H_2(g) + I_2(g) \rightleftharpoons 2HI(g)$ is 54.3 at this temperature. Calculate concentrations of H_2, I_2, and HI at equilibrium.

Answer: 0.467 mol/L, 0.258 mol/L, 2.55 mol/L

6. At 1553 K the equilibrium constant K_c for the reaction $Br_2(g) \rightleftharpoons 2Br(g)$ is 1.1×10^{-3}. If the initial concentrations are $[Br_2]=6.3 \times 10^{-2}$ mol/L and $[Br]=1.2 \times 10^{-2}$ mol/L, calculate the concentrations of these species at equilibrium.

Answer: 0.065 mol/L, 8.4×10^{-3} mol/L

7. At 703K, the equilibrium constant K_p for the reaction $2NO(g)+O_2(g) \rightleftharpoons 2NO_2(g)$ is 1.5×10^5. In one experiment, the initial pressure for NO, O_2, and NO_2 are 2.1×10^{-3} atm, 1.1×10^{-2} atm, and 0.14 atm, respectively. Calculate Q_p and predict the direction that the net reaction will shift to reach equilibrium.

Answer: 4.0×10^5, the net reaction will shift from right to left.

8. Consider the equilibrium between molecular oxygen and ozone $3O_2(g) \rightleftharpoons 2O_3(g)$, $\Delta_r H_m^{\ominus} = 284$ kJ/mol. What would be the effect of (a) increasing the pressure on the system by decreasing the volume, (b) increasing pressure by adding O_2 to the system, (c) decreasing the temperature, and (d) adding a catalyst?

Answer: The equilibrium will shift from (a) left to right, (b) left to right, (c) right to left, (d) A catalyst has no effect on the equilibrium.

Chapter 7　Solutions

  Solutions are common in our everyday life. Solutions are homogeneous and stable mixtures which are formed when one or several substances disperse in another substance in the form of molecules or ions. In a solution, the substance that is dispersed is termed solute, and the dispersing medium is called solvent. Solutions can exist as gases, solids and liquids. The air we breathe is a sample of gaseous solutions. Most natural minerals and many alloys are solid solutions. The soups and beverages we drink frequently, the tap water we use daily, and the fluids carrying nutrients, salts, dissolved gases, enzymes, hormones, wastes and other things that run through our bodies continuously are all liquid solutions.

  Solutions are extremely important in our daily lives, industry and other fields. Most chemical reactions carried out in the lab and industry take place in solution because solutions provide a condition for rapid chemical reactions to occur. "Life" is the sum of a series of complex processes occurring in solution. In clinical practice many drugs can't be used unless they are made into solutions. Most commonly encountered and used solutions are liquid solutions, in which the solvent is a liquid and the solute may be a gas, liquid, or solid. Of all the liquid solvents water is the most commonly employed and is the best of the inorganic solvents because of its ability to dissolve a variety of substances. A solution with water as the solvent is called an aqueous solution. Aqueous solutions are involved in a wide range of chemical reactions including almost all biological processes. In this chapter, we will concentrate our attention upon the aqueous solutions. Some ways to express concentration will be introduced and some properties of the aqueous solutions especially the colligative properties will be discussed in detail.

7.1　Ways of Expressing Concentration

  The concentration of a solution receives much concern while the solution is to be used. It is well known to us that the properties of a solution are often related to its concentration. For example, the properties of 98% H_2SO_4 is very different from those of 10% H_2SO_4. For chemical reactions that take place in solution we always quite concern about the concentration of the reactants because it determines how often molecules collide and thus indirectly determine the rates of reactions and the conditions at equilibrium. The concentration of a solution also takes an important role in clinical practice. Our body fluids contain a great variety of solutes (ions and molecules) including electrolytes, dissolved gases, proteins, vitamins, enzymes, hormones, wastes and so on. A healthy body can be characterized by having the concentrations of almost all these solutes within narrow, normal ranges. Doctors usually determine whether someone is sick according to these data. For instance, once the potassium ions of some case are not within normal limits, and he may get hyperpotassemia. Moreover, pharmacists in hospital are routinely involved in preparing different concentration drug solutions to meet different requirements. Thus, it is important for us to have a firm enough grasp of the concept of concentration and be able to manipulate commonly used units of concentration.

  Then, how can we express the concentration of a solution? Generally, concentration is the ratio of the amount of solute to the amount of solvent or solution. The concentration of a solution can be expressed either qualitatively or quantitatively. A solution can be qualitatively described as dilute or concentrated. A dilute solution has a relatively small concentration of solute. A concentrated solution has a relatively high concentration of solute. Various ways of expressing concentration quantitatively are in use, the choice is usually a matter of convenience in a particular application. We will only examine the most common ones used in chemistry and pharmacy, including percentage, mole fraction, mass concentration, molarity, and molality.

Percentage

The simplest statement of the concentrations of the components of a solution is in terms of their percentages by weight or volume. There are three ways that we may express this percentage, mass (weight) percent, volume percent, and mass (weight)/volume percent. They in turn can be defined as following.

$$\text{Mass(weight) percent}(w/w\%) = \frac{\text{mass of solute}}{\text{total mass of solution}} \times 100\% \tag{7.1}$$

$$\text{Volume percent}(v/v\%) = \frac{\text{volume of solute}}{\text{total volume of solution}} \times 100\% \tag{7.2}$$

$$\text{Weight/volume percent}(w/v\%) = \frac{\text{mass of solute}}{\text{total volume of solution}} \times 100\% \tag{7.3}$$

Percent concentration is widely used in commerce, medicine, and other applied fields of chemistry. Generally, mixtures of solids such as alloys of metals are specified by weight percentage concentrations, whereas mixtures of gases or liquids are usually specified by volume percentages. For example, rubbing alcohol is generally 70% by volume alcohol, which means that 100 mL of solution contains 70 mL of alcohol. When the solute is a solid and the solvent is a liquid, the concentration is often expressed as a weight percent. However, because the weight of a liquid is usually more difficult to determine than its volume, when expressing the concentration of a solid solute in a solution, it is conveniently measured by mass (weight)/volume percent. For example, normal saline is prepared by dissolving 0.9 g of NaCl in 100 mL of water and is said to be 0.9% NaCl (mass/volume).

Very low concentrations may be expressed in parts per million (ppm). For even more dilute solutions, parts per billion (ppb) is used. By definition

$$\text{Parts per million(ppm)} = \frac{\text{mass of solute}}{\text{total mass of solution}} \times 10^6 \tag{7.4}$$

$$\text{Parts per billion(ppb)} = \frac{\text{mass of solute}}{\text{total mass of solution}} \times 10^9 \tag{7.5}$$

These terms are widely employed to express the amounts of trace pollutants in the environment. For instance, the maximum allowable concentration of arsenic in drinking water is 0.010ppm.

Mole Fraction

The mole fraction (χ_i) of any component i of a solution equals the ratio of the amount of component i (in moles) to the total amount of all components (in moles).

$$\chi_i = \frac{n_i}{\sum_i n_i} = \frac{\text{amount of component } i\text{(in moles)}}{\text{total amount of all components(in moles)}} \tag{7.6}$$

Mole fractions have no units because the units in the numerator and the denominator cancel. Mole fractions run from zero (substance not present) to unity (the pure substance). Suppose we have a solution that contains a single solute B dissolved in a solvent A, the mole fraction of the solute B would be

$$\chi_B = \frac{n_B}{n_A + n_B}$$

And the mole fraction of the solvent A may be

$$\chi_A = \frac{n_A}{n_A + n_B}$$

Obviously, the sum of the mole fraction of the solute and the solvent is equal to 1, i.e. $\chi_A + \chi_B = 1$.

Sample 7.1 A solution is prepared by dissolving 18.0 g of glucose ($C_6H_{12}O_6$) and 90.0 g of water, Calculate the mole fractions of the component.

Solution:

$$n_{C_6H_{12}O_6} = \frac{18.0 \text{ g}}{180 \text{ g/mol}} = 0.100 \text{ mol}$$

$$n_{H_2O} = \frac{90.0 \text{ g}}{18 \text{ g/mol}} = 5.00 \text{ mol}$$

$$\chi_{C_6H_{12}O_6} = \frac{0.100}{5.00 + 0.100} = 0.020$$

$$\chi_{H_2O} = 1 - \chi_{C_6H_{12}O_6} = 1 - 0.020 = 0.980$$

Mass Concentration

Mass concentration is equal to the mass of the solute divided by the volume of the solution proportion. Its SI unit is kg/m³ but a frequent unit is g/L or g/mL. The symbol "ρ" is often used for mass concentration, with a subscript to indicate the solute of interest.

$$\text{Mass concentration}(\rho) = \frac{\text{mass of solute}}{\text{volume of solution}} \tag{7.7}$$

Thus, the mass concentration of normal saline which is made by dissolving 0.9 g of NaCl in 100 mL of water is about 9 g/L. A common clinical expression of concentration is milligrams per 100 mL solution (mg/100 mL). For example, the calcium ion content of a blood sample is 0.096 mg in 1.0 mL. This is most often expressed as 9.6 mg/dL, since 1 dL = 100 mL. Sometimes this is also expressed as 9.6mg%.

Molarity

Molarity (amount of substance concentration), denoted c, is by far the most important and common way to express the concentration of solutions in chemistry, which is given by

$$\text{Molarity}(c) = \frac{\text{amount of solute(in moles)}}{\text{volume of solvent(in cubic meters)}} \tag{7.8}$$

Its SI unit is mol/m³. But the unit we often used is mol/L in chemistry laboratory, and the unit mmol/L or μmol/L is frequently used in pharmacological research and clinical practice.

Molarity is a common unit of concentration because the volume of a liquid is very easy to measure. The disadvantage to molarity is that molarity can change with temperature because volume is temperature dependent.

The use of molarity in clinical practice is gradually popularized. The WHO (World Health Organization) suggests that the content of all those substance whose relative molecular weight are already known in body fluids should be specified by molarity. For example, normal fasting levels of glucose in the blood are 70 to 100 mg of glucose per 100 mL of whole blood volume. We used to express this as 70~100 mg%, and now should be expressed as 3.9~5.6 mmol/L. Yet, the mass concentration unit can still be used to express the concentration of those substance whose molar mass are unknown up to now. As for the injections, the WHO holds that both the mass concentration and molarity of the solution should be labeled in most cases. For instance, the concentration of the isotonic glucose solution which is often used in clinical practice is used to express as 5%, and now should be labeled as '50 g/L $C_6H_{12}O_6$' and '0.28 mol/L $C_6H_{12}O_6$' as well.

Molality

Molality is the amount in moles of solute per kilogram of solvent and is customarily represented by an italic lowercase m. The SI unit for molality is mol/kg.

$$\text{Molality}(m) = \frac{\text{amount of solute(in moles)}}{\text{mass of solvent(in kilograms)}} \quad (7.9)$$

Notice that it is kilograms of solvent, not solution.

Sample 7.2 Calculate the molality of an antifreeze solution that contains 545 g of ethylene glycol in 800 mL of water. The molar mass of ethylene glycol is 62.0 g/mol.

Solution:

$$m = \frac{\frac{545}{62} \text{ mol}}{800 \times 10^{-3} \times 1.0 \text{ kg}} = 11 \text{ mol/kg}$$

Molality (unlike molarity) is independent of temperature because both moles and masses do not vary with temperature. Molality is often used as the concentration unit when we discuss certain properties of solutions that involve temperature changes, e.g. the colligative properties of solutions, which we will explore in the following section.

Molarity and molality are easily confused because their definitions are similar and the two words are really close. We should pay attention to the distinction between them. Yet, in dilute aqueous solutions near room temperature and standard atmospheric pressure, the molarity and molality are nearly equal. This is because 1 kg of water nearly the same as 1 kg of solution, and 1 kg of solution roughly corresponds to a volume of 1 L at these conditions. However, in all other conditions, this is usually not the case.

Concentration Calculations

In chemical and pharmacological research, we are frequently involved in preparing solutions by diluting and mixing stock solutions in industry and clinical practice. Sometimes you may find it necessary to be able to convert from one concentration unit to another. How can we solve such problems?

1. Dilutions and Mixing

The procedure for preparing a less concentrated solution from a more concentrated one is called dilution. Dilution of stock solutions is frequently used to make solutions of any desired concentration. To dilute a solution means to add more solvent without adding more solute. Note that dilution does not change the total amount of solute, i.e., the amount of solute in the final solution is the same as the amount of solute in the initial volume of solution. You can calculate the concentration of a solution following a dilution by applying this equation

$$c_{initial} \times V_{initial} = c_{final} \times V_{final} \quad \text{(In short, } c_1 V_1 = c_2 V_2\text{)} \tag{7.10}$$

Mixing of stock solutions is another method to obtain solutions of the concentration we expect. Similarly, the amount of solute after mixing is equal to the total amount of solute before mixing, which can be expressed with the formula

$$c_1 \times V_1 + c_2 \times V_2 + \cdots = c_{final} \times V_{final}$$

("1,2" refers to the solution before mixing.)

Below are two samples for you.

Sample 7.3 Concentrated HCl is 37.0% HCl by mass and has a density of 1.19 g/mL. How many milliliters of this solution are required to prepare 125 mL of a 1.50 M HCl solution?

Solution:

Assume x mL of concentrated HCl is needed.

Because the total amount of HCl remains the same upon dilution, therefore

$$x \times 1.19 \text{ g/mL} \times 37.0\% = 1.50 \text{ mol/L} \times 0.125 \text{ L} \times 36.5 \text{ g/mol}$$
$$x = 15.5 \text{ mL}$$

Sample 7.4 A nurse need 500 mL of 25% (m/v) glucose solution in some case and she only have 500 mL of 5% (m/v) and 500 mL of 50% (m/v) stock solution. How can she make up the solution? (Assuming that the volume are additive)

Solution:

Suppose she needs x mL 5% (m/v) glucose solution.

As the mass of glucose is a constant before and after mixing, then

$$x \times 5\% + (500 - x) \times 50\% = 500 \times 25\%$$
$$x = 277.8 \text{ mL}$$

Thus she may make up the solution by mixing 277.8 mL 5% (m/v) glucose solution with 222.2 mL 50% (m/v) glucose solution.

2. Conversion of Concentration Units

As you know, there are several ways we often use to express the concentration of a solution. It is important to know how to convert between these concentration units. Usually, we may need to know the density of the solution to solve this type of problem. Then, we should learn to make an assumption to get started. Generally, we assume the quantity of solution found in the denominator unit of the concentration unit which we are trying to convert. For example, if we are trying to convert molarity to weight/weight percent, assume we have 1 liter of solution. And if we are trying to convert weight/weight percent to molarity, assume that we have 100 grams of this solution. Here is a typical sample.

Sample 7.5 The 10.0% (m/v) glucose ($C_6H_{12}O_6$) solution has several medical uses. What is the molarity, molality and mole fraction of glucose in the solution? The density of the solution is 1.04 g/mL.

Solution:

Assume we have 1L of 10.0% (m/v) glucose solution. Then

$$m_{C_6H_{12}O_6} = 100 \text{ g}$$
$$m_{solution} = 1000 \text{ mL} \times 1.04 \text{ g/L} = 1040 \text{ g}$$
$$m_{H_2O} = (1040 - 100) \text{ g} = 940 \text{ g}$$
$$n_{C_6H_{12}O_6} = \frac{100 \text{ g}}{180 \text{ g/mol}} = 0.556 \text{ mol}$$

$$n_{H_2O} = \frac{940 \text{ g}}{18 \text{ g/mol}} = 52.22 \text{ mol}$$

$$c(C_6H_{12}O_6) = \frac{0.556 \text{ mol}}{1 \text{ L}} = 0.556 \text{ mol/L}$$

$$c(C_6H_{12}O_6) = \frac{0.556 \text{ mol}}{0.94 \text{ kg}} = 0.591 \text{ mol/kg}$$

$$\chi_{C_6H_{12}O_6} = \frac{0.556 \text{ mol}}{0.556 \text{ mol} + 52.22 \text{ mol}} = 0.011$$

7.2 Colligative Properties

A solution is the combination of solute and solvent, yet the physical properties of the solution are extremely different from either the solute or the solvent used to make the solution. A simple example is a 1 mol/L solution of $CuSO_4$ is bright blue while the solute $CuSO_4$ is white and the solvent water is colorless. Another common example is pure water freezes at 0 ℃ whereas aqueous solutions freeze at lower temperatures. Generally, we divided the physical properties of solutions into two categories. One is those properties that are affected by the identity of dissolved solute particles, such as color, taste, solubility, conductivity and so on. The other is those properties that depend only on the quantity (concentration or number of the dissolved particles) but not on the kind of solute particles. Such properties are termed as colligative properties, mainly including the vapor pressure, the boiling point, the freezing point, and the osmotic pressure. All of these properties are important in physiological and natural systems. We'll explore the four colligative properties in detail respectively in the following sections, and we'll restrict our objects of study to the dilute aqueous solution in which the solute is non-volatile in order to simplify our discussion.

Vapor Pressure Lowering

1. Vapor Pressure

When a glass of water placed in a closed container, liquid water molecules at the water surface can escape to the gas phase when they have a sufficient amount of energy to break down the liquid's intermolecular forces. This process is known as evaporation. Meantime, some of the gaseous water molecules may collide with the water surface and condense back into water. When the rate of condensation of the gas becomes equal to the rate of evaporation of the liquid, the space above the liquid will eventually contain a constant amount of water vapor. The pressure exerted by the water vapor at this moment is called saturated water vapor pressure at the given temperature, or vapor pressure of water for short. Similarly, the vapor pressure of any liquid (or solid) is defined as the equilibrium pressure of a vapor above its liquid (or solid).

The vapor pressure of a liquid varies with its temperature. Table 7.1 shows the vapor pressure of pure water at different temperature. Obviously, as the temperature of a liquid or solid increases its vapor pressure also increases. It is because more molecules have enough energy to escape from the liquid or solid at a higher temperature. At a given temperature, the vapor pressure of a substance is constant.

Table 7.1 Vapor pressure of pure water at different temperature

Temperature/K	273	293	313	323	353
Vapor Pressure/kPa	0.611	2.34	7.38	12.34	47.37

The types of molecules that make up a solid or liquid determine its vapor pressure (see Table 7.2). As you know, the vapor pressure of ethanol is higher than the vapor pressure of water at a given temperature, since the intermolecular forces between ethanol molecules are relatively weak.

Table 7.2 Vapor pressure of several liquids at 293K

Liquid	Water	Alcohol	Benzene	Ether	Mercury
Vapor Pressure/kPa	2.339	5.853	9.959	57.73	1.201×10^{-3}

2. Vapor Pressure Lowering

A experiment shows the vapor pressure of pure water at 25 ℃ is 3167.7 Pa, whereas the vapor pressure of an aqueous 0.5 mol/kg sucrose solution at the same temperature is 3135.7 Pa. And as for the 1 mol/kg sucrose solution under the same condition, the vapor pressure is only 3107.7 Pa. What does these data mean? It is the presence of a nonvolatile solute lowers the vapor pressure of a solvent under the same conditions. The more concentrated the solution, the more the vapor pressure is lowered.

Have you wondered about why this happens? As you know, the vapor pressure over a liquid is the result of a dynamic equilibrium between condensation and evaporation. When we add a nonvolatile solute (e.g. the sucrose) to water, some of the water surface area will be occupied by solute particles. Thus, the amount of surface area available for the escaping water molecules will be reduced. On the other hand, the escape of the liquid water to the gas phase may be restricted because the intermolecular forces in solutions are usually greater than those in the separated substances. Therefore, a nonvolatile solute added to water reduces the capacity of the water molecules to move from the liquid phase to the vapor phase. That fact is reflected in the lower vapor pressure for a aqueous solution with a nonvolatile solute relative to the pure water. (See Figure 7.1)

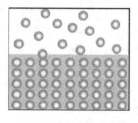
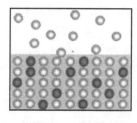

(a) Pure Solvent   (b) Solution

Figure 7.1 Vapor pressure of a solution is lower than that of the pure solvent

The French chemist Francois Raoult (10 May, 1830~1 April, 1901) discovered a law based on numerous experiments in 1887. At a given temperature, the vapor pressure of the dilute solution containing a nonvolatile nonelectrolyte solute is directly proportional to the mole fraction of solvent in the solution. This is Raoult's law, which is given by

$$p_{\text{soln}} = p^0_{\text{solvent}} \chi_{\text{solvent}} \quad (7.11)$$

Where p_{soln}-Vapor pressure of the solution;

χ_{solvent}-Mole fraction of the solvent in the solution;

p^0_{solvent}-Vapor pressure of a pure solvent at the same temperature.

Suppose we have such a solution containing only a solute and a solvent, then we have

$$\chi_{\text{solvent}} = 1 - \chi_{\text{solute}}$$

Substitute into Equation(7.11):

$$p_{\text{soln}} = p^0_{\text{solvent}}(1 - \chi_{\text{solute}})$$

$$p^0_{\text{solvent}} - p_{\text{soln}} = p^0_{\text{solvent}} \chi_{\text{solute}}$$

Therefore
$$\Delta p = p^0_{solvent} - p_{soln} = p^0_{solvent} \chi_{solute} \tag{7.12}$$

In this equation, Δp is the change in vapor pressure of solution. Equation (7.12) is another expression form of Raoult's law. It mathematically describes the vapor pressure lowering phenomenon, i. e. , at a particular temperature, the decrease in the vapor pressure of the dilute solution containing a nonvolatile nonelectrolyte solute is proportional to the mole fraction of solute and independent of the nature of the solute.

In addition, we can deduce another important and useful formula from equation(7.12), which can be given as following.
$$\Delta p = p^0_{solvent} - p_{soln} = K m_{solute} \tag{7.13}$$

Here, m_{solute} represents the molality of the solute, and K is a proportionality constant at a given temperature which is determined by the $p^0_{solvent}$ and the molar mass of the solvent. This formula indicates that at a certain temperature the decrease in the vapor pressure of the dilute solution containing a nonvolatile nonelectrolyte solute is approximately proportional to the molality of solute and independent of the nature of the solute.

Solutions that obey Raoult's law are called ideal solutions. The ideal solution could be formed when the solvent and solute have similar size, shape, and polarity in the mixture. In an ideal solution, the interactions are all the same, i. e. , solvent-solvent and solute-solute attractions are the same as solvent-solute attractions. Otherwise, solutions will show a deviation from Raoult's law. These solutions are named non-ideal solutions. Actually, very few solutions can approach ideality. Only when the solute concentration is low, the inter molecular interactions between solute and solvent is similar, the solute and solvent molecules size similar, real solutions tend to behave as ideal solutions. However, Raoult's law is a good enough approximation for the non-ideal solutions that we will continue to use Raoult's law.

Notice that in our applications of Raoult's law we have limited ourselves to the dilute solution containing a nonvolatile nonelectrolyte solute, this is because

(1) If the solute is a volatile liquid (such as ethyl alcohol), it will have some vapor pressure of its own. And it will contribute to the total vapor pressure above a solution in which it is dissolved. Therefore, the vapor pressure over a solution containing a volatile solute (or solutes) is equal to the sum of the vapor pressures of the solvent and each of the volatile solutes. The Raoult's law still applies. This will discuss in the Physical Chemistry.

(2) When the solution is a little concentrated (1 mol/kg sucrose solution, for example), the solvation effects and the interactions among the particles in the solution cannot be neglected. The actual vapor pressure lowering may not equal to that predicted by Raoult's Law. However, if the solution is sufficiently dilute, the dissolved solute will have tiny effect on the solvent. The solution will then approach ideal behaviour. That is very dilute solutions obey Raoult's Law approximately.

(3) If the solute is an electrolyte, e. g. NaCl, when dissolved in water it dissociates into one Na^+ ion and one Cl^- anion. So, if you added 0.1 moles of NaCl, there would actually be 0.2 moles of particles in the solution theoretically. That's the figure you would have to use in the mole fraction calculation. In other words, the equation for Raoult's Law should be modified in such case, as we shall see in the following section.

Sample 7.6 Estimate the vapor pressure of a 5% (m/v) glucose ($C_6H_{12}O_6$) solution which is made by dissolving 5 g of glucose in 100 mL of water. The vapor pressure of pure water at the same temperature is 3.17 kPa.

Solution:

100 mL of water has a mass of 100 g since the density of water is nearly 1 g/mL.

$$n_{C_6H_{12}O_6} = \frac{5 \text{ g}}{180 \text{ g/mol}} = 0.0278 \text{ mol}$$

$$n_{H_2O} = \frac{100 \text{ g}}{18 \text{ g/mol}} = 5.56 \text{ mol}$$

$$\chi_{H_2O} = \frac{5.56}{0.0278 + 5.56} = 0.995$$

$$p_{soln} = p^0_{solvent} \chi_{solvent} = 3.17 \text{ kPa} \times 0.995 = 3.15 \text{ kPa}$$

Boiling Point Elevation

1. Boiling Point

The boiling point of a liquid is defined as the temperature at which the vapor pressure of that liquid equals the atmospheric pressure. The normal boiling point of a liquid is the temperature at which its vapor pressure is equal to one atmosphere (101.325 kPa). For example, the normal boiling point of water is 373.15 K. Generally, the boiling point of a liquid refers to the normal boiling point when no clear indication of pressure is given.

The boiling point of a liquid is determined by the types of molecules that make up it. The stronger the intermolecular forces between molecules, the higher the boiling point, vice versa. The normal boiling point of ethyl alcohol is 351.5 K, it is lower than that of water due to relatively stronger hydrogen bonding interactions of water molecules. In addition, the boiling point of a liquid varies with external pressure. The lower the external pressure, the lower the boiling point of a liquid. The atmospheric pressure over Tibetan plateau is less than one atmosphere, the boiling point of water is lower than 373.15 K.

2. Boiling Point Elevation

It is experimentally shown that the boiling point of a solution made of a liquid solvent with a nonvolatile nonelectrolyte solute (slolutes) is higher than that of the pure solvent. This phenomenon is named boiling point elevation.

The vapor pressure of the solution which is lower than that of its solvent can be responsible for the boiling point of solution elevation. Figure 7.2 is a phase diagram for both a solution containing a nonvolatile solute (slolutes) and its pure solvent. This graph can illustrate the boiling point elevation phenomenon very well. In Figure 7.2, the line from O to A represents the vapor pressure curve of a pure liquid solvent, and the O' to A' line indicates the vapor pressure curve of a solution prepared by dissolving a nonvolatile solute (slolutes) in the solvent. As you can see from this graph, the line O' to A' is below the line O to A. This means the vapor pressure of the solution is lower than that of its pure solvent at any given temperature. For example, at T_b^0, the vapor pressure of the pure solvent is equal to 1 atm while the vapor pressure of the solution is lower than 1 atm. At this temperature, the pure solvent will boil, but the solution can't boil. To boil the solution, we must heat it to a higher temperature in order for the vapor pressure to become equal to 1 atm. Consequently, the boiling

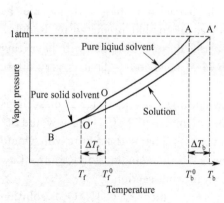

Figure 7.2 Phase diagram for a solvent and its solution with a nonvolatile solute

point of the solution goes up to T_b. Therefore, a higher temperature is required to boil the solution than the pure solvent. The difference between the boiling point of a pure solvent and that of a solution is the boiling point elevation (ΔT_b).

A large quantity of research has demonstrated that the magnitude of the boiling point elevation depends only upon the concentration of the solution, not on the nature of the solute. To be exact, for dilute solutions containing a nonvolatile nonelectrolyte solute (slolutes) the boiling point elevation is directly proportional to the molality of the solute. Hence, we can calculate that change in boiling point from the following equation(7.14).

$$\Delta T_b = T_b - T_b^0 = K_b m \qquad (7.14)$$

Where ΔT_b represents an increase in the boiling point, T_b^0 is the boiling point of the pure solvent and T_b is the boiling point of the solution. In the above formula, we use the unit molality (m) for the concentration because molality is temperature independent. The proportionality constant K_b is the molar boiling point elevation constant and has a specific value depending on the identity of the solvent being used. Some typical values for several common solvents are given in Table 7.3. For water, K_b is 0.512 K·kg/mol, therefore we can expect a 1 mol/kg aqueous solution of sucrose will boil 0.512 K higher than pure water.

Table 7.3 Molar boiling-point-elevation and freezing-point-depression constants

Solvent	T_b^0/K	K_b/(K·kg/mol)	T_f^0/K	K_f/(K·kg/mol)
Water, H_2O	373.1	0.512	273.0	1.86
Benzene, C_6H_6	353.1	2.53	278.5	5.12
Cyclohexane, C_6H_{12}	353.7	2.79	279.4	20.2
Ethanol, C_2H_5OH	351.4	1.22	158.4	1.99
Acetic acid, CH_3COOH	391.1	3.07	289.6	3.90
Acetone, CH_3COCH_3	329.2	1.71	177.6	2.40
Ether, $(C_2H_5)_2O$	307.7	2.02	156.8	1.79
Carbon tetrachloride, CCl_4	349.8	4.95	250.0	30.0
Chloroform, $CHCl_3$	334.2	3.63	209.5	4.68
Carbon disulfide, CS_2	319.2	2.37	161.0	3.80
Camphor, $C_{10}H_{16}O$	477.0	5.61	452.8	39.7
Naphthalene, $C_{10}H_8$	491.0	5.80	353.2	6.90

Sample 7.7 What is the boiling-point of a solution made by mixing 11.5 g of ethanol and 500.0 g of water?

Solution:

When the components of a solution are all liquids, the solvent is the substance in greater quantity. In this case, it is obviously water. Hence,

$$K_b = 0.512 \text{ K·kg/mol}$$

$$m = \frac{\frac{11.5 \text{ g}}{46 \text{ g/mol}}}{0.5 \text{ kg}} = 0.5 \text{ mol/kg}$$

$$\Delta T_b = K_b m = 0.512 \text{ K·kg/mol} \times 0.5 \text{ mol/kg} = 0.256 \text{ K}$$

$$T_b = T_b^0 + \Delta T_b = (373.15 + 0.256)\text{K} = 373.41 \text{ K}$$

Freezing Point Depression

1. Freezing Point

The freezing point of a substance is the temperature at which the vapor pressure of its liquid phase is equal to the vapor pressure of its solid phase at a given external pressure. In this case, the solid phase and the liquid phase can simultaneously coexist. If the vapor pressure unequal, phase transition will occur from the phase with a higher vapor pressure to the phase with a lower vapor pressure.

The freezing temperature varies slightly with the external pressure, and the normal freezing point, T_f, of a liquid is the temperature at which it freezes at 1 atm (101.325 kPa). For example, the normal freezing point of H_2O is the temperature at which the external pressure is 101.3 kPa as well as both the vapor pressure of water and the vapor pressure of ice are equal to 0.611 kPa, i.e. 273 K. At this temperature, ice and water can simultaneously coexist. If the temperature is higher than 273 K, ice will melt since the vapor pressure of ice is larger than that of water.

2. Freezing Point Depression

As it is known to all, pure water freezes at 0 ℃ while seawater freezes at a temperature is lower than 0 ℃. This is a common illustration for the freezing point depression. In fact, large number of researches have demonstrated that the freezing point of a solution of a nonvolatile solute is always lower than the pure solvent. This phenomenon is called freezing point depression.

The freezing point depression is also related to the lowering of vapor pressure by the solute. The Figure 7.2 points out that fact graphically. In Figure 7.2, for simplicity, suppose the line from O to A, the line from O' to A' and the line from O to B represents the vapor pressure curve of water, an aqueous solution and ice, respectively. As it is shown in this graph, at point O, the vapor pressure of water is equal to that of ice (0.611 kPa), hence the corresponding temperature (T_f^0) is the freezing point of pure water (273 K). As discussed earlier, the vapor pressure of the aqueous solution is lower than that of water solvent at any given temperature. At T_f^0 (or 273 K), the vapor pressure of the aqueous solution is lower than that of ice (0.611 kPa), the solution can't freeze, on the contrary, ice will melt. When the temperature continue to drop, as a result of the faster drop of the vapor pressure of ice compared to that of the aqueous solution, at a certain temperature T_f, the vapor pressure of ice and that of the aqueous solution may be equal, the aqueous solution will freeze. The temperature T_f is the freezing point of the solution. Obviously, the freezing point of the aqueous solution T_f is lower than that of pure water T_f^0. And the difference between the freezing point of a pure solvent and that of a solution is the freezing point depression (ΔT_f).

Experimentally, we know that the freezing point depression of a dilute solution containing a nonvolatile nonelectrolyte solute (slolutes) is directly proportional to the molality of the solute, and is given by the equation

$$\Delta T_f = T_f^0 - T_f = K_f m \qquad (7.15)$$

Here, ΔT_f represents a decrease in the freezing point, T_f^0 is the freezing point of the pure solvent, T_f is the freezing point of the solution, and m is molality of the solute. Note that the molar freezing point depression constant, K_f, depends only on the solvent. Some typical values for K_f for several common solvents are also given in Table 7.3. For water, K_f is 1.86 K·kg/mol, therefore we can expect a 1 mol/kg aqueous solution of sucrose will freeze at −1.86 ℃. From the above equation, we can conclude that as with all colligative properties, the magnitude of the freezing point depression depends only upon the concentration of the solution, not on the nature of the solute.

The freezing point depression is a useful property in many ways. For example, the antifreeze used in automobile heating and cooling system is a mixture of water and ethylene glycol (or propylene glycol) which has a freezing point much lower than that of pure water. The use of NaCl, $CaCl_2$, etc. on icy roads in the winter after a snowfall helps to melt the ice from the roads by lowering the melting point of the ice. In addition, measurements of freezing point depression can be used to determine the molar mass of a dissolved molecular solute, the technique is called cryoscopy. Similarly, measurements of boiling point elevation can be used to determine the molar mass of a dissolved molecular solute, a method named ebullioscopy. However, when the molar mass in need to be determined we often prefer to the cryoscopy. This is because the cryoscopy is more sensitive ($K_f > K_b$, hence $\Delta T_f > \Delta T_b$) and it can be carry out at a lower temperature whereby avoiding the biological samples to be destroy or degeneration, comparatively speaking.

Sample 7.8 Ethylene glycol is the main ingredient in antifreeze. Estimate the freezing point of a solution made by mixing 400 mL of ethylene glycol in 600 mL of water. The density of ethylene glycol is 1.11 g/mL and its molar mass is 62.01 g/mol.

Solution:

$$n(\text{ethyleneglycol}) = \frac{400 \text{ mL} \times 1.11 \text{ g/mL}}{62.01 \text{ g/mol}} = 7.16 \text{ mol}$$

$$m = \frac{7.16 \text{ mol}}{600 \text{ mL} \times 1 \text{ g/mL}} = 11.9 \text{ mol/kg}$$

$$\Delta T_f = K_f m = 1.86 \text{ K} \cdot \text{kg/mol} \times 11.9 \text{ mol/kg} = 22.1 \text{ K}$$

$$T_f = T_f^0 - \Delta T_f = 273.0 \text{ K} - 22.1 \text{ K} = 250.9 \text{ K}$$

Sample 7.9 A 1.35 g sample of an organic compound is dissolved in 10.0 g of benzene reduces the freezing point from 5.50 ℃ to 1.84 ℃. Calculate the molar mass of the organic compound. ($K_f = 5.12$ K · kg/mol).

Solution:

$$m = \frac{\Delta T}{K_f} = \frac{(5.50 - 1.84) \text{K}}{5.12 \text{ K} \cdot \text{kg/mol}} = 0.715 \text{ mol/kg}$$

Moles of the sample $= 0.715 \text{ mol/kg} \times 10.0 \times 10^{-3} \text{kg} = 7.15 \times 10^{-3} \text{ mol}$

Molar mass of the sample $= \dfrac{1.35 \text{g}}{7.15 \times 10^{-3} \text{mol}} = 189$ g/mol

Osmotic pressure

1. Osmosis and Osmotic Pressure

Another example of the colligative properties of the solution is the osmosis phenomenon. Figure 7.3 shows a typical experiment to demonstrate osmosis. In this experiment the substance in the thistle tube is a solution and in the flume is the pure solvent used in making that solution. And they are separated by a semipermeable membrane. Initially, the liquid heights of the solvent in the flume and the solution in the thistle tube are the same, as shown in Figure 7.3(a). After a while, it is found that the fluid levels in the flume and in the thistle tube become uneven, the latter is higher. The volume of the solution rise [see Figure 7.3(b)]. Eventually, as Figure 7.3(c) shows, when the height difference between the two sides becomes large enough, the ascent of the level of the solution ceases.

How can we explain these phenomena?

First, we should know the characters of the semipermeable membrane. The semipermeable membranes contain pores that permit passage of some components of a solution. They ofen allow solvent molecules such as water to pass through. However, they do not al-

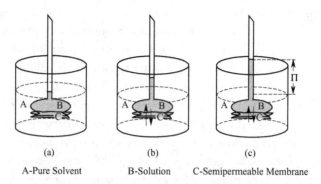

(a)   (b)   (c)

A-Pure Solvent   B-Solution   C-Semipermeable Membrane

Figure 7.3 An experiment to illustrate osmosis

low the passage of larger molecules or ions, e.g. the proteins. Important examples of semipermeable membranes are cell membranes and cellophane.

Then, let's investigate the micro process in this experiment. As we have mentioned above, the solvent can pass through the semipermeable membrane freely. Thus there is movement of solvent in both directions across a semipermeable membrane. However, the concentration of the solvent particles in the solution is relatively less, so the rate of transfer of solvent from the solution to the solvent is slower than in the opposite direction. That difference in flow rate causes the solution volume to increase, and the level of the solution inside the tube begins to rise accordingly. As the solution rises, it exerts a larger pressure to drive its solvent to flow faster to the pure solvent side. When the flow rate from both sides of the membrane are equal, equilibrium is reached. The net flow of solvent ceases and the level of the solution can't rise any longer.

Likewise, when two solutions of different concentrations are separated by a semipermeable membrane, more solvent molecules will flow out of the less concentrated solution side of the membrane than solvent flows into it from the more concentrated solution side.

In the experiments we mentioned above, the net transfer of solvent through a semipermeable membrane from a pure solvent or a dilute solution to a more concentrated solution is termed osmosis. Osmosis will occur whenever two solutions of different concentration are separated by a semipermeable membrane. The excess hydrostatic pressure on the solution compared to the pure solvent or relative dilute solution at osmotic equilibrium is called osmotic pressure, denoted by 'Π' as shown in Figure 7.3(c). Obviously, if one wishes to prevent osmosis one should apply pressure to the more concentrated solution equal to the osmotic pressure. Therefore, usually, the osmotic pressure of a solution is defined as the external pressure that must be applied to a solution to stop osmosis. When one exerts a pressure greater than the osmotic pressure on the more concentrated solution, one can force the solvent molecules from the concentrated solution to the dilute solution or pure solvent. This interesting process is known as reverse osmosis.

2. Calculation of Osmotic Pressure

As we discussed above, the osmotic pressure Π is the hydrostatic pressure required for a solution to achieve osmotic equilibrium with the pure solvent or relative dilute solution. So we can calculate osmotic pressure by using the following equation

$$\Pi = dgh \tag{7.16}$$

Where Π stands for the osmotic pressure, d is the density of the relative concentrated solution, h represents the height difference between the two liquid which we discuss, and g

is the gravitational constant.

In addition, experiments show that the osmotic pressure is related to both the concentration and the temperature of the solution. In 1886 the famous Dutch physical and organic chemist Jacobus Henricus van't Hoff (30 August, 1852~1 March, 1911) pointed out that the osmotic pressure of the dilute solution containing a nonvolatile nonelectrolyte solute (solutes) obeys a law similar in form to the ideal-gas law, which can be given by

$$\Pi V = nRT \tag{7.17}$$

or

$$\Pi = \frac{n}{V}RT = cRT \tag{7.18}$$

Where Π—the osmotic pressure of the solution, Pa;
V—the volume of the solution, dm^3;
n—the amount of solute, mol;
R—ideal gas constant, 8.314 kPa · dm^3/(mol · K);
T—thermodynamic temperature, K;
c—the molarity of the solution, mol/L.

These formulae are called van't Hoff equations. They demonstrate that the osmotic pressure of dilute solutions containing a nonvolatile nonelectrolyte solute (solutes) is proportional to the concentration of the solute particles in solution at constant temperature. In other words, the osmotic pressure, like other colligative properties, increases with the concentration of solute particles and does not depend on the type of the solute particles. Likewise, these formulae give valid result for an ideal solution or for the limiting case of a real solution containing a nonvolatile nonelectrolyte solute (solutes) that is sufficiently dilute.

As to the aqueous solution, when it is sufficiently dilute, the molarity and molality are nearly equal, therefore equation(7.18) can be rewritten as

$$\Pi = mRT \tag{7.19}$$

Where m is the molality of the solution.

Sample 7.10 The 5 g/L glucose ($C_6H_{12}O_6$) solution is an often used intravenous injection in clinical practice. What is the osmotic pressure of the solution at 310 K?

Solution:

$$\Pi = mRT = \frac{5 \text{ g/L}}{180 \text{ g/mol}} \times 8.314 \text{ kPa} \cdot \text{L/(mol} \cdot \text{K)} \times 310 \text{ K} = 71.6 \text{ kPa}$$

3. Some Applications of Osmosis and Osmotic Pressure

Osmosis widely exists in nature and organisms, and plays an essential role in many physiological processes. The basic principle of osmosis is extensively used in chemistry, biology, industry, clinical practice, and even our everyday life.

(1) Molecular Weight Determination

Like cryoscopy and ebullioscopy measurements, the osmotic pressure measurements can also be used to determine molecular mass of a dissolved substance. This technique, osmometry, is very sensitive when compared with cryoscopy and ebullioscopy because a small amount of solute will produce a much larger change. Even at extremely low concentrations (10^{-6} mol/kg, for example), the osmotic pressure of the solutions is still measurable. Taking this advantage, it is often the only practical way for determining molar masses of large molecules (such as proteins and polymers) due to their limited solubility in most solvents. However, this method can't be applied to determining molar masses of small molecules because they can pass through the semipermeable membrane.

Sample 7.11 The osmotic pressure of a solution of 0.050 g of horse hemoglobin in 10.0 mL of water solution is 1.8×10^2 Pa at 25 ℃. What is the molecular weight of the hemoglobin?

Solution:

$$c = \frac{\Pi}{RT} = \frac{1.8 \times 10^2 \times 10^{-3} \text{ kPa}}{8.314 \text{ kPa} \cdot \text{L/(mol} \cdot \text{K)} \times 298 \text{ K}} = 7.265 \times 10^{-5} \text{ mol/L}$$

$$n = 7.265 \times 10^{-5} \text{ mol/L} \times 10.0 \times 10^{-3} \text{ L} = 7.265 \times 10^{-7} \text{ mol}$$

$$\text{Molar mass} = \frac{0.050 \text{ g}}{7.265 \times 10^{-7} \text{ mol}} = 6.882 \times 10^4 \text{ g/mol}$$

(2) Application in Medicine

Osmosis is widely applied in medicine science. In this section, some typical examples of these applications will be given after some related terms been introduced.

In medicine, the term osmolarity is commonly used when we discuss the osmotic pressure of solutions. Osmolarity is defined as the total concentration of all the particles in a solution that contribute to the solution's osmotic pressure. Osmolarity is customarily represented by the symbol "c_{os}", and the frequent unit is mmol/L or mol/L. For example, a solution of 0.1 mol/L sucrose solution corresponds to an osmolarity of 0.1 mol/L, while the osmolarity of 0.1mol/L NaCl is 0.2 mol/L since a NaCl salt particle can dissociate fully in water to become two separate particles, a Na^+ ion and a Cl^- ion. Evidently, the higher the osmolarity of solution, the higher the osmotic pressure.

The osmotic pressure of different solutions may be equal and unequal. Two solutions are said to be isotonic if they have the same osmotic pressure. As for two solutions whose osmotic pressure is unequal, comparatively speaking, the solution has a higher osmotic pressure is termed hypertonic solution and the solution has a lower osmotic pressure is called hypotonic solution. In clinical practice, the classification standard of the isotonic, hypertonic and hypotonic solution is based on the total osmolarity in plasma. The osmolarity of normal human plasma is about 304 mmol/L. Therefore, it is stipulated in clinical practice that a physiological isotonic solution is a solution with an osmolarity of 280~320 mmol/L. For instance, the 9 g/L normal saline (308 mmol/L), the 50 g/L glucose solution (280 mmol/L) and the 12.5 g/L $NaHCO_3$ (298 mmol/L) are all common physiological isotonic solution. Meantime, the solution with osmolarity higher than 320 mmol/L is ascribed to the hypertonic solution, and the solution with osmolarity lower than 280 mmol/L is referred as a hypotonic solution.

In clinical practice, when a patient is given a great quantity of solutions containing electrolytes and nutrients by intravenous (Ⅳ) infusion, caution should be paid to the osmolarity of the infusion solutions, otherwise as a result, the water regulation in his body may be disorder and the red blood cells may be deformed even be destroyed. The following experiment of red blood cells in solutions with different concentration can be regard as an illustration (See Figure 7.4). Red blood cells contain water, solutes, and are surrounded by semipermeable membranes. Figure 7.4(b) shows the red blood cells in an isotonic solution (e.g. 50 g/L glucose solution), they keep their normal morphology. If red blood cells are placed in a hypertonic solution (e.g. 250 g/L glucose solution), there is a higher solute concentration in the solution than in the cell. Osmosis occurs and water passes through the membrane out of the cell. The cell shrivels up, as shown in Figure 7.4(a). This process is called crenation. When red blood cells are placed in a hypotonic solution (e.g. 10 g/L glucose solution), there is a higher solute concentration in the cell than outside the cell. Osmosis occurs and water flows into the cell. This causes the cell burst, as shown in Figure 7.4 (c), a process called hemolysis. Therefore, for intravenous feeding, the IV solutions must be isotonic with relative to the intracellular fluids of the red blood cells to prevent hemolysis or crenation.

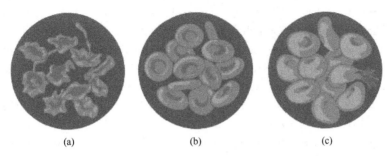

(a)   (b)   (c)

Figure 7.4   Effect of different solutions on red blood cells

Sample 7.12   The average osmotic pressure of blood is 769.9 kPa at 37 ℃. How many grams of glucose ($C_6H_{12}O_6$) are required to prepare 1 L of glucose solution isotonic with blood?

Solution:

$$\Pi = cRT$$

$$c = \frac{\Pi}{RT} = \frac{769.9 \text{ kPa}}{8.314 \text{ kPa} \cdot \text{L/mol} \cdot \text{K} \times 310 \text{ K}} = 0.2987 \text{ mol/L}$$

$$\text{Mass}_{C_6H_{12}O_6} = 0.2987 \text{ mol/L} \times 1 \text{ L} \times 180 \text{ g/mol} = 53.77 \text{ g}$$

(3) Other Typical Applications of Osmosis

Osmosis plays an important role in living systems and our everyday life. Water moves into plants, to a great extent, through osmosis. Eating large quantities of salty food causes retention of water and swelling of tissues (edema). Sometimes we use salt to preserve meats. The salt reduces the water concentration to a level below that in living organisms. Bacteria placed on the surface of the salty meat will lose water through osmosis and die of dehydration. Another use of osmosis is in reverse osmosis, a method is used in desalination of seawater and can also be used to purify fresh water for medical, industrial and domestic applications. In this process, a solvent was forced to move from a region of high solute concentration through a membrane to a region of low solute concentration by applying a pressure larger than the osmotic pressure.

Colligative Properties of Electrolyte Solutions

In our above discussion of colligative properties, we have limited the application of all the formulae only to the dilute solution containing a nonvolatile nonelectrolyte solute (solutes). And we also mentioned, all the formulae should be modified when the solute is an electrolyte since an electrolyte solute will contribute more particles per formula unit than a nonelectrolyte solute. But how can we do? Van't Hoff suggested a correction factor 'i' should be introduced into all the formulae to account for the fact that electrolyte solutes may ionize. Thus they will be rewritten as following.

$$\Delta p = iKm \tag{7.20}$$

$$\Delta T_b = iK_b m \tag{7.21}$$

$$\Delta T_f = iK_f m \tag{7.22}$$

$$\Pi = icRT \approx imRT \tag{7.23}$$

Where 'i' is called the van't Hoff factor. Theoretically, the value of 'i' is equal to the number of ions produced by a particular solute. For example, the ideal value of 'i' for NaCl is 2 because each NaCl produces one Na^+ and one Cl^-; for K_2SO_4 it is 3 since each K_2SO_4 releases two K^+ and one SO_4^{2-}. Thus, 0.100 mol/kg solution of NaCl will have a freezing point depression of 0.372 K ($\Delta T_f = 2 \times 1.86 \times 0.100 = 0.372$K), and 0.100 mol/kg solution of K_2SO_4 will have a freezing point depression of 0.558 K. However, the measured values are only 0.348 K and 0.458 K, respectively. The difference between the expected and measured value is due to electrostatic attractions of ions. These electrostatic attractions cause the number of independent particles reduced, and result in a reduction in the freezing point depression (as well as in the vapor pressure reduction, boiling point elevation, and the osmotic pressure.)

Table 7.4 gives the observed (measured) Van't Hoff factors for several substances at different concentration. Obviously, the observed value is less than the ideal value. And the more dilute the solution, the more closely it approaches the ideal value. Meantime, the lower the charges on the ions, the less it departs from the ideal value. These can be explained as, the electrostatic interaction between charged particles decreases as their separation increases and as their charges decrease. However, in dilute solutions the assumption of complete ionization will give us good results, hence when the actual value (or measured value) is absent, we will still use the ideal value in calculations.

Table 7.4 Van't Hoff factors for several substances at 298K

Compound	Concentration/(mol/kg)				Ideal Value
	0.100	0.0500	0.0100	0.00100	
Glucose	1.00	1.00	1.00	1.00	1.00
NaCl	1.87	1.89	1.94	1.97	2.00
K_2SO_4	2.32	2.59	2.70	2.84	3.00
$MgSO_4$	1.21	1.30	1.53	1.82	2.00

Sample 7.13 Blood is isotonic with 0.9% (m/v) NaCl solution at 25 ℃. What is the osmotic pressure? Assume $i = 2.0$.

Solution:

Suppose we have 100 mL of the soultion, thus there is 0.9 g of NaCl in the solution.

$$c = \frac{\frac{0.9 \text{ g}}{58.5 \text{ g/mol}}}{0.1 \text{L}} = 0.154 \text{ mol/L}$$

$$\Pi = icRT$$
$$= 2 \times 0.154 \text{ mol/L} \times 8.314 \text{ kPa} \cdot \text{L/(mol} \cdot \text{K)} \times 298 \text{ K}$$
$$= 763.1 \text{ kPa}$$

Key Words

solution [sə'lju:ʃən]　溶液，解答
solvent ['sɔlvənt]　溶剂
solute ['sɔlju:t]　溶解物，溶质
aqueous solution ['eikwiəs sə'lju:ʃən]　水溶液
concentration [ˌkɔnsen'treiʃən]　浓度
concentrated ['kɔnsentreitid]　集中的，浓缩的

percentage [pə'sentidʒ]　百分比，百分率，比例
mole fraction [məul 'frækʃən]　摩尔分数
mass concentration [mæs ˌkɔnsen'treiʃən]　质量浓度
molarity [məu'læriti]　摩尔浓度
molality [məu'læliti]　质量摩尔浓度

colligative ['kɔligətiv] （物质的物理性质）依数的

colligative properties 依数性

Raoult's law 拉乌尔定律

ideal solution 理想溶液

vapor pressure lowering 蒸气压下降

boiling point elevation 沸点升高

freezing point depression 凝固点降低

molar boiling-point-elevation constant 沸点升高常数

molar freezing-point-depression constant 凝固点降低常数

cryoscopy [krai'ɔskəpi] 冰点测定学

ebullioscopy [i,bʌli'ɔskəpi] 沸点升高测定法

osmosis [ɔz'məusis] 渗透（作用）

osmotic pressure 渗透压

semipermeable ['semi'pə;mjəbl] 半透性的

membrane ['membrein] （动物或植物体内的）薄膜，隔膜

osmometry 渗透压力测定法

osmolarity [ɔzmə'læriti] 渗透浓度

hypotonic ['haipəu'tɔnik] solution 低渗溶液

isotonic ['aisəu'tɔnik] solution 等渗溶液

hypertonic ['haipə(:)'tɔnik] solution 高渗溶液

crenation [kri'neiʃən] 圆齿状，钝齿状；（红细胞的）皱缩

hemolysis [hi:'mɔləsis] 溶血（作用），溶血现象；血球溶解

Exercises

1. A solution for intravenous injections contains 245 mg of lactate, $C_3H_5O_3^-$, in 100.0 mL. What is the molarity of the ion?

Answer: 0.0275 mol/L

2. The maximum permissible amount of cadmium (Cd^{2+}) in drinking water is 0.01 mg/L. Calculate the molarity of such a solution. (Atomic weight: Cd=112.41).

Answer: 9×10^{-8} mol/L

3. Formalin is a preservative solution used in biology laboratories. It contains 40 cm^3 of formaldehyde, CH_2O, (density 0.82 g/cm^3) per 100 cm^3 of water. Determine the molality of formalin.

Answer: 11 mol/kg

4. The disinfectant hydrogen peroxide (H_2O_2) is a 3.0% by weight solution in water. What are the molarity and the molality of the solution. Assuming a density of 1.0 g/cm^3.

Answer: 0.88 mol/L, 0.88 mol/kg

5. Hydrochloric acid (HCl) is purchased as a concentrated aqueous solution with the following composition.

Density — 1.19 g/mL   Mass percent HCl — 38%

(1) Calculate the molarity, molality and mole fraction for the solution. (Molar mass HCl=36.5 g/mol)

(2) How much concentrated HCl would be required to make 1.00 L of a 0.100 M solution of HCl?

Answer: (1) 12.4 mol/L, 16.8 mol/kg, χ_{HCl}=0.23, χ_{water}=0.77 (2) 8.06 mL

6. Urea, NH_2CONH_2, is a nonvolatile nonelectrolyte. Calculate the boiling point and the freezing point of a solution of 30.0 g of urea in 250 g of water.

Answer: 101.0 ℃, −3.72 ℃

7. A 0.350 g sample of a biomolecule is dissolved in 15.0 g of chloroform, and the freezing-point depression is determined to be 0.240 ℃. Calculate the molar mass of the biomolecule.

Answer: 455 g/mol

8. Estimate the freezing point of an antifreeze mixture is made up by combining one volume of ethylene glycol (mass weight=62, density 1.11 g/mL) with two volumes of water.

Answer: -11 ℃

9. Arrange the following solutions in the order of increasing osmotic pressure and state your reasons.
(1) 0.01 M $CaCl_2$ at 50 ℃   (2) 0.01 M glucose at 50 ℃
(3) 0.01 M NaCl at 25 ℃   (4) 0.01 M $MgSO_4$ at 25 ℃
(5) 0.01 M sucrose at 25 ℃

Answer: (1)>(3)>(4)>(2)>(5)

10. Insulin controls glucose metabolism. A solution of 2.0 g of insulin in 250.0 mL of water has an osmotic pressure of 3.48×10^3 Pa at 30 ℃. What is the molecular weight of insulin?

Answer: 5.8×10^3 g/mol

11. An aqueous solution of 10.00 g of catalase, an enzyme found in the liver has a volume of 1.00 L at 27 ℃. The solutions osmotic pressure is found to be 99.3 Pa. Calculate the molar mass of catalase.

Answer: 2.51×10^5 g/mol

12. What would be the calculated and measured freezing point for a 0.100 mol/kg solution of K_2SO_4? $K_f=1.86$ ℃ · kg/mol and $i=2.32$.

Answer: -0.56 ℃; -0.43 ℃

Chapter 8 Solubility Equilibrium

Everyday we encounter many things which are very important to our life, such as the air we breathe, the fluids that run through our body, the mineral water we drink and so on. Things above mentioned have a common character which mixture in homogeneous, that is, their component are intermingled uniformly. The homogenous mixture of two or more substances is called solution. The substance present in a larger amount is called solvent, the other things in a smaller amount is called solute. Solution may be existed as gases, liquids or solids. Water is the familiar solvent, so we will discuss only aqueous solution whose solvent is water. Chemical equilibria in solution play key roles in our daily life and chemical industry. In this chapter, we will mainly focus attention on solubility equilibrium.

8.1 Saturated Solutions and Solubility

From our everyday life, we know that different solids differ dramatically in their solubility. Salt, sugar, and ethyl alcohol are very soluble in water. Marble, salad oil, and silver chloride are very insoluble in water. How to express the solubility quantitatively?

For most substances, there is a limit to how much can be dissolved in a given volume of solvent. This limit is influenced by the nature of solute and solvent. Once the solution reaches a limit, no more solute dissolves, regardless of how much solid we add to the system. This solution is said to be saturated. The maximum quantity of solute that need be added to a given volume of solvent at a specific temperature is called solubility of the solute and denoted as s. A substance's solubility can be expressed in two other quantities. Molar solubility is the number of moles of solute per liter of a saturated solution (mol/L), solubility is grams of solute in forming a liter of a saturated solution (g/L). Note that all these solubility refer to the concentration of saturated solutions at a given temperature (usually 25 ℃) owing to the influence of temperature on solubility.

Chemists class substances as soluble, slightly soluble, or insoluble in a qualitative sense.

A soluble salt is dissolved in water to give a solution with a concentration of at least 0.1 mol/L at room temperature.

A salt is insoluble if the concentration of an aqueous solution is less than 0.001 mol/L at room temperature.

Slightly soluble salts give solutions that fall between these extremes.

Absolutely insoluble substance is not existed. When we put an iron nail in water, it does not appear to dissolve in water. If we had an instrument sensitive enough, we would find that some iron had dissolved. In the following section the insoluble substance we discussed is insoluble strong electrolyte, which means that all the dissolved dissociate completely.

8.2 Solubility Equilibria

Solubility Product Constant, K_{sp}

Consider an insoluble salt, such as AgCl, is dissolved in water. It dissociates completely into separate hydrated cations and anions by hydration, this process is called dissolution. As the dissolution proceeds, hydrated cations and anions will collide and reform the solid phase, this process is called precipitation. Once time is enough, the rate of dissolution equals to that of precipitation, an equilibrium is reached between the ions in solution and the solid in the solution. At this point no more solids will dissolve, that is, the solution becomes saturated. The equation for this equilibrium of AgCl is usually written as follow.

$$AgCl(s) \underset{precipitate}{\overset{dissolve}{\rightleftharpoons}} Ag^+(aq) + Cl^-(aq)$$

Applying the law of mass action to the above equilibrium, we may write the equilibrium constant expression for the equation at a given temperature.

$$K = \frac{[Ag^+][Cl^-]}{[AgCl(s)]} \qquad (8.1)$$

Since the concentration of pure solid is regarded as a constant, this term may be incorporated in the constant K to give another constant K_{sp}.

$$K_{sp} = [Ag^+][Cl^-] \qquad (8.2)$$

K_{sp} is called the solubility product constant or simply solubility product. It is the product of the ion concentration of the ions in the saturated solution, each raised to the power of its coefficient in balanced equation. The value of K_{sp} is the measure of solubility of a substance. The larger value of K_{sp}, the greater concentration of ions must be and more soluble the substance in water. The smaller value of K_{sp}, the more easily precipitate produce.

Each AgCl yields only one Ag^+ ion and one Cl^- ion, so it's very simple to write its solubility product expression. The following cases are more complex.

$$Ag_2CrO_4(s) \rightleftharpoons 2Ag^+ + CrO_4^{2-}$$
$$K_{sp} = [Ag^+]^2[CrO_4^{2-}] \qquad (8.3)$$
$$Ca_3(PO_4)_2(s) \rightleftharpoons 3Ca^{2+} + 2PO_4^{3-}$$
$$K_{sp} = [Ca^{2+}]^3[PO_4^{3-}]^2 \qquad (8.4)$$

A partial list of solubility product constants is given in Table 8.1 and in the Appendix E.

Table 8.1 K_{sp} value at 298K for some compounds

Compound	K_{sp}	Compound	K_{sp}
AgBr, Silver bromide	5.35×10^{-13}	CaF_2, Calcium fluorine	3.45×10^{-11}
AgCl, Silver chloride	1.77×10^{-10}	$Ca(OH)_2$, Calcium hydroxide	5.02×10^{-6}
AgI, Silver iodide	8.52×10^{-17}	CdS, Cadmium sulfide	1.40×10^{-29}
Ag_2CrO_4, Silver chromate	1.12×10^{-12}	$Cd(OH)_2$, Cadmium hydroxide	7.2×10^{-15}
$BaSO_4$, Barium sulfate	1.08×10^{-10}	CuS, Copper(II) sulfide	1.27×10^{-36}
$CaCO_3$, Calcium carbonate	3.36×10^{-9}	$Mg(OH)_2$ Magnesium hydroxide	5.61×10^{-12}
CaC_2O_4, Calcium oxalate	1.46×10^{-10}	PbS, Lead sulfide	9.04×10^{-29}
$CaSO_4$, Calcium sulfate	4.93×10^{-5}	ZnS, Zinc sulfide	2.93×10^{-25}

K_{sp} is related not only with insoluble strong electrolyte's structure, but also with temperature. Usually temperature raise, K_{sp} is increased. Because the change is feeble, we generally use the value of K_{sp} at 298K.

Sample 8.1 Give the solubility product constant expressions for the following compounds

(a) barium sulfate (b) iron (Ⅲ) hydroxide (c) bismuth sulfide
Solution:
  (a) $K_{sp}=[Ba^{2+}][SO_4^{2-}]$ (b) $K_{sp}=[Fe^{3+}][OH^-]^3$ (c) $K_{sp}=[Bi^{3+}]^2[S^{2-}]^3$

K_{sp} and Solubility

We can distinguish carefully between the solubility of a substance and its solubility product by their definition. The solubility product is an equilibrium constant and invariable for the substance at any specific temperature. In contrast solubility is the quantity that dissolves to form a saturated solution. The solubility can change considerably as the concentration of other solutes change. However, in all cases the product of ion concentration must satisfy the K_{sp} expression.

Although solubility and solubility product are distinguishing, the value of K_{sp} indicates the solubility of a substance. Solubility product and solubility are related, therefore solubility product can be calculated from the solubility, or vice versa.

Sample 8.2 The solubility of silver chloride is found experimentally to be 1.91×10^{-3} g/L in pure water at 298K. Calculate the value of K_{sp} for silver chloride.
Solution: To convert solubility to K_{sp}, we need to convert g/L to mol/L. The molar mass of silver chloride is 143.4 g/mol, so

$$s=\frac{1.91\times10^{-3}\text{ g/L}}{143.4\text{ g/mol}}=1.33\times10^{-5}\text{ mol/L}$$

The chemical equation and the K_{sp} expression are

$$AgCl(s)\rightleftharpoons Ag^++Cl^- \quad K_{sp}=[Ag^+][Cl^-]$$

Because each AgCl unit contains only one Ag^+ ion and one Cl^- ion, at equilibrium both $[Ag^+]$ and $[Cl^-]$ are equal to s. Then

$$[Ag^+]=[Cl^-]=s=1.33\times10^{-5}\text{ mol/L}$$

Now we can calculate K_{sp}

$$K_{sp}=[Ag^+][Cl^-]=s^2=(1.33\times10^{-5})^2=1.77\times10^{-10}$$

Thus, the value of K_{sp} for silver chloride is 1.77×10^{-10}.

Sample 8.3 The value of K_{sp} for Ag_2CrO_4 is 1.12×10^{-12} in pure water at 298 K. Calculate the solubility of Ag_2CrO_4 in mol/L.
Solution: Assume the solubility of Ag_2CrO_4 as s mol/L. Since each Ag_2CrO_4 yields 2 Ag^+ and CrO_4^{2-} ions

$$Ag_2CrO_4(s)\rightleftharpoons 2Ag^++CrO_4^{2-}$$
Equilibrium/(mol/L) $2s$ s

Substitute the ion concentration at equilibrium in the K_{sp} expression:

$$K_{sp}=[Ag^+]^2[CrO_4^{2-}]=(2s)^2\times s=4s^3$$

Then

$$s=\sqrt[3]{\frac{K_{sp,Ag_2CrO_4}}{4}}=\sqrt[3]{\frac{1.12\times10^{-12}}{4}}$$
$$=6.54\times10^{-5}\text{ (mol/L)}$$

Thus, the solubility of Ag_2CrO_4 is 6.54×10^{-5} mol/L.

As samples show, the relationship between molar solubility and solubility product is different according to the different chemical formula of compound. Table 8.2 summarizes the relationship among various expressions of molar solubility and solubility product.

Table 8.2 Relationship between solubility product (K_{sp}) and molar solubility (s)

Compound	Types of Compound	Relation between K_{sp} and s	Formular
$BaSO_4$	AB	$K_{sp}=s^2$	$s=\sqrt{K_{sp}}$
Ag_2CrO_4 or CaF_2	A_2B or AB_2	$K_{sp}=(2s)^2 s$	$s=\sqrt[3]{\dfrac{K_{sp}}{4}}$
$A_m B_n$	$A_m B_n$	$K_{sp}=(ms)^m(ns)^n$	$s=\sqrt[m+n]{\dfrac{K_{sp}}{m^m n^n}}$

Relative Solubility

A salt's K_{sp} value provides information about its solubility. In practice, great care must be taken in using K_{sp} values to predict the relative solubilities of a group of salts. For salts with similar formula, smaller values of K_{sp} salts with, less soluble the salt is. For example, $K_{sp,AgCl} > K_{sp,AgBr} > K_{sp,AgI}$, $s_{AgCl} > s_{AgBr} > s_{AgI}$. However, when the salts being compared produce different numbers of ions, the more K_{sp} is, the more solubility is not sure, such as $s_{Ag_2CrO_4} > s_{AgCl}$, $K_{sp,Ag_2CrO_4} < K_{sp,AgCl}$. In this case, we compare the relative solubility by calculating the solubility which using the relationship summarized in Table 8.2.

8.3 Factors Affecting Solubility

Common-Ion Effect

In the previous section we have considered insoluble salt dissolved in pure water. On the other hand, what will happen when water contains an ion in common with the dissolving salt?

Sample 8.4 what is the solubility (mol/L) of CaF_2?
(1) in pure water. (2) in 0.010 mol/L NaF solution? (CaF_2: $K_{sp}=3.45 \times 10^{-11}$)

Solution:
(1) The equilibrium equation and expression of K_{sp} are

$$CaF_2(s) \rightleftharpoons Ca^{2+} + 2F^- \qquad K_{sp}=[Ca^{2+}][F^-]^2$$

Suppose the solubility of CaF_2 in pure water as s_1 mol/L. Because 1mol CaF_2 dissociate 1mol Ca^{2+} and 2 mol F^-, we can write $[Ca^{2+}]=s_1$ mol/L and $[F^-]=2s_1$ mol/L. It follows that

$$K_{sp}=[Ca^{2+}][F^-]^2=s_1 \times (2s_1)^2=4s_1^3$$

This expression is easily solved for s_1

$$s_1=\sqrt[3]{\dfrac{K_{sp}}{4}}=\sqrt[3]{\dfrac{3.45 \times 10^{-11}}{4}}=2.05 \times 10^{-4} \text{ (mol/L)}$$

The solubility of CaF_2 in pure water is therefore 2.05×10^{-4} mol/L.

(2) Suppose s_2 (mol/L) as the solubility of CaF_2 in 0.010 mol/L NaF solution
Then
$$CaF_2(s) \rightleftharpoons Ca^{2+} + 2F^-$$
Initial/(mol/L)   0   0.010
Equilibrium/(mol/L) s_2 $2s_2+0.010$

Substituting into the expression of K_{sp} gives

$$K_{sp}=[Ca^{2+}][F^-]^2=s_2 \times (2s_2+0.010)^2=3.45 \times 10^{-11}$$

Since the K_{sp} value for CaF_2 is very small, s_2 is expected to be small compared with

0.010 mol/L, therefore, $2s_2 + 0.010 \approx 0.010$

$$s_2 \times 0.010^2 = 3.45 \times 10^{-11}$$
$$s_2 = 3.45 \times 10^{-7} \text{ mol/L}$$

Thus the solubility of CaF_2 in 0.010 mol/L NaF solution is 3.45×10^{-7} mol/L

Now we compare the solubility of CaF_2 in pure water and in 0.010 mol/L NaF solution

solubility of CaF_2 in pure water $= 2.05 \times 10^{-4}$ mol/L

solubility of CaF_2 in 0.010 mol/L NaF solution $= 3.45 \times 10^{-7}$ mol/L

Note that the solubility of CaF_2 is much less in the presence of F^- ions from NaF. Reduction in the solubility caused by the presence of a common ion is called the common-ion effect. We can use Le châtelier's principle to explain this phenomenon. If we add some soluble salt, acid or base that contains one of the same ions to a saturated solution, according to the principle the equilibrium will tend to adjust by decreasing the concentration of the added ions. On the other words, the solubility of the original salt is decreased, and it precipitates.

For example
$$CaF_2(s) \rightleftharpoons Ca^{2+} + 2F^-$$

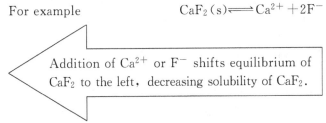

Addition of Ca^{2+} or F^- shifts equilibrium of CaF_2 to the left, decreasing solubility of CaF_2.

The common-ion effect can be useful if we wish to ensure that we have almost completely removed an ion from solution. We can conclude that the addition of excess hydroxide ion as precipitator to the wastewater supply should precipitate more of the heavy metal ions. However adding too much more precipitator is not suitable. The ions which precipitated can't remove completely from solution because of the existence of precipitation equilibrium. In practice, we consider that the ion precipitated is removed completely if its concentration is less than 1.0×10^{-5} mol/L. On the other hand, if too large an excess is used, the precipitate dissolves as a result of complex ion formation. For example, soluble $AgCl_2^-$ may be formed on the condition of excess Cl^- putting into the solution of AgCl.

$$AgCl + Cl^- \rightleftharpoons AgCl_2^-$$

Usually precipitator is added excessively 20%~30%.

Salt Effect

When dilute solution of silver nitrate ($AgNO_3$) and potassium chloride (KCl) are mixed in potassium nitrate (KNO_3) solution, less precipitation is formed. This is an example of salt effect. The increase in solubility of an insoluble strong electrolyte caused by the presence of ions in solution is called salt effect. How to explain the effect of potassium nitrate (which has no ions in common with silver chloride) on the equilibrium for saturated solution of silver chloride? In the mixed solution, the positively charged potassium ions are attracted by the chloride ions with their negative end, then cluster around the chloride ions. On the other hand, the negatively charged nitrate ions are attracted by the silver ions with one positive end, then cluster around the silver ions. Such clusters are called ionic atmosphere. The ionic atmosphere restrain combining the silver ions and chloride ions to form solid silver chloride, so the degree of collision between ions and the surface of solid is reduced. Then the rate of dissolution exceeds that of precipitation for a while, equilibrium shifts to the dissolution. When the reaction reaches a new dynamic equilibrium, the solubility of insoluble strong electrolyte is increased.

Both common-ion effect and salt effect are important factors on the solubility of substance. Common-ion effect decreases the solubility, whereas salt effect increases the solubility. In solution the common-ion effect is along with salt effect, usually common-ion effect is greater, so we neglect salt effect in the special dilute solution.

pH and Solubility

Some salt's solubility may be affected to some extent by the pH of the solution. For example, magnesium hydroxide dissolves according to the solubility equilibrium

$$Mg(OH)_2(s) \rightleftharpoons Mg^{2+}(aq) + 2OH^-(aq)$$

When we add OH^- ions which increase in pH at the same time, the common-ion effect forces the equilibrium shift to the left, decreasing the solubility of magnesium hydroxide. However if we put the H^+ ions into solution which decrease in pH at the same time, OH^- ions will react with the added H^+ ions so that the concentration of OH^- is reduced. The concentration of Mg^{2+} will have to increase to maintain the equilibrium condition, thus the solubility of magnesium hydroxide will increase in acidic solution. On base of this phenomenon, a suspension of solid $Mg(OH)_2$ known as milk of magnesia, is used to dissolve in the stomach to combat excess acidity.

This idea also applies to salts with other types of anions. For example, the solubility of calcium fluoride increases as the pH of solution decreases. Because F^- ion is a strong base, it is the conjugate base of the weak acid, HF. F^- react with H^+ to form HF and lead to the reduction of concentration for F^-. As a result, the solubility equilibrium of CaF_2 is shifted to right. Thus, the solubility of CaF_2 is enhanced.

$$CaF_2(s) \rightleftharpoons Ca^{2+} + 2F^-$$

The solubility of barium sulfate ($BaSO_4$), on the other hand, is unaffected when pH of solution changes. Because SO_4^{2-} ion is a very weak base, there is no HSO_4^- formed. That is, H^+ ions putting into a solution containing SO_4^{2-} does not react with SO_4^{2-}, so the solubility of sulfate salt is same in acid as in pure water.

These examples illustrate a general rule. if anion X^- is a strong base, on the other words, if HX is a weak acid, the solubility of salt MX is increased as the solution becomes more acidic. Salts containing effective bases, such as OH^-, S^{2-}, PO_4^{3-}, CO_3^{2-}, $C_2O_4^{2-}$ or CN^- are more soluble as $[H^+]$ increase (pH is lowered). This rule can also explain the formation of stalactites and stalagmites in cave. When carbon dioxide dissolves in groundwater, the acidity of groundwater is added. According to the rule of pH and solubility, the solubility of calcium carbonate is enhanced, then it dissolves and eventually produce huge caverns. On the other hand, as the carbon dioxide escapes to the air, the acidity of dripping water becomes weaker, then calcium carbonate precipitates, stalactites and stalagmites are formed very slowly.

8.4 Predicting Precipitation Reaction

Solubility Product Principle

Sometimes it is important to know either a precipitate will form or a precipitate will dissolve. For example we want to separate one ion from solution by precipitating. On the other hand, it may be necessary to dissolve a precipitate to identify the cation or anion in the qualitative analysis of a solution. We use the ion product Q, which is defined in a very similar way to the K_{sp} expression for a given solid. For example, solid $CaCO_3$, the ion product Q is written as follows.

$$Q = c_{Ca^{2+}} \cdot c_{CO_3^{2-}}$$

By definition, Q is the product of ions at any time, it changes during the reaction's proceeding. Only when the solubility reaction reaches equilibrium, Q equals to K_{sp}, so K_{sp} is an especial example of Q.

According to the knowledge of chemical equilibrium which Chapter 6 mentioned, we can judge the direction in which a solubility reaction will occur by comparing the ion product Q with the solubility product K_{sp}. The possible relationship between Q and K_{sp} are summarized as follows.

If $Q > K_{sp}$, solution is supersaturated. Precipitation occurs and will continue until the concentrations of ions remaining in solution satisfy K_{sp}.

If $Q = K_{sp}$, solution is saturated. The solubility equilibrium exists.

If $Q < K_{sp}$, solution is unsaturated and is capable of dissolving more of the solute until $Q = K_{sp}$.

What concluded in this section is called the solubility product principle. On base of this principle, it is applicable to control the concentration of an ion to make the reaction occur in which direction we need.

Sample 8.5 The concentration of calcium in the blood plasma is 0.0025 mol/L. If the concentration of oxalate ion is 1.0×10^{-7} mol/L, do you expect calcium oxalate ($K_{sp} = 2.3 \times 10^{-9}$) to precipitate?

Solution: The ion product for calcium oxalate is

$$Q = c_{Ca^{2+}} \cdot c_{C_2O_4^{2-}} = 0.0025 \times 1.0 \times 10^{-7} = 2.5 \times 10^{-10}$$

The value of $Q < K_{sp}$, thus we expect no precipitation occur.

Dissolving Precipitates

A precipitate will be formed on condition that the ion product Q exceeds K_{sp}, whereas a precipitate should dissolve when we decrease the ion concentrations adequately to give an ion product less than K_{sp}. If anything lowers the concentrations of ions in the solution, the equilibrium will be destroyed, solid will pass into solution to reestablish equilibrium. All solid will continue to dissolve as long as the concentrations of ions are continuously lowered. How to decrease the concentration of ions in solution may be accomplished in the following way.

(1) Addition of a reagent which forms a weak electrolyte including water, weak acid, weak base or salt which dissociated difficultly. For example, $Mg(OH)_2$ is a slightly soluble hydroxide, will dissolve by reacting with acid. The H^+ ions from the acid remove the OH^- ions by converting them into water. The decreasing of OH^- ions lead to less value of Q, then $Q < K_{sp}$ and $Mg(OH)_2$ dissolves.

$$Mg(OH)_2(s) \rightleftharpoons Mg^{2+}(aq) + 2OH^-(aq)$$
$$+$$
$$2H^+(aq)$$
$$\updownarrow$$
$$2H_2O(l)$$

The equation for the overall process is
$$Mn(OH)_2(s) + 2H^+(aq) \rightleftharpoons Mg^{2+}(aq) + 2H_2O(l)$$

Many carbonate, sulfite, and sulfide precipitates can be dissolved by adding acid because the anions react with the acid to form slightly dissociating acid.

$$CaCO_3(s) \rightleftharpoons Ca^{2+} + CO_3^{2-}$$
$$+$$
$$H^+ + Cl^- \longleftarrow HCl$$
$$\Updownarrow$$
$$HCO_3^- \xrightleftharpoons{H^+} CO_2 + H_2O$$

As CO_3^{2-} ions are removed by combination with H^+ ions, more $CaCO_3$ passes into solution to replenish the supply of CO_3^{2-} ions.

The equation for the overall process is
$$CaCO_3(s) + 2HCl(aq) \rightleftharpoons Ca^{2+}(aq) + H_2CO_3 + 2Cl^-(aq)$$
$$\longrightarrow CO_2(g) + H_2O(l)$$

Phenomena that the appearance of many historic marble and limestone monuments are damaged by acid rain are typical examples of the dissolution of carbonates by acid.

(2) We can remove an ion from solution by using oxidation-reduction reactions to changing its oxidation state. For example, the sulfide ions in very insoluble heavy metal sulfide precipitates can be oxidized to elemental sulfur. The concentration of the sulfide ions are decreased and $Q < K_{sp}$, then very insoluble heavy metal sulfide precipitates will dissolve. For example, copper(II) sulfide will dissolve when nitric acid is added.

$$CuS(s) \rightleftharpoons Cu^{2+}(aq) + S^{2-}(aq)$$
$$+$$
$$HNO_3(aq)$$
$$\Updownarrow$$
$$S(s) + NO(g)$$

The equation for the overall process is
$$3CuS(s) + 8HNO_3(aq) \rightleftharpoons 3Cu(NO_3)_2(aq) + 3S(s) + 2NO(g) + 4H_2O(l)$$

(3) Addition of a reagent which reacts with solid to form complex ions. The formation of a complex can also remove an ion and shift the equilibrium to dissolve solids. For example, a precipitate of silver chloride dissolves when enough ammonia is added.

$$AgCl(s) \rightleftharpoons Ag^+ + Cl^-$$
$$+$$
$$2NH_3$$
$$\Updownarrow$$
$$[Ag(NH_3)_2]^+$$

The equation for the overall process is
$$AgCl(s) + 2NH_3 \rightleftharpoons [Ag(NH_3)_2]^+ + Cl^-$$

8.5　Selective Precipitation

If there are several ions in a solution, when we add a precipitator which can react with all of them, what will happen? Usually we focus our attention on two problems. Which ion precipitate first (the order of precipitation)? What concentration of the ion precipitated first will be in solution when another ion precipitate (whether these ions are separated completely)? For example, assume that we have a solution containing both 0.010 mol/L CrO_4^{2-} ions and 0.010 mol/L Cl^- ions. If a source of Ag^+ is added gradually to the solution, AgCl precipitates first as a white solid. After most of the Cl^- ions have precipitated, Ag_2CrO_4 precipitation begins to form. Thus, separating CrO_4^{2-} and Cl^- is feasible.

Such a technique of separating several ions is called selective precipitation by which adding a reactant that precipitates these ions in turn. To understand why CrO_4^{2-} and Cl^- can be separated this way, let's see the Sample 8.6.

Sample 8.6 A solution containing 0.010 mol/L CrO_4^{2-} and 0.010 mol/L Cl^-. As $AgNO_3$ solution is added drop by drop (assume no volume change on addition of the $AgNO_3$). Calculate

(1) Which ion precipitates first?

(2) What concentration of the ion precipitated first remain in solution when the second ion begins to precipitate?

Solution:

(1) For AgCl, the K_{sp} expression is
$$K_{sp}=[Ag^+][Cl^-]=1.77\times10^{-10}$$
Therefore, $[Ag^+]=\dfrac{K_{sp,AgCl}}{c_{Cl^-}}=\dfrac{1.77\times10^{-10}}{0.010}=1.77\times10^{-8}\text{ mol/L}$

In the other words, any Ag^+ in excess of 1.77×10^{-8} mol/L will cause precipitation AgCl to form.

Similarly, for Ag_2CrO_4 the K_{sp} expression is
$$K_{sp}=[Ag^+]^2[CrO_4^{2-}]=1.12\times10^{-12}$$
Therefore,
$$[Ag^+]=\sqrt{\dfrac{K_{sp,Ag_2CrO_4}}{c_{CrO_4^{2-}}}}=\sqrt{\dfrac{1.12\times10^{-12}}{0.010}}=1.06\times10^{-5}\text{ mol/L}$$

The Ag^+ ion concentration of 1.06×10^{-5} mol/L is required for Ag_2CrO_4 to precipiate.

Although the solubility product of Ag_2CrO_4 is greater than the solubility product of AgCl, AgCl will precipitate first because the $[Ag^+]$ required is less.

(2) When the Ag_2CrO_4 precipitation begins to form, the concentration of Ag^+ in solution is 1.06×10^{-5} mol/L. At this moment, the solubility equilibrium of AgCl is still existed in solution.
$$AgCl(s)\rightleftharpoons Ag^+(aq)+Cl^-(aq)$$
So the concentration of Ag^+ remaining in solution satisfy the K_{sp} expression of AgCl, $[Ag^+][Cl^-]=K_{sp}(AgCl)$

Therefore, $[Cl^-]=\dfrac{K_{sp,AgCl}}{[Ag^+]}=\dfrac{1.77\times10^{-10}}{1.06\times10^{-5}}=1.67\times10^{-5}\text{ mol/L}$

We can see the concentration of Cl^- left in solution when Ag_2CrO_4 begins to precipitate is lower than 1.67×10^{-5} mol/L.

Since solubilities of metal sulfide salts span a wide range, the sulfide ion is often used as precipitating reagent to separate metal ions by selective precipitation. For example, suppose a solution mixed with Ni^{2+} and Mn^{2+}. Because of 10^6 fold difference between the solubility product for NiS and MnS, careful addition of S^{2-} to the solution will precipitate Ni^{2+} as NiS, whereas Mn^{2+} is still in solution.

$$MnS: K_{sp}=4.65\times10^{-14}$$
$$NiS: K_{sp}=1.07\times10^{-21}$$

Because the sulfide ion is the acid radical of weak acid hydrogen sulfide, its concentration is related directly with the concentration of H^+. In practice, we put H_2S to solution until it to be saturated, then adjust the pH of solution, consequently the concentration of S^{2-} can be controlled.

Hydrogen sulfide is a diprotic acid that dissociates in two steps.

$$H_2S+H_2O \rightleftharpoons H_3O^+ + HS^- \quad K_{a1}=9.1\times10^{-8}$$
$$HS^- + H_2O \rightleftharpoons H_3O^+ + S^{2-} \quad K_{a2}=1.1\times10^{-12}$$

According to Le chatelier's principle, the dissociation of this acid can be changed by adjusting the pH of the solution. If we add a strong acid, the concentration of S^{2-} will be relatively small since the dissociation equilibrium is driven toward the left. On the other hand, If we add a strong base, the concentration of S^{2-} will be relative large, since the equilibrium is shifted to the right, producing more S^{2-}.

When we put either a strong acid or a strong base into the solution of hydrogen sulfide, the equilibria are shifted very much in one direction or the other, then we can treat H_2S as if it dissociates by losing two protons in one step.

$$H_2S+2H_2O \rightleftharpoons 2H_3O^+ + S^{2-}$$

The equilibrium constant expression for this overall reaction is equal to the product of equilibrium constant expressions for these individual steps by multiple equilibrium principle.

$$\frac{[H^+]^2[S^{2-}]}{[H_2S]}=K_{a1}K_{a2}=1.0\times10^{-19}$$

A saturated solution of hydrogen sulfide has an initial concentration of 0.1 mol/L at room temperature. Because hydrogen sulfide is a weak acid, we can assume that the concentration of this acid at equilibrium is approximately equal to its initial concentration.

$$\frac{[H^+]^2[S^{2-}]}{[0.10]}\approx 1.0\times10^{-19} \quad \text{or} \quad [H^+]\approx\sqrt{\frac{1.0\times10^{-20}}{[S^{2-}]}}$$

By the expression we know that $[S^{2-}]$ and $[H^+]$ are inverse ratio. That is, $[S^{2-}]$ is sensitive to the change of $[H^+]$, it may be changed remarkably according to the little change of $[H^+]$.

Sample 8.7 A solution contains the following concentration of cations, 0.10 mol/L Pb^{2+} and 0.10 mol/L Mn^{2+}. Add gas of hydrogen sulfide until the solution of this gas is saturated. How to control the pH of solution in order to separate Pb^{2+} and Mn^{2+} completely? $K_{sp}(PbS)=9.04\times10^{-29}$, $K_{sp}(MnS)=4.65\times10^{-14}$.

Solution:

Because the same concentration of ions and similar formular, lead sulfide (PbS) with which less K_{sp} will precipitate first.

When Pb^{2+} precipitate as PbS completely, the concentration of Pb^{2+} is less than 1.0×10^{-5} mol/L, at this point, $[Pb^{2+}]$ and $[S^{2-}]$ should satisfy the expression of K_{sp} for PbS.

$$K_{sp}=[Pb^{2+}][S^{2-}]=1.0\times10^{-5}\times[S^{2-}]=9.04\times10^{-29}$$

So
$$[S^{2-}]=\frac{K_{sp}}{1.0\times10^{-5}}=\frac{9.04\times10^{-29}}{1.0\times10^{-5}}=9.04\times10^{-24}\,\text{mol/L}$$

According to the expression between $[S^{2-}]$ and $[H^+]$, then

$$[H^+]\approx\sqrt{\frac{1.0\times10^{-20}}{[S^{2-}]}}=\sqrt{\frac{1.0\times10^{-20}}{9.04\times10^{-24}}}=33.3 \text{ mol/L}$$

That is, when $[H^+]$ is less then 33.3 mol/L, the Pb^{2+} will precipitate as PbS completely.

The solubility product expression for MnS.

$$K_{sp}=[Mn^{2+}][S^{2-}]=4.65\times10^{-14}$$

Substituting the concentration of Mn^{2+} into the expression and calculate $[S^{2-}]$ when MnS just starts to precipitate.

$$[0.10][S^{2-}]=4.65\times10^{-14}$$
$$[S^{2-}]=4.65\times10^{-13} \text{ mol/L}$$

According to the expression between $[S^{2-}]$ and $[H^+]$, then

$$[H^+]\approx\sqrt{\frac{1.0\times10^{-20}}{[S^{2-}]}}=\sqrt{\frac{1.0\times10^{-20}}{4.65\times10^{-13}}}=1.47\times10^{-4} \text{ mol/L}$$

Therefore, Mn^{2+} can be kept from precipitation if we retain the H^+ ion concentration more than 1.47×10^{-4} mol/L.

Summarizing what have calculated, if the H^+ ion concentration can be controlled between $1.47\times10^{-4}\sim33.3$ mol/L, we can separate Pb^{2+} and Mn^{2+} completely.

8.6 Application of Precipitate on Medicine

Barium Sulfate's Use on Digestive tract

Barium sulfate, an insoluble compound is also used to diagnose ailments of the digestive tract. After the patient has drunk a suspension of $BaSO_4(s)$ in $Na_2SO_4(aq)$, X-rays are taken for the $BaSO_4(s)$ to reach the digestive tract. Because $BaSO_4(s)$ is opaque to X-rays, the stomach and intestines are white and can be seen clearly in the image. Although barium ion is very toxic, the very low solubility of barium sulfate makes the ingestion of the compound safe. According to the common-ion effect, sulfate ions from sodium sulfate decrease the solubility of Barium sulfate.

Pure barium sulfate is also made by precipitation.

$$BaCl_2(aq)+Na_2SO_4(aq)\rightleftharpoons BaSO_4(s)+2NaCl(aq)$$

After treated with filtration, washing, desiccation, barium sulfate precipitation is checked the limitation of quantity for impurity and content. Only when precipitation satisfies the quality standards of Chinese codex, it can be used as medicine.

Analysis on the Limitation of Quantity

For example, the concentration of Cl^- in officinal water can be determined by precipitating the Cl^- on addition of an excess of $AgNO_3(aq)$.

An excess of $AgNO_3(aq)$ shifts the equilibrium to the left.

$$AgCl(s)\rightleftharpoons Cl^-+Ag^+$$

All but a negligible amount of the Cl^- is precipitated by the common-ion effect. The $AgCl(s)$ is collected by filtration, washed to remove excess Ag^+ and NO_3^-, and dried. From the mass of $AgCl(s)$ formed, the amount of Cl^- in the sample can be calculated.

Some impurity may be intermingled inevitably during the process of medicine's production. For the sake of medicine quality, check is necessary according to the standard of medicine quality prescribed by countries. One of the checks is analysis on the limitation of quantity by the principle of precipitation.

In practice, first add precipitate reagent to the solution, observe whether precipitation produce. Then compare with the standard solution on parallel condition, consequently judge whether the impurity ion exceed the limitation. Another way is calculating the amount of impurity according to K_{sp}, then judge the amount whether satisfy the limitation prescribed.

Sample 8.8 The limitation of Cl^- in officinal water can be determined in such way.

5 drops of dilute nitric acid are putted to the 50 mL sample solution in order to avoid the interferer of CO_3^{2-} and OH^-. Then add 0.10 mol/L $AgNO_3$(aq) 1 mL, the reacting solution deposited semi-minute should be clear. At this time

$$c_{Ag^+} = 0.10 \times \frac{1}{50+1} = 2.0 \times 10^{-3} \text{ mol/L}$$

According to Solubility Product Principle, if there is not precipitation, $Q_c = K_{sp}$,

Thus $$c(Cl^-) \leqslant K_{sp}/c(Ag^+) \leqslant \frac{1.77 \times 10^{-10}}{2.0 \times 10^{-3}} \leqslant 9.0 \times 10^{-8} \text{ mol/L}$$

Tooth Decay

Tooth enamel which is composed of the mineral hydroxyapatite, is the hardest substance in the body. When people don't clean mouth, acids will be created by bacteria as they metabolize carbohydrates in food. These acids will dissolve the mineral hydroxyapatite.

$$Ca_{10}(PO_4)_6(OH)_2(s) + 8H^+(aq) = 10Ca^{2+}(aq) + 6HPO_4^{2-}(aq) + 2H_2O(l)$$

The resultant Ca^{2+} and HPO_4^{2-} ions diffuse out of the tooth enamel and are washed away by saliva. Parts of tooth surface become porous and spongy, and develop Swiss-cheese-like holes. With enough time, holes eventually turn into tooth cavities.

Toothpastes containing fluoride ion decrease the incidence of tooth decay. If the affected tooth is bathed in a solution containing appropriate amount of fluoride ion, the fluoride ion can react with hydroxyapatite to form fluoroapatite, $Ca_{10}(PO_4)_6F_2$.

$$Ca_{10}(PO_4)_6(OH)_2(s) + 2F^- = Ca_{10}(PO_4)_6F_2(s) + 2OH^-$$

This mineral, in which F^- has replaced OH^-, is much more resistant to further decay. Because the fluoride ion is a much weaker base than the hydroxide ion, the fluoroapatite, $Ca_{10}(PO_4)_6F_2$ react with acid more difficultly than hydroxyapatite, $Ca_{10}(PO_4)_6(OH)_2$.

Formation of Kidney Stone

The precipitation of certain salts in our kidneys produce kidney stones. Kidney stones, which can be painful in the extreme, consist largely of calcium oxalate, $Ca_2C_2O_4$. Kidney stone patient should eat not or less vegetables such as amaranth and spinach, which containing oxalic acid abundantly. Because oxalate ions ($C_2O_4^{2-}$) supplied from oxalic acid, react with the calcium ions in blood plasma, and produce insoluble calcium oxalate, then kidney stone can gradually build up in the kidneys. Proper adjustment of a patient's diet can help to reduce precipitation formation. Another way of prevention and cure is drinking more water. Drinking more water not only quicken emiction to decrease the urine's remaining time, but also reduce the concentration of calcium ions and oxalate ions.

Key Words

solubility equilibrium [ˌsɔljuˈbiliti ˌiːkwiˈlibriəm]　沉淀溶解平衡
homogenous [həˈmɔdʒinəs]　单相的

saturated solution [ˌsætʃəreitid səˈljuːʃən]　饱和溶液
solubility [ˌsɔljuˈbiliti]　溶解度

molar solubility ['məulə ˌsɔlju'biliti] 摩尔溶解度
soluble ['sɔljubl] 易溶的
slightly soluble ['slaitli 'məulə] 微溶的
insoluble [in'sɔljubl] 难溶的
hydrated ['haidreitid] 水合的
hydration [hai'dreiʃən] 水合作用
dissolution [disə'ljuːʃən] 溶解
precipitation [priˌsipi'teiʃən] 沉淀
solubility product constant [ˌsɔlju'biliti 'prɔdəkt 'kɔnstənt] 溶度积常数
precipitate [pri'sipiteit] 沉淀

common-ion effect ['kɔmən 'aiən i'fekt] 同离子效应
salt effect [sɔːlt i'fekt] 盐效应
dilute solution [dai'ljuːt sə'ljuːʃən] 稀溶液
solubility product principle ['sɔlju'biliti 'prɔdəkt 'prinsəpl] 溶度积规则
ion product ['aiən product] 离子积
supersaturated [sjuːpə'sætʃəreitid] 过饱和的
selective precipitation [si'lektiv priˌsipi'teiʃən] 分步沉淀
precipitator [pri'sipiteitə] 沉淀剂

Exercises

1. The addition of $AgNO_3$ to a saturated solution of AgCl would (　　)
(A) cause more AgCl to precipitate;
(B) increase the solubility of AgCl due to the interionic attraction of NO_3^- and Ag^+;
(C) lower the value of K_{sp} for AgCl;
(D) shift the equilibrium to the right $AgCl(s) \rightleftharpoons Ag^+(aq) + Cl^-(aq)$.

Answer: A

2. In order to remove 90% of the Ag^+ from a solution originally 0.10 mol/L in Ag^+, the $[CrO_4^{2-}]$ must be (　　)
(A) 1.12×10^{-12}　　(B) 1.12×10^{-11}　　(C) 1.12×10^{-10}　　(D) 1.12×10^{-8}

Answer: C

3. Give the solubility product constant expressions for the following compounds.
(1) silver chromate, Ag_2CrO_4; (2) calcium carbonate, $CaCO_3$; (3) magnesium hydroxide, $Mg(OH)_2$; (4) mercurous chloride, Hg_2Cl_2; (5) bismuth sulfide, Bi_2S_3.

Answer: skipped

4. The solubility of calcium sulfate ($CaSO_4$) is found experimentally to be 0.67 g/L. Calculate the value of K_{sp} for calcium sulfate.

Answer: 2.4×10^{-5}

5. The value of K_{sp} for magnesium hydroxide is 5.61×10^{-12}. Calculate
(1) the solubility of magnesium hydroxide in mole per liter.
(2) the pH of a saturated solution for magnesium hydroxide.

Answer: (1) 1.12×10^{-4} mol/L　(2) 10.4

6. If CaF_2 (s) is precipitated from a solution that contains Ca^{2+} by adding F^-, what $[Ca^{2+}]$ will remain in solution when $[F^-]$ reaches 1.0×10^{-4}? Assume at temperature of 298K. (CaF_2, $K_{sp} = 3.45 \times 10^{-11}$)

Answer: 3.45×10^{-3} mol/L

7. Will a precipitate form at the following situation:
(1) If 25.0 mL of 2.0×10^{-4} mol/L $(NH_4)_2C_2O_4$ is mixed with 75.0 mL of 3×10^{-5} mol/L $AgNO_3$ at 298K.
(2) Exactly 200 mL of 0.0040 mol/L $BaCl_2$ are added to exactly 600 mL of 0.0080 mol/L K_2SO_4.
(3) If 2.00 mL of 0.20 mol/L NaOH are added to 1.00 L of 0.100 mol/L $CaCl_2$.

Answer: (1) precipitate occur (2) precipitate occur (3) no precipitate occur

8. Explain the following phenomena.

(1) calcium carbonate will dissolve in solution of oxalic acid.

(2) barium sulfate is insoluble in hydrochloric acid.

(3) silver chloride dissolves in concentrated aqueous ammonia solution, precipitate again when the nitric acid is added.

(4) solubility of barium sulfate in officinal brine is greater than in pure water, however the solubility of silver chloride in officinal brine less than in pure water.

9. Which of the following slightly soluble compounds are more soluble in dilute nitric acid than they are in pure water?

(1) $MgCO_3$ (2) ZnS   (3) AgCl   (4) $Ca(OH)_2$

Answer: (1), (2), (4) are more soluble in dilute nitric acid than they are in pure water.

10. The pH of a saturated solution of a metal hydroxide MOH is 9.68. Calculate the K_{sp} for the compound.

Answer: 2.29×10^{-9}

11. A solution contains 0.10 mol/L Ag^+ and 0.10 mol/L Pb^{2+}. Can Ag^+ and Pb^{2+} be separated completely by selective precipitation with Cl^-? If "completely" means that 99.9% of the Ag^+ can be precipitated without precipitating Pb^{2+}.

Answer: can be achievable

12. Calculate the pH of solution when 0.10 mol/L Cd^{2+} begins to precipitate and precipitate completely. [$Cd(OH)_2$: $K_{sp} = 7.2 \times 10^{-15}$]

Answer: 7.43~9.43

13. The recovery of magnesium from seawater depends on the position of the following equilibrium:

$$Ca(OH)_2(s) + Mg^{2+} \rightleftharpoons Mg(OH)_2(s) + Ca^{2+}$$

What is the value of K for this equilibrium? At 298 K, the value of K_{sp} for $Mg(OH)_2$ is 5.61×10^{-12}, $Ca(OH)_2$ is 5.02×10^{-6}.

Answer: 8.95×10^5

14. Calculate the solubility of silver chloride (in g/L) in a 6.5×10^{-3} mol/L silver nitrate solution.

Answer: 3.8×10^{-6} g/L

15. A solution containing a mixture of 10^{-3} mol/L Fe^{2+} and 10^{-3} mol/L Mn^{2+}. Add gas of hydrogen sulfide until the solution of this gas is saturated. How to control the pH of solution in order to separate Fe^{2+} and Mn^{2+} completely? $K_{sp}(FeS) = 1.3 \times 10^{-18}$, $K_{sp}(MnS) = 4.65 \times 10^{-14}$

Answer: 3.56~4.83

Chapter 9 Acid-Base Equilibria

Acids and bases are great important in industry daily life. Many chemical processes involve acids or bases as reactants or catalysts, many reactions that occur in solution are closely tied to acid-base levels, the usage of many medicine, and even the ability of an aquatic environment to support fish and plant life, are all critically dependent upon the acidity or basicity of solutions. But what makes a substance behave as an acid or as a base? In this chapter, we will examine acids and bases, taking a closer look at their definitions, how they are identified and characterized, and what is common ion effect, how buffer solution resist the change of pH. In doing so, we will consider their behavior not only in terms of their structure but also in terms of the chemical equilibria in which they participate.

9.1 Acids and Bases

From the earliest days of experimental chemistry, acids were recognized first because of their sour taste (for example, citric acid in lemon juice) and cause certain dyes to change color (for example, litmus turns red on contact with acids). Indeed, the word "acid" comes from the Latin word acidus, meaning sour. Bases were noticed at a later date as bitter taste compounds (soap is a good example). The word "base" comes from an old English meaning of the word, which is "to bring low". When bases are added to acids, they lower the amount of acid.

Arrhenius Acids and Bases

Acids and bases were first defined in 1887 by a Swedish chemist Svante Arrhenius (1859~1927). He defined acids as substances that produce H^+ in water, and bases as substances that dissolve in water to produce OH^-.

HCl is an example of an Arrhenius acid. HCl gas is highly soluble in water because of its chemical reaction with water, which produces H^+ and Cl^- ions.

$$HCl(g) \xrightarrow{H_2O} H^+(aq) + Cl^-(aq)$$

The aqueous solution of HCl comes to be known as hydrochloric acid. Concentrated hycrochloric acid is about 37% HCl by mass and is 12 mol/L.

Sodium hydroxide (NaOH) is an example of an Arrhenius base. Because NaOH is an ionic compound, it dissociates into Na^+ and OH^- ions when dissolving in water, thereby releasing OH^- ions into the solution.

Brønsted-Lowry Acids and Bases

The Arrhenius theory of acids and bases, while useful, has limitations (for example, it is restricted to aqueous solutions). In 1923, the Danish chemist Johannes Brønsted (1879~1947) and the English chemist Thomas Lowry (1874~1936) independently proposed a more general theory of acids and bases. According to Brønsted-Lowry theory, an acid is a substance that can donate a proton to another substance, and a base is a substance that can accept a proton (remember that a proton is simply a H^+ ion). For example, HCl, HAc, NH_4^+ and H_2SO_3 can donate one or two protons, so they are acids. But OH^-, Ac^-, NH_3 and SO_3^{2-} are bases because of their abilities to accept a proton. Some substances can act as an acid in one reaction and as a base in another, we called them amphoteric substances (for example H_2O, HSO_3^-). You can get more information from the following Table 9.1.

Table 9.1 Examples for Brønsted-Lowry Acids and Bases

Acids	⇌	H^+ (proton)	Bases	Acids	⇌	H^+ (proton)	Bases
HCl	⇌	H^+	Cl^-	NH_4^+	⇌	H^+	NH_3
HAc	⇌	H^+	Ac^-	H_3O^+	⇌	H^+	H_2O
H_2SO_3	⇌	H^+	HSO_3^-	H_2O	⇌	H^+	OH^-
HSO_3^-	⇌	H^+	SO_3^{2-}				

From the table we know that acids can be either molecules, cations or anions, and the same to the bases. The Brønsted-Lowry theory doesn't require that a base produce OH^-, a base simply has to be able to accept a proton from a donor. Therefore, the Brønsted-Lowry theory applies to a greater variety of situations than does the Arrhenius theory.

The Brønsted-Lowry theory thinks of acid-base reactions as the transfer of H^+ ions from one substance to another, and it does not restrict acids and bases to aqueous solutions. For example, in the reaction between HCl and NH_3, a proton is transferred from the acid HCl to the base NH_3.

$$HCl(aq) + NH_3(aq) \rightleftharpoons Cl^-(aq) + NH_4^+(aq)$$

This reaction can occur in the gas phase. The hazy film that forms on the windows of general chemistry laboratories is largely solid NH_4Cl formed by the gas-phase reaction of HCl and NH_3.

Another example, when HCl(g) dissolves in water, the transfer of H^+ occurs.

$$HCl(g) + H_2O(l) \rightleftharpoons H_3O^+(aq) + Cl^-(aq)$$
$$\text{acid}_1 \quad \text{base}_2 \quad\quad \text{acid}_2 \quad\quad \text{base}_1$$

The acid, HCl has actually transferred a H^+ ion (a proton) to the base, H_2O, and the products are also an acid (H_3O^+) and a base (Cl^-). That is after H_2O accepts a proton, H_3O^+ then becomes an acid, because it can donate proton, and Cl^- can accept a proton. Thus, acids and bases act in pairs. we can write a general equation to indicate this reciprocal relationship.

$$\text{acid}_1 + \text{base}_2 \rightleftharpoons \text{acid}_2 + \text{base}_1 \tag{9.1}$$

The acid-base pairs (acid_1-base_1, acid_2-base_2) are said to be conjugate of each other. Thus, Cl^- is the conjugate base of the acid HCl, and H_3O^+ is the conjugate acid of H_2O. An acid and a base such as HCl and Cl^- that differ only in the presence or absence of a proton are called a conjugate acid-base pair. Every acid can form its conjugate base by losing a proton. Similarly, every base can form its conjugate acid by gaining a proton. Thus, H_3O^+ is the conjugate acid of H_2O, and HCl is the conjugate acid of Cl^-.

Lewis Acids and Bases

Another definition of acids and bases was formulated by an American chemist G. N. Lewis. According to Lewis's definition, an acid is a substance that can accept a pair of electrons, and a base is a substance that can donate a pair of electrons. For example, in the reaction of ammonia and H^+, NH_3 acts as a Lewis base because it donates a pair of electrons to the proton H^+, which acts as a Lewis acid by accepting the pair of electrons. Therefore, a Lewis acid-base reaction is one that involves the donation of a pair of electrons from one substance to another. Such a reaction does not produce a salt and water.

Several categories of substances can be considered Lewis acids, positive ions, having less than a full octet in the valence shell, polar double bonds (one end), expandable valence shells. And several categories of substances can be considered Lewis bases, negative ions, one of more unshared pairs in the valence shell, polar double bonds (the other end).

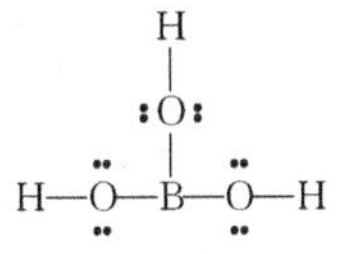

Figure 9.1 The structure of boric acid

The significance of the Lewis concept is that it is much more general than other definitions. For example, boric acid (H_3BO_3, a weak acid used in eyewash) is an oxoacid with the structure shown in figure 9.1.

Note that boric acid does not ionize in water to produce an H^+ ion. Instead, its reaction with water is

$$B(OH)_3(aq) + H_2O(l) \rightleftharpoons B(OH)_4^-(aq) + H^+(aq)$$

In this Lewis acid-base reaction, boric acid accepts a pair of electrons from the hydroxide ion that is derived from the H_2O molecule.

9.2 Ion Product of Water

Water, as we know, is a unique solvent. It is a poor conductor of electricity because it is a very weak electrolyte, but it can transfer a proton to another water molecule.

$$H_2O(l) + H_2O(l) \rightleftharpoons H_3O^+(aq) + OH^-(aq)$$

This reaction is sometimes called the autoprotolysis equilibrium of water. We can write expression of the equilibrium constant.

$$K = \frac{[H_3O^+][OH^-]}{[H_2O][H_2O]} \quad \text{or} \quad K = \frac{[H^+][OH^-]}{[H_2O][H_2O]}$$

Because a very small fraction of water molecules are reacted, the concentration of water, $[H_2O]$, remains virtually unchanged. Therefore

$$K[H_2O]^2 = K_w = [H^+][OH^-] \tag{9.2}$$

The equilibrium constant K_w is called the ion-product constant of water, which is the product of the molar concentrations of H^+ and OH^- ions at a particular temperature. Table 9.2 shows some values of K_w at different temperatures. These values were determined by careful measurements of the conductivity of pure water. As we can see, the value of K_w increases as temperature increases. At the temperature of 298K the value of K_w is 1.0×10^{-14}.

$$K_w = [H^+][OH^-] = 1.0 \times 10^{-14}$$

Then

$$[H^+] = [OH^-] = \sqrt{K_w} = \sqrt{1.0 \times 10^{-14}} = 1 \times 10^{-7} \text{mol/L}$$

This value should be memorized because we will be using it often.

Table 9.2 Ion product of water at different temperatures

T/K	273	283	298	323	373
K_w	1.139×10^{-15}	1.290×10^{-15}	1.008×10^{-14}	5.474×10^{-14}	5.5×10^{-13}

If $[H^+] = [OH^-]$, the aqueous solution is said to be neutral. In an acidic solution, $[H^+] > [OH^-]$. In a basic solution, $[H^+] < [OH^-]$. In practice we can change the concentration of either H^+ or OH^- ions in solution, but we cannot vary them independently. If we adjust the solution so that $[OH^-] = 1.0 \times 10^{-4}$ mol/L, the concentration of H^+ must change to

$$[H^+] = \frac{K_w}{[OH^-]} = \frac{10 \times 10^{-14}}{1.0 \times 10^{-4}} = 1 \times 10^{-10} \text{ mol/L}$$

Because the concentrations of H^+ and OH^- ions in aqueous solutions are frequently very small numbers, a Danish biochemist Sorensen proposed a more practical measure called pH. The pH of a solution is defined as the negative logarithm of the $[H^+]$ (in moles per liter).

$$\text{pH} = -\lg[H^+] \tag{9.3}$$

Keep in mind that equation 9.3 is simply a definition designed to give us convenient numbers to work with. Note that the term $[H^+]$ in equation 9.3 belongs only to the value for H^+ concentration, for we cannot take the logarithm of units. Thus, the pH of a solution is a dimensionless quantity. Acidic and basic solutions at 298K can be identified by their pH values.

Acidic solutions $[H^+] > 1.0 \times 10^{-7}$ mol/L, pH < 7.00
Neutral solutions $[H^+] = 1.0 \times 10^{-7}$ mol/L, pH = 7.00
Basic solutions $[H^+] < 1.0 \times 10^{-7}$ mol/L, pH > 7.00

Notice that pH increases as $[H^+]$ decreases. Table 9.3 lists the pH for a number common fluids.

Table 9.3   The pH of some common fluids

Sample	pH Value	Sample	pH Value
Gastric fluid	1.5	Saliva	6.4~6.9
Lemon juice	2.3	Milk	6.5
Vinegar	3.0	Pure water	7.0
Orange juice	3.5	Blood	7.35~7.45
Coffee	5.2	Tears	7.4
Urine	4.8~7.5	Ammonia	11.5

Similar to pH scale, a pOH scale can be defined the negative logarithm of the OH^- concentration.

$$\text{pOH} = -\lg[OH^-] \tag{9.4}$$

Now consider again the ion product constant for water.

$$[H^+][OH^-] = K_w = 1.0 \times 10^{-14}$$

Taking the negative logarithm of both sides, we obtain

$$-\lg[H^+] - \lg[OH^-] = -\lg K_w = -\lg(1.0 \times 10^{-14})$$

From the definitions of pH and pOH we obtain

$$\text{pH} + \text{pOH} = K_w = 14 \tag{9.5}$$

Equation 9.5 provides us with another way to express the relationship between the concentrations of H^+ and OH^- ions in aqueous solution.

9.3 Strength of Weak Acids and Bases

As we know, numerous acids are weak acids. Consider a weak monoprotic acid HB. Its dissociation in water is represented by

$$HB(aq) + H_2O(l) \rightleftharpoons H_3O^+(aq) + B^-(aq)$$

Or

$$HB(aq) \rightleftharpoons H^+(aq) + B^-(aq)$$

The equilibrium constant for this acid dissociation is called the acidity constant or acid dissociation constant, K_a, is given by

$$K_a = \frac{[H^+][B^-]}{[HB]} \quad (9.6)$$

At a given temperature, the strength of the acid HB is measured quantitatively by the value of K_a. The larger K_a, the stronger the acid, the greater the concentration of H^+ ions at equilibrium because of its dissociation.

Weak bases are treated like weak acids. When a weak monoprotic base B^- dissolves in water, it undergoes the reaction

$$B^-(aq) + H_2O(aq) \rightleftharpoons HB(aq) + OH^-(aq)$$

The production of OH^- ions in this base dissociation reaction means that, in this solution at 25 °C, pH>7.

Compared with the total concentration of water, very few water molecules are consumed by the reaction, so we can treat $[H_2O]$ as a constant. Thus we can write expression of the equilibrium constant:

$$K_b = \frac{[HB][OH^-]}{[B^-]} \quad (9.7)$$

The equilibrium constant for base dissociation, K_b, is called the basicity constant or base dissociation constant. At a given temperature, the strength of the base B^- is measured quantitatively by the magnitude of K_b. The larger K_b, the stronger the base that is, the greater the concentration of OH^- ions at equilibrium because of its dissociation.

Table 9.4 lists a number of common weak acids and bases and their dissociation constants. We can see if an acid is strong, its conjugate base is weak.

Table 9.4 Conjugate acid-base pairs in water and K_a, K_b

Conjugate Acid	K_a	Conjugate Base	K_b	Conjugate Acid	K_a	Conjugate Base	K_b
H_2SO_3	1.54×10^{-2}	HSO_3^-	6.49×10^{-13}	$H_2PO_4^-$	6.23×10^{-8}	HPO_4^{2-}	1.61×10^{-7}
HSO_4^-	1.20×10^{-2}	SO_4^{2-}	8.33×10^{-13}	NH_4^+	5.68×10^{-10}	NH_3	1.76×10^{-5}
H_3PO_4	7.52×10^{-3}	$H_2PO_4^-$	1.33×10^{-12}	HCO_3^-	5.61×10^{-11}	CO_3^{2-}	1.78×10^{-4}
HAc	1.76×10^{-5}	Ac^-	5.68×10^{-10}	H_2O_2	2.4×10^{-12}	HO_2^-	4.17×10^{-3}
HSO_3^-	1.02×10^{-7}	SO_3^{2-}	9.80×10^{-8}	HS^-	1.2×10^{-13}	S^{2-}	9.09×10^{-3}
H_2CO_3	4.30×10^{-7}	HCO_3^-	2.33×10^{-8}	HPO_4^{2-}	2.2×10^{-13}	PO_4^{3-}	4.55×10^{-2}
H_2S	8.9×10^{-8}	HS^-	1.1×10^{-7}	H_2O	1.0×10^{-14}	OH^-	1.0

An important relationship between the acidity constant and the basicity constant of its conjugate base can be derived as follows. Using acetic acid (HAc) and Ac^- Conjugate acid-base as an example, each of these species reacts with water.

(1) $HAc(aq) \rightleftharpoons H^+(aq) + Ac^-(aq)$ $\quad K_a = \dfrac{[H^+][Ac^-]}{[HAc]}$

(2) $Ac^-(aq) + H_2O(aq) \rightleftharpoons HAc(aq) + OH^-(aq)$ $\quad K_b = \dfrac{[HAc][OH^-]}{[Ac^-]}$

(3) $H_2O(aq) \rightleftharpoons H^+(aq) + OH^-(aq)$ $\quad K_w = [H^+][OH^-]$

The product of K_a and K_b is given by

$$K_a K_b = \frac{[H^+][Ac^-]}{[HAc]} \times \frac{[HAc][OH^-]}{[Ac^-]}$$
$$= [H^+][OH^-]$$
$$= K_w$$

Now notice something very interesting and important. The sum of reactions (1) and (2) here is reactions (3), that is the autoionization of water. In order to this example we make use of a rule for chemical equilibrium, the rule of multiple equilibrium. When two reactions are added to give a third reaction, the equilibrium constant for the third reaction is equal to the product of the equilibrium constants for the two added reactions. Thus for any conjugate acid-base pair it is always true that

$$K_a K_b = K_w \tag{9.8}$$

Equation (9.8) means the product of K_a for an acid and K_b for its conjugate base is the ion product constant for water. The stronger the acid (the larger K_a), the weaker its conjugate base (the smaller K_b), and vice versa (see Table 9.4).

Taking the negative logarithm of both sides of Equation 9.8, we obtain

$$-\lg K_a - \lg K_b = -\lg K_w$$

Referring to the definitions of pH and pOH we obtain

$$pK_a + pK_b = 14 \tag{9.9}$$

We can use Equation (9.8) to calculate the K_b of the conjugate base of HAc as follows.

$$K_b = \frac{K_w}{K_a} = \frac{1.0 \times 10^{-14}}{1.76 \times 10^{-5}} = 5.68 \times 10^{-10}$$

9.4 The pH of Weak Acids and Bases

We can use K_a or K_b and the initial concentration of an acid or base to calculate equilibrium concentrations of all the species and pH of the solution. In calculating the equilibrium concentrations of a weak acid (or a weak base), we follow essentially a procedure of three basic steps.

① Express the equilibrium concentrations of all substances in terms of the initial concentrations and a single unknown x, which represents the change in concentration.

② Write the expression of acidity constant (or basicity constant) in terms of the equilibrium concentrations. We can solve for x if knowing the value of K_a or K_b.

③ Calculate the equilibrium concentrations of all substances and the pH of the solution.

Consider a weak acid HB, its acidity constant and initial concentration are K_a and c respectively, calculate the concentration of H^+.

Solution:
Step 1 Because HB is monoprotic, one HB molecule ionizes to give one H^+ ion and one B^- ion. Let x mol/L be the equilibrium concentration of H^+, then the equilibrium concentration of B^- is x too. We can now summarize the changes in concentrations as follows.

$$HB(aq) \rightleftharpoons H^+(aq) + B^-(aq)$$

Initial concentrations:	c	0.0	0.0	(mol/L)
Change:	x	x	x	(mol/L)
Equilibrium concentrations:	$c-x$	x	x	(mol/L)

Step 2 According to the principle of chemical equilibrium.

$$K_a = \frac{[H^+][B^-]}{[HB]} = \frac{x^2}{c-x}$$

This equation can be rewritten as

$$x^2 + xK_a - cK_a = 0$$

$$[H^+]=x=-\frac{K_a}{2}+\sqrt{\frac{K_a^2}{4}+cK_a} \qquad (9.10)$$

Equation 9.10 is just the formula for calculating $[H^+]$ that a weak monoprotic acid dissociates. We can often apply a simplifying approximation to this type of problem. Because HB is a weak acid, the extent of its dissociation must be small. Therefore x is small compared with c. As a general rule, if $c/K_a \geqslant 500$, $x/c = 5\%$. So we can assume $c - x \approx c$, then

$$K_a = \frac{[H^+][B^-]}{[HB]} = \frac{x^2}{c-x} = \frac{x^2}{c}$$

$$x^2 = K_a c$$

Taking the square root of both sides, we obtain

$$x = \sqrt{K_a c}$$

Step 3 At equilibrium, therefore

$$[H^+] = \sqrt{K_a c} \qquad (9.11)$$

Equation 9.11 is the approximate formula for calculating $[H^+]$ that a weak monoprotic acid dissociates.

Similarly, to a weak monoprotic base, we can get

$$[OH^-] = -\frac{K_b}{2} + \sqrt{\frac{K_b^2}{4} + cK_b} \qquad (9.12)$$

When $c/K_b \geqslant 500$, $c - x \approx c$.

$$[OH^-] = \sqrt{K_b c} \qquad (9.13)$$

Sample 9.1 Calculate the pH, [HAc] and [Ac$^-$] in a 0.10 mol/L HAc solution at equilibrium. $K_a = 1.76 \times 10^{-5}$

Solution:

Step 1 let x be the equilibrium concentration of H^+.

$$\text{HAc(aq)} \rightleftharpoons H^+(aq) + Ac^-(aq)$$

Initial concentrations	0.10	0.0	0.0	(mol/L)
Change	x	x	x	(mol/L)
Equilibrium concentrations	0.10 $-x$	x	x	(mol/L)

Step 2 According to the principle of chemical equilibrium

$$K_a = \frac{[H^+][Ac^-]}{[HAc]} = \frac{x^2}{0.10 - x}$$

Because $c/K_a = 0.10/(1.76 \times 10^{-5}) = 5.68 \times 10^3 > 500$, then $0.10 - x \approx 0.10$

$$x = \sqrt{K_a \cdot c} = \sqrt{1.76 \times 10^{-5} \times 0.10} = 1.33 \times 10^{-3} \, \text{mol/L}$$

Step 3 At equilibrium

$$[H^+] = [Ac^-] = 1.33 \times 10^{-3} \, \text{mol/L}$$
$$[HAc] = 0.10 - 1.33 \times 10^{-3} = 9.99 \times 10^{-2} \, \text{mol/L}$$

and

$$pH = -\lg[H^+] = -\lg(1.33 \times 10^{-3}) = 2.88$$

Sample 9.2 What is the pH of a 0.30 mol/L ammonia solution? $K_b = 1.80 \times 10^{-5}$

Solution:

Step 1 Let x be the equilibrium concentration of OH^-.

$$NH_3(aq) + H_2O(aq) \rightleftharpoons NH_4^+(aq) + OH^-(aq)$$

Initial concentrations	0.30	0.0	0.0	(mol/L)

Change	x	x	x	(mol/L)
Equilibrium concentrations	$0.30-x$	x	x	(mol/L)

Step 2 According to the principle of chemical equilibrium.

$$K_b = \frac{[NH_4^+][OH^-]}{[NH_3]} = \frac{x^2}{0.30-x}$$

Because $c/K_b = 0.30/(1.80\times10^{-5}) = 1.67\times10^4 > 500$, then $0.30-x \approx 0.30$

$$x = \sqrt{K_b \cdot c} = \sqrt{1.80\times10^{-5}\times0.30} = 4.90\times10^{-3}\,\text{mol/L}$$

Step 3 At equilibrium

$$[OH^-] = 4.90\times10^{-3}\,\text{mol/L}$$
$$pH = 14 - pOH = 14 - \lg(4.90\times10^{-3}) = 10.31$$

Sample 9.3 Calculate the pH of a 0.15 mol/L solution of sodium acetate (NaAc). $K_b = 5.68\times10^{-10}$.

Solution:

Note that the salt first dissociates into cation and anion in solution. We then examine separately the possible reaction of each ion with water.

Step 1 The major species in a sodium acetate solution are Na^+ and Ac^- ions. Because we started with a 0.15 mol/L NaAc, the initial concentrations of the ions are also equal to 0.15 mol/L.

$$NaAc(aq) \rightleftharpoons Na^+(aq) + Ac^-(aq)$$
$$\text{0.15 mol/L} \quad \text{0.15 mol/L}$$

Of these ions, only the Ac^- ion is a weak base. We ignore the contribution to H^+ and OH^- ions by water.

Step 2 Let x be the equilibrium concentration of OH^- ions in mol/L, we summarize.

$$Ac^-(aq) + H_2O(aq) \rightleftharpoons HAc(aq) + OH^-(aq)$$

Initial concentrations	0.15	0.0	0.0	(mol/L)
Change	x	x	x	(mol/L)
Equilibrium concentrations	$0.15-x$	x	x	(mol/L)

$$K_b = \frac{[HAc][OH^-]}{[Ac^-]} = \frac{x^2}{0.15-x} = 5.68\times10^{-10}$$

Because K_b is very small and the initial concentration of the base is large, and $c/K_b = 0.15/(5.68\times10^{-10}) = 2.64\times10^8 > 500$, we can apply the approximation $0.15-x \approx 0.15$.

$$\frac{x^2}{0.15-x} \approx \frac{x^2}{0.15} = 5.68\times10^{-10}$$

or $\quad x = \sqrt{K_b c} = \sqrt{5.68\times10^{-10}\times0.15} = 9.23\times10^{-6}\,\text{mol/L}$

Step 3 At equilibrium

$$[OH^-] = 9.23\times10^{-6}\,\text{mol/L}$$
$$pOH = -\lg[OH^-] = -\lg(9.23\times10^{-6}) = 5.04$$
$$pH = 14 - pOH = 14 - 5.04 = 8.96$$

9.5 The Common-Ion Effect

We have examined the equilibrium concentrations of ions in solutions containing a weak acid or a weak base. We now consider such solutions that contain not only a weak acid, such as acetic acid (HAc), but also a soluble salt of that acid, such as NaAc. What happens when NaAc is added to a solution of HAc? Because Ac^- is a weak base, it isn't surprising that $[H^+]$ decreases, that is, the pH of the solution increases according to Le Chatelier's principle.

HAc is a weak acid. In contrast, NaAc dissociates completely in aqueous solution to form Na^+ and Ac^- ions because NaAc is a strong electrolyte.

$$HAc(aq) \rightleftharpoons H^+(aq) + Ac^-(aq)$$
$$NaAc(aq) \rightleftharpoons Na^+(aq) + Ac^-(aq)$$

The addition of Ac^-, causes this equilibrium to shift to the left, thereby decreasing the equilibrium concentration of $H^+(aq)$.

The dissociation of the weak acid HAc decreases when add the strong electrolyte NaAc, which has an ion in common with it. We call the observation common-ion effect. The dissociation of a weak electrolyte is decreased by adding to the solution, a strong electrolyte that has an ion in common with the weak electrolyte. For example, the addition of NH_4^+ (as from the strong electrolyte NH_4Cl) causes the base-dissociation equilibrium of NH_3 to shift the left, decreasing the equilibrium concentration of OH^- and increasing the pH.

$$NH_3(aq) + H_2O(aq) \rightleftharpoons NH_4^+(aq) + OH^-(aq)$$
$$NH_4Cl(s) \rightleftharpoons NH_4^+(aq) + Cl^-(aq)$$

The following samples illustrate how equilibrium concentrations may be calculated when a solution contains a mixture of a weak electrolyte and a strong electrolyte that have a common ion. You will see that the procedure is similar to those encountered for weak acids and weak bases.

Sample 9.4 Calculate the pH of a solution made by adding 0.10 mol of acetic acid, HAc, and 0.10 mol of sodium acetate, NaAc, to enough water to make 1.0 L of solution. $K_a = 1.76 \times 10^{-5}$.

Solution:

Step 1 Identify the major substances in solution and consider their acidity or basicity. Because HAc is a weak electrolyte and NaAc is a strong electrolyte, the major substances in the solution are HAc (a weak acid), Na^+ (neither acidic nor basic), Ac^- (the conjugate base of HAc) and H_2O (solvent).

Then, identify the important equilibrium reaction. The pH of the solution will be controlled by the dissociation equilibrium of HAc because H_2O is a much weaker acid than HAc.

$$HAc(aq) \rightleftharpoons H^+(aq) + Ac^-(aq)$$

Note that because NaAc was added to the solution, the values of $[H^+]$ and $[Ac^-]$ are not the same. The Na^+ ion is merely a spectator ion and will have no influence on the pH of the solution.

Step 2 Calculate the initial and equilibrium concentrations of each of the substances that participate in the equilibrium.

Let x be the equilibrium concentration of H^+, we summarize

$$HAc(aq) \rightleftharpoons H^+(aq) + Ac^-(aq)$$

Initial concentrations	0.10	0.0	0.10	(mol/L)
Change	x	x	x	(mol/L)
Equilibrium concentrations	$0.10-x$	x	$0.10+x$	(mol/L)

Notice that the equilibrium concentration of Ac^- (the common ion) is the initial concentration due to the NaAc (0.10 mol/L) plus the change in concentration (x) due to the dissociation of HAc.

The expression of the acidity constant is

$$K_a = \frac{[H^+][Ac^-]}{[HAc]} = 1.76 \times 10^{-5}$$

The addition of NaAc does not change the value of K_a. Substituting the equilibrium concentrations into K_a gives

$$K_a = \frac{x(0.10+x)}{0.10-x} = 1.76 \times 10^{-5}$$

Because K_a is small, we assume that x is small compared to the original concentrations of HAc and Ac⁻ (0.10 mol/L each). Therefore, $0.10-x \approx 0.10$, $0.10+x \approx 0.10$. We can simplify our equation.

$$\frac{x(0.10+x)}{0.10-x} \approx \frac{x(0.10)}{0.10} = 1.76 \times 10^{-5}$$

$$x = 1.76 \times 10^{-5} \text{ mol/L}$$

The value of x is indeed small relative to 0.10, justifying the approximation made in simplifying the problem.

Step 3 At equilibrium

$$[H^+] = 1.76 \times 10^{-5} \text{ mol/L}$$

And

$$pH = -\lg[H^+] = -\lg(1.76 \times 10^{-5}) = 4.74$$

In Sample 9.1, we calculated that a 0.10 mol/L solution of HAc has a pH of 2.88, corresponding to $[H^+] = 1.33 \times 10^{-3}$ mol/L. Thus the addition of NaAc has substantially decreased $[H^+]$, as expected from Le Chatelier's principle.

9.6 Buffer Solutions

The most interesting thing about solutions that contain a weak conjugate acid-base pair is that they resist drastic changes in pH. Neither addition of small amounts of acids or bases nor dilution results in much change in pH. Solutions that resist a change in pH upon addition of small amounts of acids or bases are called buffer solutions. It occurs frequently in nature. Human blood is an important example of a complex aqueous medium with a pH buffered at about 7.4, variation in the pH of blood is greater than 0.6 pH unit for more than a few seconds results in death. Seawater is buffered at about 8.1 to 8.3 near the surface. Culture media used in bacteriology are usually buffed at the pH needed for the growth of the bacteria being studied. Buffer solutions are very important in the laboratory and in medicine.

Composition of Buffer Solutions and Buffer Mechanism

Buffer solutions generally are composed of a weak conjugate acid-base pair such as NH_4^+-NH_3, $H_2PO_4^-$-HPO_4^{2-}, HAc-Ac⁻ etc. The acidic species neutralize OH⁻ ions and the basic ones neutralize H^+ ions. The two species of the buffer don't consume each other through a neutralization reaction. The conjugate acid-base pair is also called buffer pair. By choosing appropriate components and adjusting their relative concentrations, we can buffer a solution at virtually any pH.

How does a buffer solution keep the pH from changing? Let's consider a buffer solution composed of HAc and Ac⁻. The acid dissociation equilibrium is

$$HAc(aq) \rightleftharpoons H^+(aq) + Ac^-(aq)$$

The corresponding acidity constant expression is

$$K_a = \frac{[H^+][Ac^-]}{[HAc]}$$

Solving this expression for $[H^+]$, we have

$$[H^+] = K_a \frac{[HAc]}{[Ac^-]} \tag{9.14}$$

From this expression $[H^+]$, and thus the pH, is determined by two factors, the value of K_a of HAc and the ratio of the concentrations of the conjugate acid-base pair, $[HAc]/[Ac^-]$.

(1) If an acid is added to the buffer solution, the base component of the buffer, Ac^- will react with the added H^+ ions.

$$H^+(aq) + Ac^-(aq) \rightleftharpoons HAc(aq)$$

We see that the reaction causes $[Ac^-]$ to decrease and $[HAc]$ to increase. However, as long as the quantities of HAc and Ac^- in the buffer are large compared to the amount of H^+ added, the ratio $[HAc]/[Ac^-]$ doesn't change much, so the $[H^+]$ does not increase very much.

(2) If a base is added to the buffer solution, the acid component of the buffer, HAc will react with the added OH^- ions.

$$OH^- + HAc(aq) \rightleftharpoons H_2O(aq) + Ac^-(aq)$$

This reaction causes $[HAc]$ to decrease and $[Ac^-]$ to increase. As long as the change in the ratio $[HAc]/[Ac^-]$ is small, the $[H^+]$ does not decrease very much.

The pH of Buffer Solutions

Buffer capacity and pH are the most important characteristics of a buffer. Buffer capacity is the quantity of acid or base that can be added before the pH changes significantly. Buffer capacity depends on the amount of the conjugate acid-base pair. The pH of the buffer depends on the K_a for the acid, on the relative concentrations of the acid and base that comprise the buffer. For example, we can see from Equation 9.14 that $[H^+]$ for a 1 L solution that is 1.0 mol/L in HAc and 1.0 mol/L in NaAc will be the same as for a 1 L solution that is 0.10 mol/L in HAc and 0.10 mol/L in NaAc. However, the first solution has a greater buffer capacity because it contains more HAc and Ac^-. The greater the quantities of the conjugate acid-base pair, the more resistant the ratio of their concentrations, and hence the pH, is to change.

How to calculate the pH of a buffer solution? Let's continue to use the equation 9.14. Taking the negative logarithm of both sides of equation 9.14, we have

$$-\lg[H^+] = -\lg\left(K_a \frac{[HAc]}{[Ac^-]}\right) = -\lg K_a - \lg \frac{[HAc]}{[Ac^-]}$$

Because $-\lg[H^+] = pH$ and $-\lg K_a = pK_a$, we have

$$pH = pK_a - \lg \frac{[HAc]}{[Ac^-]} = pK_a + \lg \frac{[Ac^-]}{[HAc]}$$

In general,

$$pH = pK_a + \lg \frac{[base]}{[acid]} \tag{9.15}$$

Note that In equation 9.15, [acid] and [base] refer to the equilibrium concentrations of the conjugate acid-base pair. And when [acid]=[base], $pH = pK_a$.

Equation 9.15 is called the Henderson-Hasselbalch equation. We can use this equation to calculate the pH of buffer solutions. In doing equilibrium calculations, we have seen that we can normally neglect the amounts of the acid and base that dissociate. Therefore, we can usually use the initial concentrations of the acid and base components of the buffer solution directly, that is $[acid] \approx c_{acid}$, $[base] \approx c_{base}$.

Equation 9.15 can be expressed by the following equation.

$$pH = pK_a + \lg \frac{c_{base}}{c_{acid}} \tag{9.16}$$

Sample 9.5

① Calculate the pH of a buffer system containing 0.10 mol/L HAc and 0.10 mol/L NaAc.

② What is the pH of the buffer system after the addition of 0.01 mol of HCl to 1 L of the solution? Assume that the volume of the solution does not change when the HCl is added.

③ What is the pH of the buffer system after the addition of 0.01 mol of NaOH to 1 L of the solution? $K_a = 1.76 \times 10^{-5}$.

Solution:

① $c_{HAc} = c_{Ac^-} = 0.10$ mol/L, $K_a = 1.76 \times 10^{-5}$

Write the Henderson-Hasselbalch equation

$$pH = pK_a + \lg \frac{c_{Ac^-}}{c_{HAc}}$$

$$pH = 4.75 + \lg \frac{0.10}{0.10} = 4.75$$

② After the addition of HCl, complete dissociation of HCl acid occurs

$$HCl(aq) = H^+(aq) + Cl^-(aq)$$

0.01mol   0.01mol   0.01mol

Originally, 0.1 mol HAc and 0.10 mol Ac⁻ were present in 1 L of the solution. After neutralization of the

HCl acid by Ac⁻, which we write as

$$Ac^-(aq) + H^+(aq) \rightleftharpoons HAc(aq)$$

0.01mol   0.01mol   0.01mol

The concentrations of HAc and Ac⁻ present are

$$c'_{HAc} \approx 0.10 + 0.01 = 0.11 \text{ mol/L}$$
$$c'_{Ac^-} \approx 0.10 - 0.01 = 0.09 \text{ mol/L}$$

Next we calculate the pH

$$pH = pK_a + \lg \frac{c'_{Ac^-}}{c'_{HAc}} = 4.75 + \lg \frac{0.09}{0.11}$$

$$pH = 4.66$$

③ This problem is treated like the problem ②.

After the addition of NaOH, complete ionization of NaOH occurs

$$NaOH(aq) \rightleftharpoons Na^+(aq) + OH^-(aq)$$

0.01mol   0.01mol   0.01mol

After neutralization of the NaOH by HAc, which we write as

$$OH^-(aq) + HAc(aq) \rightleftharpoons H_2O(aq) + Ac^-(aq)$$

0.01 mol   0.01 mol   0.01 mol

The number of moles of HAc and the number of moles of Ac⁻ present are

$$c''_{HAc} \approx 0.10 - 0.01 = 0.09 \text{ mol/L}$$

$$c''_{Ac^-} \approx 0.10 + 0.01 = 0.11 \text{ mol/L}$$

Next we calculate the pH

$$\text{pH} = \text{p}K_a + \lg \frac{c''_{Ac^-}}{c''_{HAc}} = 4.75 + \lg \frac{0.11}{0.09}$$

$$\text{pH} = 4.84$$

The buffer equation indicates that the pH of a buffer solution is close to the pK_a of the acid in a buffer solution (see equation 9.15). And we have known that a buffer solution must contain a weak acid and its conjugate base as major species. When the ratio of weak base to weak acid is between 0.1 and 10, the buffer solution will have sufficient capacity. The buffer equation translates this restriction into a pH range (buffer range) over which the acid and its conjugate base can serve as an effective buffer:

$$\text{pH}_{\text{low}} = \text{p}K_a + \lg 0.1 = \text{p}K_a - 1$$

$$\text{pH}_{\text{high}} = \text{p}K_a + \lg 10 = \text{p}K_a + 1$$

$$\text{pH range} = \text{p}K_a \pm 1$$

With a given weak acid, a buffer solution can be prepared at any pH within about ± 1 pH unit of its pK_a value. For example, we need a buffer solution to maintain the pH of a solution close to 5.0. How do we select buffer pairs? According to the previous analysis, we must choose a weak acid with a pK_a between 4.0 and 6.0. A buffer has maximum capacity, therefore, when its acid has its pK_a as close as possible to the target pH. Table 9.5 lists some acid-base pairs often used as buffer solutions. For a pH = 5.0 buffer, HAc-Ac$^-$ (pK_a = 4.75) would be a good choice.

Table 9.5 Common Buffer Systems

Conjugate Acid-Base Pairs	pK_a	pH Range	Conjugate Acid-Base Pairs	pK_a	pH Range
H_3PO_4-$H_2PO_4^-$	2.12	1.12~3.12	HCOOH-HCOO$^-$	3.74	2.74~4.74
HAc-Ac$^-$	4.75	3.75~5.75	$H_2PO_4^-$-HPO_4^{2-}	7.21	6.21~8.21
H_3BO_3-$H_2BO_3^-$	9.24	8.24~10.24	NH_4^+-NH_3	9.25	8.25~10.25
HCO_3^--CO_3^{2-}	10.25	9.25~11.25	HPO_4^{2-}-PO_4^{3-}	12.66	11.66~13.66

Key Words

acid-base equilibria [ˈæsid beis ˌiːkwiˈlibriə] 酸碱平衡
acid [ˈæsid] 酸
base [beis] 碱
Brønsted-Lowry theory　Brønsted-Lowry 理论
conjugate acid-base pair [ˈkɔndʒugitˈæsid beis pɛə] 共轭酸碱对
ion-product constant of water　水的离子积常数

acidity constant [əˈsiditi ˈkɔnstənt] 酸度常数
basicity constant [bəˈsisiti ˈkɔnstənt] 碱度常数
buffer solution [ˈbʌfə səˈljuːʃən] 缓冲溶液
buffer pair [ˈbʌfə pɛə] 缓冲对
buffer system [ˈbʌfə ˈsistəm] 缓冲系统
buffer capacity [ˈbʌfə kəˈpæsiti] 缓冲容量
Henderson-Hasselbalch equation　Henderson-Hasselbalch 方程
buffer range [ˈbʌfə reindʒ] 缓冲范围

Exercises

1. What's the definitions of Brønsted acids and bases?
2. Give the conjugate acid of each of the following bases:

SO_4^{2-} ; S^{2-} ; HPO_4^{2-} ; NH_3 ; H_2O ; $[Cr(H_2O)_5(OH)]^{2+}$

3. Give the conjugate base of each of the following acids:

NH_4^+ ; $H_2PO_4^-$; H_2S ; HCO_3^- ; H_2SO_4 ; H_2O ; $[Cu(H_2O)_4]^{2+}$

4. Classify each of the following species as a Brønsted acid or base, amphoteric:

HS^- ; CO_3^{2-} ; $H_2PO_4^-$; NH_3 ; H_2S ; NO_2^- ; HCl ; Ac^- ; OH^- ; H_2O

5. HCO_3^- is amphoteric. (a) Write an equation for the reaction of HCO_3^- with water, in which the ion acts as an acid. (b) Write an equation for the reaction of HCO_3^- with water, in which the ion acts as a base. In both cases identify the conjugate acid-base pairs.

6. In a sample of lemon juice $[H^+]$ is 3.8×10^{-4} mol/L, what is the pH?

Answer: 3.42

7. A solution has a $[H^+]$ of 5.3×10^{-9} mol/L, what is the pH?

Answer: 8.28

8. Calculate the pH of a 0.052 mol/L nitrous acid (HNO_2) solution.

Answer: 2.34

9. What is the concentration of a solution of KOH for which the pH is 11.89?

Answer: 7.8×10^{-3} mol/L

10. Calculate the pH of a 0.20 mol/L solution of HCN?

Answer: 5.0

11. What is the pH of a 0.40 mol/L NH_3?

Answer: 11.43

12. The pH of a 0.10 mol/L benzoic acid (C_6H_5COOH) solution is 2.60. What is the K_a of the acid?

Answer: 6.4×10^{-5}

13. The pH of a 0.30 mol/L KA solution is 9.50, what is the K_a of the acid HA?

Answer: 3.0×10^{-6}

14. Calculate the pH of a solution containing 0.085 mol/L nitrous acid HNO_2 ($K_a = 4.5 \times 10^{-4}$) and 0.10 mol/L potassium nitrite, KNO_2.

Answer: 3.42

15. How many moles of NH_4Cl must be added to 2.0 L of 0.10 mol/L NH_3 to form a buffer solution whose pH is 9.0? (Assume that the addition of NH_4Cl does not change the volume of the solution.)

Answer: 0.36 mol

Chapter 10 An Introduction to Electrochemistry

Oxidation-reduction reactions or redox reactions for short are very much a part of the world around us. Examples range from the rusting of iron, the metabolic breakdown of carbohydrates to the bleaching of hair. They are characterized by electron transfer between chemical species. When a redox reaction is coupled with electron flow through a circuit, the process is electrochemical. And electrochemistry deals with the interchange of chemical energy and electrical energy.

In this chapter, we will learn the principles of oxidation-reduction reaction and an introduction to the principles of electrochemistry.

10.1 Oxidation-Reduction Reactions

Oxidizing Agent and Reducing Agent

Oxidation-reduction or redox reactions are electron transfer reactions. In a redox reaction, electrons are transferred from one specie to another. The specie that loses electrons is oxidized and acts as a reducing agent. The specie that gains electrons is reduced and acts as an oxidizing agent. That is, electrons are conserved in all chemical processes. They can be neither created nor destroyed.

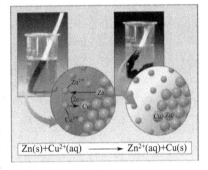

Figure 10.1 The reaction of zinc metal with $CuSO_4$ aqueous solution

The reaction of zinc metal with $CuSO_4$ aqueous solution, which is shown in Figure 10.1, illustrates the fundamental principles of redox reaction. When a piece of zinc is dropped into 1 mol/L $CuSO_4$ solution, the zinc metal dissolves and the copper plates out immediately. As more copper plates out, the blue color of the $CuSO_4$ solution due to the copper (II) ion fades. The molecular and net ionic equation for this reaction are, respectively,

$Zn(s) + CuSO_4(aq) \longrightarrow Cu(s) + ZnSO_4(aq)$ and $Zn(s) + Cu^{2+}(aq) \longrightarrow Cu(s) + Zn^{2+}(aq)$

Solid zinc has been transformed into Zn^{2+} ions by losing two electrons and Cu^{2+} ions into copper metal by gaining two electrons. Because electrons must be conserved in every chemical process, the electrons lost by zinc must be gained by Cu^{2+} ions. Gains and losses of electrons always occur together. Therefore, the reaction of zinc metal with $CuSO_4$ aqueous solution is a redox reaction. Zinc atoms are oxidized and act as a reducing agent because it donates electrons to Cu^{2+} ions and causes Cu^{2+} ions to be reduced. Copper ions are reduced and act as an oxidizing agent because it accepts electrons from zinc and causes Zn atoms to be oxidized.

Oxidation Number

Although electrons are gained and lost in a redox reaction, the reaction equation itself does not show the electrons that are being transferred. In order to tell whether a redox reaction has occurred or not, the best way to keep track of these electrons is by assigning oxidation numbers to the atoms or ions involved in the reaction.

The oxidation number (or oxidation state) is the apparent charge an atom has if all bonds between atoms of different elements are assumed to be completely ionic. The shared electrons in a bond are assigned to the more electronegative atom. Consequently, for elements and monatomic ions, the oxidation number is the same as the charge, while the oxidation number of a covalently bound element is the charge the element would carry if all the shared pairs of electrons were transferred to the more electronegative atom. Therefore oxidation number is hypothetical number assigned to an individual atom or ion present in a substance. It should be remembered that the oxidation number of an atom does not represent the "real" charge on that atom. For example, here are the oxidation numbers of the atoms on the species as follows.

$$\overset{0}{S} \quad \overset{1}{N}\overset{1}{Cl} \quad \overset{2-2}{CaO} \quad \overset{1}{K^+} \quad \overset{2}{Zn^{2+}} \quad \overset{1-2}{H_2O} \quad \overset{1-1}{HCl} \quad \overset{0}{F_2} \quad \overset{0}{H_2}$$

The numbers above the element symbols are the oxidation numbers. There is no charge on the atoms in S, F_2 and H_2. Thus each atom in S, F_2 and H_2 has a zero oxidation number. Therefore the oxidation numbers of atoms in pure neutral elements are zero. The oxidation numbers of monatomic ions equal their net charges. Thus the oxidation numbers of K^+ and Zn^{2+} are $+1$ and $+2$ respectively. To determine the oxidation numbers of atoms in ionic compounds NaCl and CaO, we divide them into their individual ions, and there are monatomic ions present, such as Na^+, Cl^- for NaCl and Ca^{2+}, O^{2-} for CaO. Thus the oxidation numbers of Na and Cl in NaCl are $+1$ and -1, whereas the oxidation numbers of Ca and O in CaO are $+2$ are -2. If the shared pair of electrons in polar H—Cl bond was transferred to the more electronegative chlorine atom, the chlorine atom would have apparent charge of -1 and the hydrogen atom would have apparent charge of $+1$. Thus the oxidation numbers of H and Cl in HCl are $+1$ and -1 whereas the oxidation numbers of H and O in H_2O are $+1$ are -2.

Oxidation numbers enable us to describe oxidation-reduction reactions, and balancing redox chemical reactions. It is helpful and important to assign oxidation numbers (or oxidation states) to a variety of compounds and ions using a set of rules.

(1) An atom of any element in the free state has an oxidation number of 0.

Examples S_8 The oxidation number of S is 0.

Fe The oxidation number of Fe is 0.

(2) The oxidation number of a monatomic ion equals to the charge of the monatomic ion.

Examples S^{2-} The oxidation number of S is -2.

Al^{3+} The oxidation number of Al^{3+} is $+3$.

(3) Hydrogen is assigned an oxidation of $+1$ in most compounds, except for metal hydrides, such as NaH, LiH, etc., in which the oxidation state for H is -1.

Examples HCl The oxidation number of H is $+1$.

HNO_3 The oxidation number of H is $+1$.

(4) Oxygen is assigned an oxidation number of -2 in most compounds.

Examples H_2O The oxidation number of oxygen is -2.

CuO The oxidation number of oxygen is -2.

Except

(a) Peroxides oxidation number of oxygen is -1.

Examples　H_2O_2 The oxidation number of oxygen is -1.

(b) Superoxides　oxidation number of oxygen is $-1/2$.

Examples　KO_2 The oxidation number of oxygen is $-1/2$.

(c) Oxygen fluorides.

Examples　OF_2 The oxidation number of oxygen is $+2$.
　　　　　O_2F_2 The oxidation number of oxygen is $+1$.

(5) Fluorine is assigned an oxidation number of -1 in all its compounds.

Examples　HF The oxidation number of fluorine is -1.
　　　　　NaF The oxidation number of fluorine is -1.

(6) Group I elements (Alkali metals) are assigned an oxidation number of $+1$ in compounds.

Examples　$NaNO_3$ The oxidation number of sodium is $+1$.
　　　　　$Na_2C_2O_4$ The oxidation number of sodium is $+1$.

(7) Group II elements (Alkaline-earth metals) are assigned an oxidation number of $+2$ in compounds.

Examples　$MgBr_2$ The oxidation number of magnesium is $+2$.
　　　　　$BaCO_3$ The oxidation number of barium is $+2$.

(8) The sum of the oxidation numbers of all the atoms in a formula equals the electrical charge shown with the formula.

(a) The sum of the oxidation numbers of all the atoms in a neutral compound is 0.

Example　SO_2 let x be the oxidation number of sulfur
$$0 = x + 2 \times (-2)$$
So x(oxidation number of S)$= 0 + 4 = +4$

(b) The sum of the oxidation numbers of all the atoms in a polyatomic ion or complex ion equals the electrical charge on the ion.

Example　$Cr_2O_7^{2-}$ let x be the oxidation number of chromium
$$-2 = 2x + 7 \times (-2)$$
x (oxidation number of Cr)$= +6$

These rules are wordy. You may simply remember the oxidation states for H and O are $+1$ and -2 respectively in a compound. Oxidation of other elements can be assigned by making the algebraic sum of the oxidation states equal to the net charge on the molecule or ion.

Now we can observe that oxidation numbers can be positive, negative, or zero and are always reported for one individual atom or ion and not for groups of atoms or ions.

Sample 10.1　What is the oxidation number of the S atom in $Na_2S_4O_6$?

Solution:

According to rule 8, the sum of the oxidation numbers of sodium, sulfur and oxygen in the $Na_2S_4O_6$ molecule must be zero. By rule 6, the oxidation numbers of sodium is $+1$ and by rule 4, the oxidation number of oxygen is -2. Let x be the unknown oxidation number of sulfur. Therefore, for $Na_2S_4O_6$
$$2 \times (+1) + 4x + 6 \times (-2) = 0$$
$$x = +2.5$$

Here, the oxidation number of sulfur is a fraction. In fact, the structure of $S_4O_6^{2-}$ is

$$^-O-\overset{\overset{O}{\uparrow}}{\underset{\underset{O}{\downarrow}}{S}}-S-S-\overset{\overset{O}{\uparrow}}{\underset{\underset{O}{\downarrow}}{S}}-O^-$$

The oxidation number of the two sulfurs bonded to three oxygen atoms and each sulfur atom is $+5$, while the oxidation number of the two sulfurs bonded only to sulfur atom is 0 (zero). The total oxidation states of the four sulfur atoms are $5+5+0+0=10$, or an average of $+2.5$.

When there are several same atoms in a structure or a molecule, fractional oxidation number is often used to represent the average oxidation number of each atom. Another example is KO_2. In KO_2, oxygen has an average oxidation state of $-1/2$, which results from having one oxygen atom with oxidation number 0 and one with oxidation number -1.

In a compound, when an atom's oxidation number needs to be specified, we generally write it as a Roman numeral (or 0). The oxidation number is placed either as a right superscript to the element symbol, e.g. Fe^{III}, or in parentheses after the name of the element, e.g. Fe(III), in the latter case, there is no space between the element name and the oxidation number. The negative sign is used for negative oxidation numbers, while the plus sign can be omitted for positive oxidation numbers.

10.2 Balancing Oxidation-Reduction Equations

Although some redox equations are relatively easy to balance by inspection, other more complex redox equations require systematic balancing procedures. Here we will discuss two such procedures, called the oxidation number method and ion-electron method. Whatever method you employed, any balanced equation must meet the following requirements.

(1) Atoms of each element must be conserved. In other words, the amount of each element must be the same on both sides of the equation.

(2) Electrons must be conserved. In other words, the gains and losses of electrons must be balanced.

Oxidation Number Method

Oxidation Number method (also called Oxidation State method) is relatively an easy way to balance redox (including ionic redox) equations. The key principle is that the gain in the oxidation number in one reactant must equal to the loss in the oxidation number of the other reactant. Suppose we are asked to balance the equation as following.

$$KMnO_4 + HCl \longrightarrow MnCl_2 + Cl_2 + KCl$$

The method involves six simple steps.

Step 1. Assign oxidation numbers to all atoms in the reactants and products according to the rules for assigning oxidation states.

$$\overset{1\ \ 7-2}{KMnO_4} + \overset{1-1}{HCl} \longrightarrow \overset{2-1}{MnCl_2} + \overset{0}{Cl_2} + \overset{1-1}{KCl}$$

Step 2. Identify which elements undergo a change in the oxidation state.

$$\overset{7}{K}\overset{}{MnO_4} + 2\overset{-1}{H}Cl \longrightarrow \overset{2}{Mn}Cl_2 + \overset{0}{Cl_2} + KCl$$

Step 3. Note the change in oxidation state of each element on each side of the equation.

$$KMnO_4 + 2HCl \longrightarrow MnCl_2 + Cl_2 + KCl$$
(change of -5 for Mn; change of $+2$ for Cl)

Draw a large bracket from the element in reactant side to that in product side, and write down the increase (+) and decrease (−) in oxidation number at the middle of the bracket.

Step 4. Multiply these changes by appropriate factors so that the total increase in oxidation number equals the total decrease in oxidation number.

$$\overset{(-5)\times 2}{2KMnO_4 + 10HCl \longrightarrow 2MnCl_2 + 5Cl_2 + KCl}\underset{(+2)\times 5}{}$$

Add coefficients to the formulas so as to obtain the correct ratio of the atoms whose oxidation numbers are changing. (These coefficients are usually placed in front of the formulas.)

Step 5. Balance the rest of the equation by final inspection.

$$2KMnO_4 + 16HCl = 2MnCl_2 + 5Cl_2 + 2KCl + 8H_2O$$

Firstly, balance all other atoms in the equation except those of hydrogen and oxygen. In doing this, do not alter the coefficients determined in the previous step. Then balance the oxygen atoms by adding H_2O molecules to the side deficient to oxygen atoms. In this case, $8H_2O$ are needed on the products side to balance oxygen atoms. For reactions in acidic or basic solution, add H^+/H_2O (acidic) or OH^-/H_2O (basic) to balance the charges.

Sample 10.2 Use the Oxidation Number Method to balance the equation for the reaction

$$Cr_2O_7^{2-} + HNO_2 \longrightarrow Cr^{3+} + NO_3^-$$

Solution:

According to the rules for assigning oxidation states, we can find out Cr and N atoms undergo a change in oxidation number. And the increase and decrease in oxidation number are $+2$ and -6, respectively.

$$\overset{(-6)}{Cr_2O_7^{2-} + HNO_2 \longrightarrow 2Cr^{3+} + NO_3^-}\underset{(+2)}{}$$

Multiply these changes by appropriate factors,

$$\overset{(-6)\times 1}{Cr_2O_7^{2-} + HNO_2 \longrightarrow 2Cr^{3+} + NO_3^-}\underset{(+2)\times 3}{}$$

Add coefficients to the formulas,

$$Cr_2O_7^{2-} + 3HNO_2 \longrightarrow 2Cr^{3+} + 3NO_3^-$$

Balance the rest of the equation by adding H^+ and H_2O.

$$Cr_2O_7^{2-} + 3HNO_2 + 5H^+ = 2Cr^{3+} + 3NO_3^- + 4H_2O$$

Ion-Electron Method (or Half-Reaction Method)

Ion-Electron Method (or Half-Reaction Method) is applied the best for redox reactions with ions and in aqueous solution. This method is to split an unbalanced redox equation into two parts called half-reactions. One half-reaction describes the oxidation, and the other describes the reduction. When zinc reduces copper (Ⅱ) sulfate to copper, the ionic equation for the reaction is

$$Zn + Cu^{2+} \longrightarrow Zn^{2+} + Cu$$

To determine how to split the redox reaction into two half-reactions, it is useful to look at it from the point of view of the zinc and of the copper (II) ions separately. Zinc is losing electrons to become a positive ion. This is the process of oxidation. The reaction we can write for this process is called the oxidation half-reaction (increase in oxidation number going from reactant to product).

$$Zn(s) - 2e^- \text{(electrons)} \longrightarrow Zn^{2+}(aq)$$

Copper (II) ion is gaining electrons to become a copper. This is the process of reduction. The reaction we write for this process is called the reduction half-reaction (decrease in oxidation number from left to right).

$$Cu^{2+}(aq) + 2e^- \longrightarrow Cu(s)$$

Both half-reactions are then balanced in according to the change of electrons and finally added to obtain the final balanced equation.

The half-reaction method can be applied to more complex redox reactions, such as the reaction of permanganate ion, MnO_4^-, with I^- in acidic solution.

$$MnO_4^- + I^- \longrightarrow Mn^{2+}(aq) + I_2 \text{ (unbalanced equation)}$$

The method involves eight simple steps

Step 1. Write the unbalanced equation in ionic form.

$$MnO_4^- + I^- \longrightarrow Mn^{2+}(aq) + I_2$$

Step 2. Assign oxidation numbers to each element and determine which substances are changing their oxidation states.

$$\overset{7-2}{MnO_4^-} + \overset{-1}{I^-} \longrightarrow \overset{2}{Mn^{2+}} + \overset{0}{I_2}$$

It's obvious that I underwent an increase in oxidation number (-1 to 0), so it was oxidized, and Mn underwent a decrease in oxidation number (7 to 2), so it was reduced.

Step 3. Separate the equation into two half-reactions.

Oxidation: $I^- \longrightarrow I_2$

Reduction: $MnO_4^- \longrightarrow Mn^{2+}$

Step 4. Balance each half reaction by

a. Balance all atoms other than H and O by inspection.

$$2I^- \longrightarrow I_2$$
$$MnO_4^- \longrightarrow Mn^{2+}$$

b. Balance the O and H atoms. The way this is done depends on whether the solution is acidic or basic

Acidic Solution: add H_2O to the appropriate side that needs more oxygen atoms, add the appropriate number of H^+ to the other side deficient in H.

$$2I^- \longrightarrow I_2$$
$$MnO_4^- + 8H^+ \longrightarrow Mn^{2+} + 4H_2O$$

Basic Solution: add one H_2O molecule to the side deficient in H and one OH^- ion to the opposite side, for each H atom needed. You may need to cancel out H_2O molecules duplicated on each side at this point.

c. Balance the charges by adding electrons (e^-) to the side deficient in negative charge.

$$2I^- \longrightarrow I_2 + 2e^-$$
$$MnO_4^- + 8H^+ + 5e^- \longrightarrow Mn^{2+} + 4H_2O$$

Step 5. Multiply each half-reaction by a coefficient so that the number of electrons lost equals the number of electrons gained.
$$5\times(2I^- \longrightarrow I_2+2e^-)$$
$$2\times(MnO_4^- +8H^+ +5e^- \longrightarrow Mn^{2+} +4H_2O)$$

Step 6. Add the two half reactions together. Keeping all of the reactants together on the left side of the reaction arrow and all of the products together on the right side of the reaction arrow.
$$10I^- +16H^+ +2MnO_4^- +10e^- \longrightarrow 5I_2 +10e^- +2Mn^{2+} +8H_2O$$

Step 7. Eliminate anything that appears in identical form on both sides of the equation.
$$10I^- +16H^+ +2MnO_4^- \longrightarrow 5I_2 +2Mn^{2+} +8H_2O$$

Step 8. Check that the charges balance on both sides of the equation.
Balanced equation $10I^- +16H^+ +2MnO_4^- = 5I_2 +2Mn^{2+} +8H_2O$

Sample 10.3 Balance the following equation by Ion-Electron Method.
$$CrO_2^- +Br_2 +OH^- \longrightarrow CrO_4^{2-} +Br^- +H_2O$$

Solution:

It's obvious that Cr underwent an increase in oxidation number (+3 to +6), while Br underwent a decrease in oxidation number (0 to -1) in basic-solution.

Separate the equation into two half-reactions.
Oxidation $CrO_2^- \longrightarrow CrO_4^{2-}$
Reduction $Br_2 \longrightarrow Br^-$
Balance each half reaction.
$$CrO_2^- +4OH^- \longrightarrow CrO_4^{2-} +2H_2O+3e^-$$
$$Br_2 +2e^- \longrightarrow 2Br^-$$
Multiply each half-reaction by a coefficient, then add the two half reactions together.
$$2\times(CrO_2^- +4OH^- \longrightarrow CrO_4^{2-} +2H_2O+3e^-)$$
$$+3\times(Br_2 +2e^- \longrightarrow 2Br^-)$$
$$=2CrO_2^- +3Br_2 +8OH^- \longrightarrow 2CrO_4^{2-} +6Br^- +4H_2O$$

10.3 Voltaic Cells

As mentioned above (Figure 10.1), when a strip of Zn is placed in a solution of copper (Ⅱ) sulfate, the zinc slowly dissolves and copper metal immediately begins to plate out on the zinc strip.
$$Zn(s)+Cu^{2+}(aq)\longrightarrow Cu(s)+Zn^{2+}(aq)$$

Each zinc atom transfers two electrons to a copper cation directly, and the energy gives off heats the water and its surroundings.

However, when carried out under appropriate conditions, a redox reaction can generate electrical energy if it takes place spontaneously. Here conditions mean proper device, concentration, temperature or pressure. It is possible to build such an apparatus where the two half-reactions take place at separate sites, with electrons being transferred indirectly. Such an apparatus which generates electricity through the use of a spontaneous redox reaction is known as a voltaic cell or galvanic cell. A wide variety of such cells may be constructed. The Cu-Zn or Daniell cell will illustrates the general principles involved.

The Device of Daniell Cell

The Daniel cell, also called the gravity cell was invented in 1836 by John Frederic

Daniell and the chemical reaction is

$$Zn(s) + Cu^{2+}(aq) \longrightarrow Cu(s) + Zn^{2+}(aq)$$

This reaction can also occur by indirect electron transfer. In the device (Daniell cell) shown in Figure 10.2, the left hand compartment consists of a zinc strip or rod partly immersed in 1.0 mol/L $ZnSO_4$ solution. The right hand compartment consists of a copper strip or rod partly immersed in 1.0 mol/L $CuSO_4$ solution. A wire connects the two metal strips to allow indirect electron transfer.

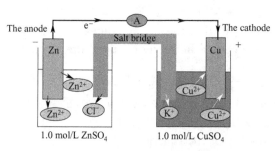

Figure 10.2 The conventional notation for representing galvanic cells

As illustrated, the compartment on the left is the site for the oxidation half-reaction. Here zinc loses two electrons and releases Zn^{2+} ions into the solution.

$$Zn(s) - 2e^- \longrightarrow Zn^{2+}(aq)$$

The zinc strip is called the anode electrode where oxidation takes place, and the oxidation reaction that takes place at anode electrode is called Zn^{2+}/Zn half-cell reaction. The compartment on the right is the site for the reduction half-reaction. Here Cu^{2+} gains two electrons when they collide with the surface of the copper strip.

$$Cu^{2+}(aq) + 2e^- \longrightarrow Cu(s)$$

The copper strip is called the cathode electrode where reduction takes place, and the reduction reaction that takes place at cathode electrode is called Cu^{2+}/Cu half-cell reaction.

The electrons lost by the zinc electrode flow in the wire to the copper electrode where they are picked up by the Cu^{2+} ions. The overall cell reaction is the sum of the two half-cell reactions.

$$Zn(s) + Cu^{2+} \longrightarrow Cu(s) + Zn^{2+}$$

The two half-cell reactions result in zinc cations (Zn^{2+}) being produced in the anode compartment, and sulphate anions (SO_4^{2-}) being left in the cathode compartment. A salt bridge (a glass tube filled with a jelly containing a suitable electrolyte such as saturated KCl or KNO_3) placed between the two half-cells (just like "liquid wire") connects the two solutions and allows the flow of cations and anions to maintain charge balance in each solution. Cations such as Zn^{2+} and K^+ move from anode to cathode, anions such as Cl^- and SO_4^{2-} move from cathode to anode along the salt bridge.

The characteristics of Voltaic or galvanic cells can be summarized as follows.

1. Voltaic cells use spontaneous redox reactions to produce electrical energy

Here, a spontaneous redox reaction means that the reaction proceeds spontaneously in the forward direction or shifts to the right to reach equilibrium.

2. Voltaic cells are devices in which electron transfer occurs through an external circuit rather than directly between reactants.

The essential components for Voltaic or galvanic cells are

(1) Two half-cells or electrodes (cathode and anode)

For Cu-Zn or Daniell cell

Cathode: A metallic Cu strip immersed in copper (Ⅱ) sulfate solution.

Anode: A metallic Zn strip immersed in zinc (Ⅱ) sulfate solution.

(2) A salt bridge to connect the oxidation and reduction half-cells and prevent solutions from mixing.

(3) A wire to complete circuit.

Shorthand Notation (or Cell Diagram) for Voltaic or Galvanic Cells

Rather than drawing out the entire voltaic cell, chemists have devised a shorthand line notation for describing voltaic cells.

The cathode (oxidation) is typically written on the right side, the anode on the left side. A single vertical line " | " indicates a phase boundary between different phases such as metal and electrolyte solution, gas and metal. A double vertical line " ‖ " indicates the salt bridge between solutions.

Generally

Anode(Oxidation) | Electrolyte 1 ‖ Electrolyte 2 | Cathode(Reduction)

Using this notation, for the Cu/Zn battery, it would be described as follows.

$$Zn(s) | Zn^{2+} (1\ mol/L) \| Cu^{2+} (1\ mol/L) | Cu(s)$$

Unlike the Cu/Zn battery which uses Zn and Cu strips as anode and cathode, the cell derived from the redox reaction: $H_2(g) + 2Fe^{3+}(aq) \longrightarrow 2H^+(aq) + 2Fe^{2+}(aq)$ should use inert electrodes to conduct electrons to and from the external circuit. Platinum metal or carbon is often used as such inert electrodes without participating chemically in the half-cell reactions, but simply provides a return path for the current.

If all the substances on one side are aqueous or involve a gas, an inert electrode such as platinum or carbon is used and should be indicated in cell diagram. The concentrations of aqueous substances or the pressure of a gas should be specified except for 1 mol/L for aqueous substance and 101.3 kPa (1 atm) for gas. Using this notation, for the battery derived from,

$$H_2(g) + 2Fe^{3+}(c_2) \longrightarrow 2H^+(c_1) + 2Fe^{2+}(c_3)$$

it would be described as follows.

$$Pt(s) | H_2(g) | H^+(c_1) \| Fe^{3+}(c_2), Fe^{2+}(c_3) | Pt(s)$$

Types of Electrodes

Electrodes are usually classified as four types according to the structure of the electrode and the characteristics of the reaction.

1. Metal-Metal Ion Electrodes

Metal-metal ion electrode is usually made by a metal electrode dipped into an aqueous solution of a salt of that metal. Electron transfer occurs between the metal atoms of the electrode and the metal ions in solution.

For example, Cu-Cu^{2+} ion electrode is made by dipping a strip of Cu into a solution containing Cu^{2+} ion.

Electrode reaction $Cu^{2+}(c) + 2e^- \rightleftharpoons Cu(s)$

Electrode notation $Cu(s) | Cu^{2+}(c)$

2. Gas Electrodes

Gas electrode typically consists of a platinum electrode (or carbon electrode) dipped into a solution containing the anion or cation of the gas. The gas is bubbled slowly over the surface of the platinum electrode. Platinum is a noble metal which is inert and which only serves to establish the electric contact with the solution. Electron transfer occurs between the gas and the ions in solution.

For example, hydrogen electrode is made when hydrogen gas is bubbled over metallic platinum immersed in an aqueous solution containing hydronium cations.

Electrode reaction $2H^+(c)+2e^- \rightleftharpoons H_2(p)$

Electrode notation $Pt(s) \mid H_2(p) \mid H^+(c)$

3. Metal-Insoluble Salt Electrodes

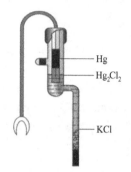

Figure 10.3 The saturated calomel electrode

Metal-insoluble salt electrode normally consists of a metal electrode in contact with (partially coated with a thin layer of) an insoluble salt of the metal, and then immersed a solution containing the anion of the insoluble salt. Electron transfer occurs between the metal atoms of the electrode and the metal ions in the insoluble salt.

For example, saturated calomel electrode (SCE) is made by solid calomel (mercury chloride) with mercury paste and potassium chloride solution (Figure 10.3).

Electrode reaction $Hg_2Cl_2(s)+2e^- \rightleftharpoons 2Hg(l)+2Cl^-(c)$

Electrode notation $Hg(l) \mid Hg_2Cl_2(s) \mid Cl^-(c)$

4. Redox electrodes

Redox electrode consists of a platinum electrode (or carbon electrode) dipped into a solution containing ions of a metal in two different oxidation states. The platinum electrode (or carbon electrode) is an inert electrode to act as a source or acceptor of electrons and allow the electrode system to be linked to another electrode to create a complete cell.

For example, a platinum electrode is dipped into a solution containing both Fe^{3+} and Fe^{2+} ions.

Electrode reaction $Fe^{3+}(c_1)+e^- \rightleftharpoons Fe^{2+}(c_2)$

Electrode notation $Pt(s) \mid Fe^{3+}(c_1), Fe^{2+}(c_2)$

10.4 Standard Cell Electromotive Force

To understand how a voltaic cell generates electrical current that consists of electrons spontaneously flowing from one place to another, we can compare the electron flow generated by a voltaic cell to the flow of water in a waterfall. Water always flows downhill, from higher altitude to lower altitude, because of a difference in gravitational potential energy between the top of the fall and the stream below. In a similar fashion, electrons flow from anode to cathode in a voltaic cell because of a difference in electrical potential energy between the anode and cathode.

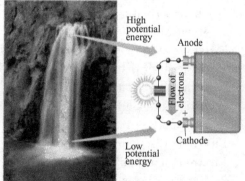

Figure 10.4 Gravitational potential and electrical potential have common features

The difference in gravitational potential energy between two places can be measured by their difference in altitude, and similarly a difference in electrical potential energy between the anode and cathode is measured as a voltage (Figure 10.4).

In a voltaic cell, the difference in electrical potential between two electrodes is called the electromotive force (emf), symbolized by the E_{cell}, of the cell or the cell potential or cell voltage. And the emf of a voltaic cell provides the driving force that pushes electrons through the external circuit.

Standard Electrode Potential

The potential difference between an anode and a cathode of a voltaic cell can be measured by a voltage measuring device. Each half-reaction or each electrode has an electrode potential. But the absolute potential of the anode and cathode cannot be measured directly. Measurement of the potential of a single half-cell is impossible. Only potential differences between two half-cells can be measured. So two main hurdles must be overcome to determine a single electrode potential.

(1) The electrode potential cannot be determined in isolation, but in a reaction with some other electrode.

(2) The electrode potential varies with the concentrations of the substances, the temperature, and the pressure in the case of a gas electrode.

In practice, the first hurdle is overcome by defining a standard electrode, and all other potential measurements can be made against this standard electrode. The standard hydrogen electrode (SHE) that contains one molar hydrogen ion and hydrogen gas at a partial pressure of 1 atm serves as a reference and is arbitrarily given a potential of 0.00 volts. This reference half-cell employs the reaction.

$$2H^+(aq) + 2e^- \rightleftharpoons H_2(g)$$

The basic construction of the standard hydrogen electrode is shown in Figure 10.5. The platinum electrode has two functions.

It provides a surface on which the dissociation of H_2 molecules can take place, it acts as an electrical conductor to the external circuit.

The second hurdle is overcome by choosing standard thermodynamic conditions (standard state) for the measurement of the potentials. The standard thermodynamic conditions are customarily determined at solute concentrations of 1 molar, gas pressures of 1 atmosphere, and a standard temperature which is usually 25 ℃. The standard potential is denoted by a degree sign as a superscript.

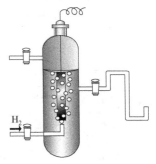

Figure 10.5 The basic construction of the standard hydrogen electrode

In addition to defining a standard electrode and standard thermodynamic conditions, we also need to specify a reference direction for half-reactions since the half-cell reactions are reversible. Depending on the conditions, any half-cell reaction can act as either as an anode or as a cathode. This means that any half-cell reaction can run as an oxidation that supplies electrons or as a reduction that gains electrons. An oxidation half-reaction is the reverse of the corresponding reduction half-reaction. If the potential of an electrode is a measure of its ability to gain electrons with the half-reaction written as a reduction, the potential will be called its "reduction potential". If the potential of an electrode is a measure of its ability to lose electrons with the half-reaction written as an oxidation, the potential will be called its "oxidation potential". For example, the reduction half-reaction

$Zn^{2+}(aq) + 2e^- \rightleftharpoons Zn(s)$ has, a reduction potential of -0.76 volts.

When the half-reaction is written as its reverse,
$$Zn(s) - 2e^- \rightleftharpoons Zn^{2+}(aq)$$
The oxidation half-reaction has an oxidation potential of 0.76 volts.

Historically, many countries, including the United States and Canada, used standard oxidation potentials rather than reduction potentials in their calculations. Since the oxidation potential of a half-reaction is the negative of the corresponding reduction potential, it is sufficient to calculate either one of the potentials. Therefore, standard electrode potential is commonly written as standard reduction potential. The conventional reference direction is reduction. Thus the standard reduction potential, denoted $E^\ominus$, is the voltage associated with a reduction reaction at an electrode at standard state.

By convention, the standard emf of the cell, $E^\ominus_{cell}$ is the difference between the standard reduction potentials of the two half-cells that make up the cell.
$$E^\ominus_{cell} = E^\ominus_{cathode} - E^\ominus_{anode} \tag{10.1}$$
In the equation above,

$E^\ominus_{cathode}$ means the reduction potential for the cathode reaction; $E^\ominus_{anode}$ means the reduction potential for the anode reaction

When we use this equation, we are emphasizing that $E^\ominus_{cell}$ depends on the difference between the reducing strengths of the two half-cells and that $E^\ominus_{cell}$ is the difference between two reduction potentials.

Now we can use the SHE and the following strategy to assign the $E^\ominus$ values to other half-reactions.

(1) Connect another half-cell to the SHE.

(2) Measure the cell potential (i.e. voltage) and determine the direction of electron flow.

(3) Assign that voltage to the half-cell reaction using a "+" sign or a "−" sign.

For example, connect a standard zinc electrode to a SHE and the voltaic cell is shown in Figure 10.6. In this case the zinc electrode is the anode and the SHE is the cathode. We deduce this fact from the observation that the zinc electrode decreases in mass during the operation of the cell, which is consistent with the loss of zinc to the solution as Zn^{2+} ions.

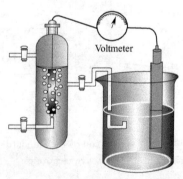

Figure 10.6 Schematic diagram of a cell for measuring the standard reduction potential of Zn^{2+}/Zn half-cell relative to the standard hydrogen electrode (SHE)

The cell diagram is

$Zn(s) | Zn^{2+}(1mol/L) \| H^+(1mol/L) | H_2(1atm) | Pt(s)$

An experimental measurement of the cell potential under standard state gives $E^\ominus_{cell} = 0.76$ V.

We can write the half-cell reactions as reductions.

Cathode $\quad 2H^+ + 2e^- \rightleftharpoons H_2$

Anode $\quad Zn^{2+} + 2e^- \rightleftharpoons Zn$

Overall $\quad Zn + 2H^+ \rightleftharpoons Zn^{2+} + H_2$

Using the equation, we can determine the standard reduction potential for the Zn^{2+}/Zn half-reaction.

$$E^\ominus_{cell} = E^\ominus_{H^+/H_2} - E^\ominus_{Zn^{2+}/Zn}$$
$$+0.76 \text{ V} = 0 \text{ V} - E^\ominus_{Zn^{2+}/Zn}$$
$$E^\ominus_{Zn^{2+}/Zn} = -0.76 \text{ V}$$

Thus, a standard reduction potential of -0.76 V can be assigned to the Zn^{2+}/Zn half-reaction.

$$Zn^{2+} + 2e^- \rightleftharpoons Zn \qquad E^{\ominus}_{Zn^{2+}/Zn} = -0.76 \text{ V}$$

The standard reduction potential of Cu^{2+}/Cu half-cell can be assigned in a similar fashion, by connecting a standard copper electrode to a SHE. The voltaic cell is shown in Figure 10.7. In this case the copper electrode is the cathode and the SHE is the anode, because the copper electrode increases in mass during the operation of the cell, which is consistent with the deposit of copper from the Cu^{2+} ions.

The cell diagram is

$Pt(s) | H_2(1atm) | H^+(1mol/L) \| Cu^{2+}(1mol/L) | Cu(s)$

An experimental measurement of the cell potential under standard state gives $E^{\ominus}_{cell} = 0.34$ V.

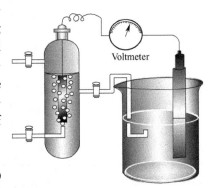

Figure 10.7 Schematic diagram of a cell for measuring the standard reduction potential of Cu^{2+}/Cu half-cell relative to the standard hydrogen electrode (SHE)

We can write the half-cell reactions as reductions.

Cathode $Cu^{2+} + 2e^- \rightleftharpoons Cu$
Anode $2H^+ + 2e^- \rightleftharpoons H_2$
Overall $Cu^{2+} + H_2 \rightleftharpoons Cu^{2+} + 2H^+$

Using the equation, we can determine the standard reduction potential for the Cu^{2+}/Cu half-reaction.

$$E^{\ominus}_{cell} = E^{\ominus}_{Cu^{2+}/Cu} - E^{\ominus}_{H^+/H_2}$$
$$+0.34 \text{ V} = E^{\ominus}_{Cu^{2+}/Cu} - 0 \text{ V}$$
$$E^{\ominus}_{Cu^{2+}/Cu} = +0.34 \text{ V}$$

Each standard reduction potential has a specific sign, a "+" sign or a "−" sign. When a species is easier to reduce than hydronium ions under standard conditions, its $E^{\ominus}$ is positive. When a species is more difficult to reduce than hydronium ions under standard conditions, its $E^{\ominus}$ is negative. The more positive the reduction potential is, the greater the species' affinity for electrons and tendency to be reduced.

Table of Standard Reduction Potentials

Over the years, chemists have carried out many measurements. As a result, numerous standard reduction potentials are tabulated in reference sources such as the Handbook of Chemistry and Physics. Table 10.1 lists some representative values, a more complete list appears in Appendix F. These available standard reduction potentials, often called half-cell potentials, could help us predict the cell potentials of voltaic cells created from any pair of electrodes, and predict comparative strengths of oxidizing and reducing agents.

Table 10.1 Standard reduction potentials of half-cells

Half-Reaction					Standard Potential $E^{\ominus}$/V
Oxidizing Agents				Reducing Agents	
$Li^+(aq)$	+	e^-	$\rightarrow$	$Li(s)$	-3.04
$SO_4^{2-}(aq) + H_2O(l)$	+	$2e^-$	$\rightarrow$	$SO_3^{2-}(aq) + 2OH^-(aq)$	-0.93
$2H_2O(l)$	+	$2e^-$	$\rightarrow$	$H_2(g) + 2OH^-(aq)$	-0.83
$Zn^{2+}(aq)$	+	$2e^-$	$\rightarrow$	$Zn(s)$	-0.76
$Fe^{2+}(aq)$	+	$2e^-$	$\rightarrow$	$Fe(s)$	-0.45

Table 10.1 (Continued)

Oxidizing Agents		Half-Reaction		Reducing Agents	Standard Potential $E^\ominus$/V
Co^{2+} (aq)	+	$2e^-$	$\longrightarrow$	Co(s)	−0.28
Pb^{2+} (aq)	+	$2e^-$	$\longrightarrow$	Pb(s)	−0.13
$2H^+$ (aq)	+	$2e^-$	$\longrightarrow$	H_2(g)	0.00
Cu^{2+} (aq)	+	$2e^-$	$\longrightarrow$	Cu(s)	+0.34
O_2(g)$+2H_2O$(l)	+	$4e^-$	$\longrightarrow$	$4OH^-$ (aq)	+0.40
I_2(s)	+	$2e^-$	$\longrightarrow$	$2I^-$ (aq)	+0.54
O_2(g)$+2H^+$ (aq)	+	$2e^-$	$\longrightarrow$	H_2O_2(l)	+0.70
Br_2(aq)	+	$2e^-$	$\longrightarrow$	$2Br^-$ (aq)	+1.08
O_2(g)$+4H^+$ (aq)	+	$4e^-$	$\longrightarrow$	$2H_2O$(l)	+1.23
$Cr_2O_7^{2-}$ (aq)$+14H^+$ (aq)	+	$6e^-$	$\longrightarrow$	$2Cr^{3+}$ (aq)$+7H_2O$(l)	+1.33
Cl_2(g)	+	$2e^-$	$\longrightarrow$	$2Cl^-$ (aq)	+1.36
MnO_4^- (aq)$+8H^+$ (aq)	+	$5e^-$	$\longrightarrow$	Mn^{2+} (aq)$+4H_2O$(l)	+1.51
F_2(g)	+	$2e^-$	$\longrightarrow$	$2F^-$ (aq)	+2.87

The values for the table entries are reduction potentials and listed in order of increasing value. It is important to know these points when using the values in the table.

(1) All of the half-reactions in the table are written as reductions. If we reverse a half-reaction, then $E^\ominus$ changes sign.

(2) Electrode potentials are intensive properties. If we multiply a half-reaction by "n", $E^\ominus$ is unchanged. The following two half-reactions have the same potential values.

$$Cl_2 + 2e \rightleftharpoons 2Cl^- \qquad E^\ominus = 1.358 \text{ V}$$
$$1/2Cl_2 + e \rightleftharpoons Cl^- \qquad E^\ominus = 1.358 \text{ V}$$

(3) The bigger the standard reduction potential is, the easier the specie to the left of the arrow is to be reduced. In other words, the specie is a stronger oxidizing agent, while the specie to the right of the arrow is a weaker reducing agent. The smaller the standard reduction potential is, the easier the specie to the right of the arrow is to be oxidized. In other words, the specie is a stronger reducing agent, while the specie to the left of the arrow is a weaker oxidizing agent.

Since lithium at the top of the list has the most negative value, indicating that it is the strongest reducing agent and would rather undergo oxidation. Fluorine with the largest positive value for standard electrode potential is the most easily reduced species and the strongest oxidizing agent.

(4) When "coupling" two half-reactions to form a Voltaic cell, the half-reaction with higher reduction potential $E^\ominus$ value has a greater tendency to occur as a reduction and should be cathode. The half-reaction with lower reduction potential $E^\ominus$ value has a greater tendency to occur as a oxidation and should be anode. The overall reaction should be that the half-reaction with higher reduction potential $E^\ominus$ value minus the half-reaction with lower reduction potential $E^\ominus$ value. The $E^\ominus_{cell}$ should be that the higher reduction potential $E^\ominus$ value minus the lower reduction potential $E^\ominus$ value.

For example, when "coupling" the half-reaction
$$Sn^{4+} (aq) + 2e^- \rightleftharpoons Sn^{2+} (aq) \qquad E^\ominus = 0.14 \text{ V}$$
with
$$Fe^{3+} (aq) + e^- \rightleftharpoons Fe^{2+} (aq) \qquad E^\ominus = 0.77 \text{ V}$$

The overall reaction is

$$2 \times Fe^{3+} (aq) + e^- \rightleftharpoons Fe^{2+} (aq)$$
$$-) Sn^{4+} (aq) + 2e^- \rightleftharpoons Sn^{2+} (aq)$$
$$\overline{= 2Fe^{3+} (aq) + Sn^{2+} (aq) \rightleftharpoons 2Fe^{2+} (aq) + Sn^{4+} (aq)}$$

and $E^\ominus_{cell} = 0.77 \text{ V} - 0.14 \text{ V} = 0.63 \text{ V}$

Applications of Standard Reduction Potentials

(1) Calculation of Standard EMF of Cell

As mentioned above, Voltaic cells use spontaneous redox reactions to produce electrical energy. The difference in electrical potential between two electrodes is called the electromotive force (emf), symbolized by the E_{cell}, of the cell or the cell potential or cell voltage. Standard cell potentials, symbolized by the $E_{cell}^{\ominus}$, are measured under standard conditions.

In principle, any redox reaction could be utilized to make an electrochemical cell. If the values of the standard reduction potentials of the two half-reactions which make up the cell are available, the emf of any cell could be easily calculated using the formula.

$$E_{cell}^{\ominus} = E_{cathode}^{\ominus} - E_{anode}^{\ominus} \qquad (10.1)$$

In the meantime, we can use the known values of the standard-state reduction potentials to determine which electrode is the anode and which is the cathode. The half-reaction with higher reduction potential $E^{\ominus}$ value has a greater tendency to occur as a reduction and should be cathode. The half-reaction with lower reduction potential $E^{\ominus}$ value has a greater tendency to occur as a oxidation and should be anode.

We saw in a previous section that the redox reaction

$$Zn(s) + Cu^{2+} \rightleftharpoons Cu(s) + Zn^{2+}$$

Consider the two half-reactions which contribute to the cell, written down as reductions (gain of electrons).

Cathode $\quad Cu^{2+}(aq) + 2e^- \rightleftharpoons Cu(s) \quad E_{Cu^{2+}/Cu}^{\ominus} = +0.34$ V

Anode $\quad Zn^{2+}(aq) + 2e^- \rightleftharpoons Zn(s) \quad E_{Zn^{2+}/Zn}^{\ominus} = -0.76$ V

It can be exploited to create a voltaic cell which is written conventionally as

$$Zn(s) | Zn^{2+}(1mol/L) \| Cu^{2+}(1mol/L) | Cu(s)$$

We could therefore calculate the standard emf, $E_{cell}^{\ominus}$, of the above cell (under standard conditions) as.

emf of cell = potential of Cu − potential of Zn

or conventionally

$$E_{cell}^{\ominus} = E_{Cu^{2+}/Cu}^{\ominus} - E_{Zn^{2+}/Zn}^{\ominus}$$
$$E_{cell}^{\ominus} = +0.34 \text{ V} - (-0.76) = 1.10 \text{ V}$$

Another example, one could use the reaction

$$Fe(s) + 2Ag^+(aq) \rightleftharpoons 2Ag(s) + Fe^{2+}(aq)$$

Write down the two half-reactions as reductions:

Cathode $\quad Ag^+(aq) + e^- \rightleftharpoons Ag(s) \quad E_{Ag^+/Ag}^{\ominus} = +0.80$ V

Anode $\quad Fe^{2+}(aq) + 2e^- \rightleftharpoons Fe(s) \quad E_{Fe^{2+}/Fe}^{\ominus} = -0.45$ V

Then construct the cell

$$Fe(s) | Fe^{2+}(1 \text{ mol/L}) \| Ag^+(1 \text{ mol/L}) | Ag(s)$$

Therefore the standard emf, $E_{cell}^{\ominus}$, of the above cell (under standard conditions) as follows.

emf of cell = potential of Ag − potential of Fe

or conventionally

$$E_{cell}^{\ominus} = E_{Ag^+/Ag}^{\ominus} - E_{Fe^{2+}/Fe}^{\ominus}$$
$$E_{cell}^{\ominus} = +0.80 \text{ V} - (-0.45) = 1.25 \text{ V}$$

We can calculate the value of $E_{cell}^{\ominus}$ for any oxidation-reduction reaction provided we know the reduction potentials $E^{\ominus}$ for the two half-reactions.

(2) Spontaneity of Redox Reaction

As we know, not all reactions are spontaneous. How can we tell which redox reactions will be spontaneous and which will be nonspontaneous?

Voltaic cells use spontaneous redox reactions to produce electrical energy, and we have observed that any reaction that spontaneously occurs in a voltaic cell must produce a positive emf. The fact that $E^{\ominus}_{cell}$ is positive tells us that when the reaction is present at standard-state conditions, it is spontaneous and shifts to the right to reach equilibrium. The more positive $E^{\ominus}_{cell}$, the more the reaction will proceed to the right.

Consequently, it is possible to predict whether a redox reaction will be spontaneous by using the potentials of half-reactions to calculate the emf associated with it.

To calculate the emf values of general redox reactions at standard-state conditions, not just spontaneous redox reactions, we will first suppose the forward direction is spontaneous. We can then break the redox reaction down into oxidation and reduction half-reactions, or identify the half-cell of reduction and the half-cell of oxidation according to the redox reaction. Finally, we use the reduction potentials for the two half-reactions to determine the emf. According to the sign of the emf, we can determine whether the reaction is spontaneous or not. If the sign is positive, it confirms that the redox reaction is spontaneous. If the sign is negative, it means that the redox reaction is nonspontaneous, in other words, the reverse reaction should be occur in a voltaic cell.

Consider the reaction between silver and copper, for example,
$$Cu(s) + 2Ag^+ \rightleftharpoons 2Ag(s) + Cu^{2+}$$

In this reaction, suppose the forward direction is spontaneous, then Cu is oxidized to Cu^{2+} and Ag^+ is reduced to Ag. Thus it can be utilized to form a voltaic cell
$$Cu(s)|Cu^{2+}(1\ mol/L)\ \|\ Ag^+(1\ mol/L)|Ag(s)$$

And it can be formally divided into separate oxidation and reduction half-reactions.

Reduction $\quad Ag^+ + e^- \rightleftharpoons Ag(s)$

Oxidation $Cu^{2+} + 2e^- \rightleftharpoons Cu(s)$

Then $\quad E^{\ominus}_{cell} = E^{\ominus}_{half\text{-}cell\ of\ reduction} - E^{\ominus}_{half\text{-}cell\ of\ oxidation}$ \hfill (10.2)

$\quad E^{\ominus}_{cell} = E^{\ominus}_{Ag^+/Ag} - E^{\ominus}_{Cu^{2+}/Cu}$

From Appendix F, $E^{\ominus}_{Cu^{2+}/Cu} = +0.34\ V$ and $E^{\ominus}_{Ag^+/Ag} = +0.80\ V$. Substitution of these values gives $E^{\ominus}_{cell} = +0.80\ V - (+0.34\ V) = +0.46\ V$

The positive voltage indicates that the chemical reaction is spontaneous.

What happens to the cell potential when we reverse the equation's direction?
$$2Ag(s) + Cu^{2+} \longrightarrow Cu(s) + 2Ag^+$$

In this reaction, suppose the forward direction is spontaneous, then Ag is oxidized to Ag^+ and Cu^{2+} is reduced to Cu. Thus it can be utilized to form a voltaic cell.
$$Ag(s)|Ag^+(1mol/L)\ \|\ Cu^{2+}(1mol/L)|Cu(s)$$

Write the oxidation and reduction half-reactions for this reaction.

Reduction $\quad Cu^{2+} + 2e^- \longrightarrow Cu(s)$

Oxidation $\quad Ag^+ + e^- \longrightarrow Ag(s)$

Then $\quad E^{\ominus}_{cell} = E^{\ominus}_{half\text{-}cell\ of\ reduction} - E^{\ominus}_{half\text{-}cell\ of\ oxidation}$

$\quad E^{\ominus}_{cell} = E^{\ominus}_{Cu^{2+}/Cu} - E^{\ominus}_{Ag^+/Ag}$

$\quad E^{\ominus}_{cell} = +0.34\ V - (+0.80\ V) = -0.46\ V$

Turning the equation around changes the sign of the cell potential, and the negative voltage indicates that the reverse chemical reaction is spontaneous. In this case, the electrode of the voltaic cell should be written in a reversed order.

We can now make a general statement.

For a given reaction, the $E_{cell}^{\ominus}$ value calculated by using two half-cell potentials can be positive or negative, and the sign of the $E_{cell}^{\ominus}$ value has the following interpretation.

$E_{cell}^{\ominus} > 0$ Under standard conditions, the reaction proceeds spontaneously in the forward direction.

$E_{cell}^{\ominus} < 0$ Under standard conditions, the reaction proceeds spontaneously in the reverse direction.

$E_{cell}^{\ominus} = 0$ Under standard conditions, the reaction is at equilibrium.

10.5   Electrochemistry and Thermodynamics

Cell EMF and Free Energy

Recall that spontaneous redox reactions have negative free energy changes. And now we know that spontaneous redox reactions have positive emf. Thus a reaction that has a negative free energy generates a positive emf. The linkage between free energy and cell potentials can be described by the following equation.

$$\Delta G = -nFE_{cell} \tag{10.3}$$

In this equation n is the number of moles of electrons transferred in the reaction and F is Faraday's constant, named after Michael Faraday. Faraday's constant is the charge carried by one mole of electrons and has the currently accepted value.

$$F = 96485.3399 \text{ coulomb/mole}$$

For reactions in which reactants and products are in their standard states, the equation above can be modified to give

$$\Delta G^{\ominus} = -nFE_{cell}^{\ominus} \tag{10.4}$$

Both n and F are positive quantities. A positive $E_{cell}^{\ominus}$ value will lead to a negative value of $\Delta G^{\ominus}$ which is the condition for spontaneity. Equation 10.4 relates the standard free energy change $\Delta G^{\ominus}$ for a reaction to its standard cell potential, $E_{cell}^{\ominus}$. It confirms that spontaneous redox reactions have positive electric potentials and negative free energy changes.

Standard EMF Cell and Chemical Equilibrium

We have known that the standard free-energy change $\Delta G^{\ominus}$ for a reaction is also related to its standard equilibrium constant, $K^{\ominus}$ [Equation (6.7) $\Delta G^{\ominus} = -RT\ln K^{\ominus}$].

Combining these two equations and grouping the constants, we obtain

$$-nFE_{cell}^{\ominus} = -RT\ln K^{\ominus}$$

$$E_{cell}^{\ominus} = \frac{RT}{nF}\ln K^{\ominus} \tag{10.5}$$

$$E_{cell}^{\ominus} = \frac{2.303RT}{nF}\lg K^{\ominus} \tag{10.6}$$

Substitution of values for the constants R and F at 298K gives

$$E_{cell}^{\ominus} = \frac{0.0592}{n}\lg K^{\ominus} \tag{10.7}$$

Thus, we build the relationships among $\Delta G^{\ominus}$, $E_{cell}^{\ominus}$, and $K^{\ominus}$ (Table 10.2).

If any one of the three quantities is known, the other two can be calculated using equation 10.4 or equation 10.7.

Table 10.2 Relationships among $\Delta G^\ominus$, $E^\ominus_{cell}$, and $K^\ominus$

$\Delta G^\ominus$	$K^\ominus$	$E^\ominus_{cell}$	Reaction at Standard-State Conditions
Negative	>1	Positive	Favors formation of products.
0	=1	0	Reactants and products are equally favored.
Positive	<1	Negative	Favors formation of reactants

Sample 10.4 Use standard reduction potentials to calculate the equilibrium constant for the reaction below.

$$2Cr^{3+}(aq) + 3Ni(s) \rightleftharpoons 2Cr(s) + 3Ni^{2+}(aq)$$

Solution:

The corresponding half-reactions and associated standard reduction potentials are

Reduction $Cr^{3+} + 3e^- \rightleftharpoons Cr(s)$ $E^\ominus_{Cr^{3+}/Cr} = -0.744$ V (1)

Oxidation $Ni^{2+} + 2e^- \rightleftharpoons Ni(s)$ $E^\ominus_{Ni^{2+}/Ni} = -0.26$ V (2)

The overall reaction can be obtained by $2 \times (1) - 3 \times (2)$.

The balanced overall reaction shows that $n = 6$.

We could therefore calculate the standard emf, $E^\ominus_{cell}$, of the above cell as follows.

$$E^\ominus_{cell} = E^\ominus_{Cr^{3+}/Cr} - E^\ominus_{Ni^{2+}/Ni}$$

$$E^\ominus_{cell} = -0.744 - (-0.26) = -0.484 \text{ V}$$

Use equation 10.7, we can calculate the equilibrium constant.

$$E^\ominus_{cell} = \frac{0.0592}{n} \lg K^\ominus$$

$$\lg K^\ominus = \frac{nE^\ominus_{cell}}{0.0592} = \frac{6 \times (-0.484)}{0.0592} = -49.05$$

$$K^\ominus = 8.83 \times 10^{-50}$$

10.6 Effect of Concentration on Cell EMF

In general, a real voltaic cell will differ from the standard conditions (concentrations not 1 molar and/or pressures not 1 atmosphere), so we need to be able to adjust the calculated cell potential at the non-standard state conditions to account for the differences. This can be done with the application of the Nernst equation described by the following equation.

$$E_{cell} = E^\ominus_{cell} - \frac{RT}{nF} \ln Q \quad (10.8)$$

In the Nernst equation, E_{cell} is the cell potential at the non-standard state conditions, $E^\ominus_{cell}$ is the cell potential when the reaction is at standard-state conditions; R is the ideal gas constant in units of joules per mole and has the value of 8.314 J/(mol·K); T is the temperature in kelvin; n is the number of moles of electrons transferred in the balanced equation for the reaction, F is Faraday's constant, and Q is the thermodynamic reaction quotient.

Substitution of values for the constants R and F at 298K gives

$$E_{cell} = E^\ominus_{cell} - \frac{0.0592}{n} \lg Q \quad (10.9)$$

The thermodynamic reaction quotient, Q, has the same equation as the equilibrium constant expression for that reaction, but Q is calculated using the current concentrations or pressures, not the equilibrium ones.

For a reaction

$$aA+bB \rightleftharpoons cC+dD$$
$$\text{reactants} \quad \text{products}$$

the reaction quotient has the form

$$Q=\frac{(c_C)^c(c_D)^d}{(c_A)^a(c_B)^b}$$

where (c_C) is understood to be the molar concentration of product C, or the numerical value of the partial pressure in atm if it is a gas.

So the reaction quotient is defined as the concentrations of the products over the reactants raised to the power of their respective coefficients in the balanced reaction at any moment in time.

Some examples are given to illustrate how to write the expression for the reaction quotient.

For a gas-phase reaction, the thermo-dynamic reaction quotient Q is identical to Q_p.

$$N_2(g)+3H_2(g) \rightleftharpoons 2NH_3(g)$$

$$Q=Q_p=\frac{p_{NH_3}^2}{p_{N_2} p_{H_2}^3}$$

For a reaction that occurs in aqueous solution, Q is identical to Q_c.

$$Zn(s)+Cu^{2+}(aq) \rightleftharpoons Zn^{2+}(aq)+Cu(s)$$

$$Q=Q_c=\frac{c(Zn^{2+})}{c(Cu^{2+})}$$

Q is different from Q_p or Q_c when the reaction involves gases and aqueous solutions.

$$H_2(g)+Cu^{2+}(aq) \rightleftharpoons 2H^+(aq)+Cu(s)$$

$$Q=\frac{(c_{H^+})^2}{p_{H_2} c_{Cu^{2+}}}$$

When write the expression for the reaction quotient, we need to notice that

(1) The partial pressures or concentrations are raised to the power of the stoichiometric coefficient for the balanced reaction.

(2) Substances in a solid or pure liquid state are not included.

(3) Express gas concentrations as partial pressure in atm, and dissolved species in molar concentration.

The Nernst equation relates the cell potential to its standard cell potential. It allows us to calculate the cell potential of any votaic cell for any concentrations.

Sample 10.5 Predict whether the following reaction would proceed spontaneously as written at 298K.

$$2H^+(aq)+Zn(s) \longrightarrow H_2(g)+Zn^{2+}(aq)$$

given that $[Zn^{2+}]=1.0$ mol/L, $[H^+]=6\times10^{-6}$ mol/L, and $p_{H_2}=1.0$ atm.

Solution:

We can use the Nernst equation to calculate the value of E for the redox reaction to determine the spontaneity of the reaction.

The corresponding half-reactions and associated standard reduction potentials are

Reduction $\quad 2H^+(aq)+2e^- \longrightarrow H_2(g) \qquad E_{H^+/H_2}^{\ominus}=+0.00$ V

Oxidation $\quad Zn^{2+}(aq)+2e^- \longrightarrow Zn(s) \qquad E_{Zn^{2+}/Zn}^{\ominus}=-0.76$ V

We could therefore calculate the standard emf, $E_{cell}^{\ominus}$, of the above cell as follows.

$$E_{cell}^{\ominus}=E_{H^+/H_2}^{\ominus}-E_{Zn^{2+}/Zn}^{\ominus}$$

$$E_{cell}^{\ominus}=+0.00 \text{ V}-(-0.76)=+0.76 \text{ V}$$

The balanced half-reactions show that $n=2$. Now evaluate Q

$$Q=\frac{p_{H_2}\times c_{Zn^{2+}}}{(c_{H^+})^2}=\frac{(1.0\text{ atm})(1.0\text{M})}{(6\times10^{-6})^2}=2.78\times10^{10}$$

Now apply the Nernst equation

$$E_{cell}=E^\ominus_{cell}-\frac{RT}{nF}\ln Q$$

$$E_{cell}=0.76-\frac{0.0592}{2}\lg(2.78\times10^{10})$$

$$E_{cell}=0.45\text{ V}$$

Because E_{cell} is positive, the reaction is spontaneous in the direction written.

Sample 10.6 Predict the cell potential for the following reaction when the pressure of the oxygen gas is 2.50 atm, the hydrogen ion concentration is 0.10 mol/L, and the bromide ion concentration is 0.25 mol/L.

$$O_2(g)+4H^+(aq)+4Br^-(aq)\longrightarrow 2H_2O(l)+2Br_2(l)$$

Solution:

The corresponding half-reactions and associated standard reduction potentials are

Reduction $O_2(g)+4H^+(aq)+4e^-\longrightarrow 2H_2O(l)$ $E^\ominus_{O_2,H^+/H_2O}=+1.229$ V

Oxidation $2Br_2(l)+4e^-\longrightarrow 4Br^-(aq)$ $E^\ominus_{Br_2/Br^-}=+1.087$ V

We could therefore calculate the standard emf, $E^\ominus_{cell}$, of the above cell as follows.

$$E^\ominus_{cell}=E^\ominus_{O_2,H^+/H_2O}-E^\ominus_{Br_2/Br^-}$$

$$E^\ominus_{cell}=+1.229-(+1.087)=+0.142\text{ V}$$

The balanced half-reactions show that $n=2$. Water and bromine are both liquids, therefore they are not included in the calculation of Q. Now evaluate Q

$$Q=\frac{1}{p_{O_2}(c_{H^+})^4\times(c_{Br^-})^4}=\frac{1}{(2.50\text{ atm})(0.10\text{ M})^4(0.25\text{ M})^4}=1.02\times10^6$$

Now apply the Nernst equation

$$E_{cell}=E^\ominus_{cell}-\frac{RT}{nF}\ln Q$$

$$E_{cell}=0.142-\frac{0.05916}{4}\lg(1.02\times10^6)$$

$$E_{cell}=+0.063\text{ V}$$

10.7 Reduction Potential Diagrams of Elements (Latimer Diagrams)

Structure of Latimer Diagrams

Reduction potential diagrams (or Latimer diagrams) show the standard reduction potentials connecting various oxidation states of an element. This is a convenient method of conveying large amounts of reduction potential values for half reactions involving all the oxidation states of a given element. In a Latimer diagram for an element, the formulas of the species that represent each oxidation state of the element are written from left to right in order of decreasing oxidation number. That is the species in the highest oxidation state is written on the left end, the species in the lowest oxidation state on the extreme right. The standard electrode potential value for the reduction half reaction involving any pair of the species connected by a horizontal line is shown above the line.

Since electrode potential differs between an acidic and a basic solution, different diagrams are required depending on the pH of the solution. Conventionally, Latimer diagrams are constructed for an element in two popular forms: (a) "In acidic medium" and (b) "In alkaline medium" for the two extremes of pH=0 and pH=14 (effective hydrogen ion concentrations of 1mol/L and 10^{-14} mol/L, respectively).

As examples, diagrams for iron and chlorine are shown below.

acid medium $\quad FeO_4^{2-} \xrightarrow{+1.9} Fe^{3+} \xrightarrow{+0.771} Fe^{2+} \xrightarrow{-0.440} Fe$

alkaline medium $\quad FeO_4^{2-} \xrightarrow{+0.9} Fe(OH)_3 \xrightarrow{-0.56} Fe(OH)_2 \xrightarrow{-0.89} Fe$

acid medium

$$ClO_4^- \xrightarrow{+1.19} ClO_3^- \xrightarrow{+1.21} ClO_2^- \xrightarrow{+1.64} HClO \xrightarrow{+1.63} Cl_2 \xrightarrow{+1.35} Cl^-$$

alkaline medium

$$ClO_4^- \xrightarrow{+0.36} ClO_3^- \xrightarrow{+0.33} ClO_2^- \xrightarrow{+0.66} ClO^- \xrightarrow{+0.40} Cl_2 \xrightarrow{+1.35} Cl^-$$

Any pair of the species in diagram could easily be converted back to the redox reaction by writing first the predominant species present and then adding the other characteristic species in the acidic (H^+ and H_2O) and in alkaline (OH^- and H_2O) solutions, respectively. Charge balancing is affected using the appropriate number of electrons. Thus, the redox reaction for the ClO_4^-/ClO_3^- couple in acidic solution is

$$ClO_4^-(aq) + 2H^+(aq) + 2e^- \rightleftharpoons ClO_3^-(aq) + H_2O(l) \quad E^{\ominus} = +1.19 \text{ V}$$

And in basic solution will be

$$ClO_4^-(aq) + H_2O(l) + e^- \rightleftharpoons ClO_3^-(aq) + OH^-(aq) \quad E^{\ominus} = +0.36 \text{ V}$$

Variations arise in $E^{\ominus}$ values in the two media, due to the involvement of H^+ or OH^- in the steps. If the involvement is not there, the values remain the same ($E^{\ominus}_{Cl_2/Cl^-}$ is almost the same in the two diagrams).

Application of Latimer Diagrams

(1) Determine whether a specie in an intermediate oxidation state will disproportionate

A disproportionation reaction is one in which a single element is simultaneously oxidized and reduced.

Consider the Latimer diagram for copper in acid solution shown below

$$Cu^{3+} \xrightarrow{+1.8 \text{ V}} Cu^{2+} \xrightarrow{+0.159} Cu^+ \xrightarrow{+0.520} Cu$$

Determine whether Cu^+ will disproportionate (convert to the more oxidized species Cu^{2+} to its right and the more reduced species Cu to its left).

Let us consider both of these processes.

$$Cu^+(aq) + e^- \rightleftharpoons Cu(s) \qquad E^{\ominus}_{Cu^+/Cu} = +0.521 \text{ V}$$
$$-)Cu^{2+}(aq) + e^- \rightleftharpoons Cu^+(aq) \qquad E^{\ominus}_{Cu^{2+}/Cu^+} = +0.153 \text{ V}$$
$$2Cu^+(aq) \rightleftharpoons Cu^{2+} + Cu(s) \qquad E^{\ominus}_{cell} = +0.368 \text{ V}$$

A positive $E^{\ominus}_{cell}$ value indicates that the final reaction is spontaneous, i.e., the Cu^+ is unstable towards disproportionation.

Sample 10.7 Determine whether Fe^{2+} will disproportionate in acid solution?

Solution: The Latimer diagram for iron in acid solution shown below

$$Fe^{3+} \xrightarrow{+0.771 \text{ V}} Fe^{2+} \xrightarrow{-0.440} Fe$$

Let us consider both of these processes.

$$Fe^{2+}(aq) + 2e^- \longrightarrow Fe(aq) \qquad E^{\ominus}_{Fe^{2+}/Fe} = -0.447 \text{ V}$$

$$-)2Fe^{3+}(aq)+2e^- \longrightarrow 2Fe^{2+}(aq) \quad E^{\ominus}_{Fe^{3+}/Fe^{2+}}=+0.771 \text{ V}$$

$$3Fe^{2+}(aq)=2Fe^{3+}(aq)+Fe(s) \quad E^{\ominus}_{cell}=-1.218 \text{ V}$$

A negative $E^{\ominus}_{cell}$ value indicates that the final reaction is nonspontaneous, i.e., Fe^{2+} will not disproportionate in acid solution.

One can determine whether a specie in an intermediate oxidation state will disproportionation simply by comparing the redox potential for its formation vs. the redox potential for its reduction. Disproportionation occurs when the $E^{\ominus}$ value on left side is less than that on right.

(2) Calculate the standard reduction potential between remote oxidation states

If we're given

$$Fe^{3+} \xrightarrow{+0.771 \text{ V}} Fe^{2+} \xrightarrow{-0.447} Fe$$
$$\underbrace{\qquad\qquad\qquad\qquad}_{E^{\ominus}_{Fe^{3+}/Fe}}$$

Then, how do we calculate $E^{\ominus}_{Fe^{3+}/Fe}$, the so-called skip potential?

Remember the equation
$$\Delta G^{\ominus}=-nFE^{\ominus}_{cell} \qquad (10.4)$$

Which describes the linkage between free energy and cell potentials?

We then remember that the free energies of chemical reactions, i.e. ΔG, obey the Hess' Law of Heat Summation. This allows us to combine the two equations to obtain the skip potential to an element. The calculations are summarized.

	$E^{\ominus}$	$\Delta G^{\ominus}$
$Fe^{3+}(aq)+e^- \longrightarrow Fe^{2+}(aq)$	$+0.771$ V	$\Delta G^{\ominus}_1=-n_1FE^{\ominus}_1$
$+)Fe^{2+}(aq)+2e^- \longrightarrow Fe(aq)$	-0.447 V	$\Delta G^{\ominus}_2=-n_2FE^{\ominus}_2$
$Fe^{3+}(aq)+3e^- \longrightarrow Fe(aq)$	$E^{\ominus}=?$	$\Delta G^{\ominus}=-nFE^{\ominus}$

$$\Delta G^{\ominus}=\Delta G^{\ominus}_1+\Delta G^{\ominus}_2$$
$$-nFE^{\ominus}=-n_1FE^{\ominus}_1+(-n_2FE^{\ominus}_2)$$

Where n_1, n_2, n are the number of transferred electrons, and $n=n_1+n_2$, Namely,

$$E^{\ominus}=\frac{n_1E^{\ominus}_1+n_2E^{\ominus}_2}{n_1+n_2}$$

Substitution of values,

$$E^{\ominus}_{Fe^{3+}/Fe}=\frac{1\times 0.771+2\times(-0.447)}{1+2}=-0.036 \text{ V}$$

Sample 10.8 Calculate the $E^{\ominus}_{ClO_3^-/ClO_2^-}$ according to the diagram shown below for Cl element.

$$ClO_3^- \xrightarrow{?} ClO_2^- \xrightarrow{+0.66} ClO^-$$
$$\underbrace{\qquad\qquad\qquad\qquad}_{+0.05}$$

Solution:

Two electrons are transferred to form ClO_2^- via ClO_3^- and two electrons are transferred to form ClO^- via ClO_2^-, therefore four electrons are totally transferred to form ClO^- via ClO_3^-.

$$-nFE^{\ominus}_{ClO_3^-/ClO^-}=-n_1FE^{\ominus}_{ClO_3^-/ClO_2^-}+(-n_2FE^{\ominus}_{ClO_2^-/ClO^-})$$

the potential becomes

$$E^{\ominus}_{ClO_3^-/ClO_2^-} = \frac{nE^{\ominus}_{ClO_3^-/ClO^-} - n_2 E^{\ominus}_{ClO_2^-/ClO^-}}{n - n_2} = \frac{4 \times 0.05 - 2 \times 0.66}{2} = 0.34 \text{ V}$$

Key Words

electrochemistry ［i:lektrəu'kemistri］ 电化学
anode ['ænəud] 阴极，负极
cathode ['kæθəud] 阳极，正极
redox reaction ［ri'dɔks ri'ækʃən］ 氧化还原反应
oxidizing agent ['ɔksidaiziŋ 'eidʒənt] 氧化剂
reducing agent ［ri'dju:siŋ 'eidʒənt］ 还原剂
oxidation number ['ɔksideiʃən 'nʌmbə]（or oxidation state）氧化数
monoatomic ['mɔnəuə'tɔmik] 单原子的
peroxides ［pə'rɔksaid］ 过氧化物
superoxide ［,sju:pə'ɔksaid］ 超氧化物
voltaic cell ［vɔl'teiik sel］ 原电池
galvanic cell ［gæl'vænik sel］ 原电池
electrode potential ［i'lektrəud pə'tenʃ(ə)l］ 电极电势
electrolytic cell ［i,lektrəu'litik sel］ 电解池
electromotive force ［əilektrəu'məutiv fɔ:s］ 电动势
potential difference ［pə'tenʃ(ə)l 'difərəns］ 电势差
Standard Hydrogen Electrode ［'stændəd 'haidrəudʒən i'lektrəud］ 标准氢电极

standard reduction potential ['stændəd ri'dʌkʃən pə'tenʃ(ə)l] 标准还原电势
Nernst Equation [nεənst i'kweiʃən] 能斯特方程
Latimer diagram [læ'timə 'daiəgræm] 莱铁莫尔电势图
disproportionation [,disprə,pɔ:ʃə'neiʃən] 歧化
copper sulfate ['kɔpə 'sʌlfeit] $CuSO_4$
zinc sulfate [ziŋk 'sʌlfeit] $ZnSO_4$
ferrous sulfate ['ferəs 'sʌlfeit] $FeSO_4$
hydrogen peroxide ['haidrəudʒən pə'rɔksaid] H_2O_2
manganese dioxide [,mæŋgə'ni:z, 'mæŋgəni:z dai'ɔksaid] MnO_2
potassium dichromate [pə'tæsjəm dai'krəumeit] $K_2Cr_2O_7$
potassium permanganate [pə'tæsjəm pə:'mæŋgənit, -neit] $KMnO_4$
sodium bromide ['səudjəm, -diəm 'brəumaid] $NaBr$
sodium iodide ['səudjəm 'aiədaid] NaI
sodium sulfite ['səudjəm 'sʌlfait] Na_2SO_3

Exercises

1. Determine the oxidation numbers of all elements in each of the following substances.

(1) $Na_2S_2O_3$; (2) C_2H_4; (3) C_2H_2; (4) Fe_3O_4; (5) $C_6H_8O_7$.

2. Use the oxidation Number method to balance the following equations.

* in basic solution, * * in neutral solution, others in acidic solution.

(1) $BrO_3^- + I^- = Br^- + I_2$

(2) $MnO_4^- + C_2O_4^{2-} = Mn^{2+} + CO_2$

(3) * * $MnO_4^- + Mn^{2+} = MnO_2$

(4) * * $Cl_2 + I_2 = IO_3^- + Cl^-$

(5) $PbO_2 + Mn^{2+} = MnO_4^- + Pb^{2+}$

(6) * $CrO_2^- + H_2O_2 = CrO_4^{2-}$

(7) $MnO_4^- + H_2O_2 = Mn^{2+} + O_2$

(8) * $MnO_4^- + SO_3^{2-} = MnO_4^{2-} + SO_4^{2-}$

(9) $NO_2^- + I^- = NO + I_2$

(10) ** $CH_2\text{—}CH\text{—}CH_2 + MnO_4^- = CO_3^{2-} + MnO_4^{2-}$
$\quad\quad\quad |\quad\quad |\quad\quad |$
$\quad\quad OH\ \ OH\ \ OH$

3. Balance the following equation by Ion-Electron Method.

(1) $BiO_3^- + Mn^{2+} + H^+ \longrightarrow Bi^{3+} + MnO_4^- + H_2O$

(2) $S_2O_3^{2-} + I_2 \longrightarrow S_4O_6^{2-} + I^-$

(3) $H_2O_2 + H^+ + I^- \longrightarrow I_2 + H_2O$

(4) $S_2O_8^{2-} + Mn^{2+} + H_2O \longrightarrow MnO_4^- + SO_4^{2-} + H^+$

(5) $Cr^{3+} + MnO_4^- + H_2O \longrightarrow Cr_2O_7^{2-} + Mn^{2+} + H^+$

(6) $MnO_4^- + H_2S + H^+ \longrightarrow Mn^{2+} + S + H_2O$

(7) $Mn^{3+} + H_2O \longrightarrow MnO_2 + Mn^{2+} + H^+$

(8) $HCOO^- + MnO_4^- + OH^- \longrightarrow CO_3^{2-} + MnO_4^{2-} + H_2O$

(9) $I_2 + OH^- \longrightarrow I^- + IO_3^- + H_2O$

(10) $MnO_4^{2-} + H_2O \longrightarrow MnO_4^- + MnO_2 + OH^-$

4. According to the following cell notations, write the corresponding half-cell reactions and overall cell reactions.

(1) $Pt|H_2(100\ kPa)|OH^-(1.0\ mol/L)\ ||\ H^+(1.0\ mol/L)|H_2(100\ kPa)|Pt$

(2) $Ag|AgCl|Cl^-(1.0\ mol/L)\ ||\ Ag^+(1.0\ mol/L)|Ag$

(3) $Pt|Cu^{2+}(1.0\ mol/L),\ Cu^+(1.0\ mol/L)\ ||\ Cu^{2+}(1.0\ mol/L),\ Cl^-(1.0\ mol/L)|CuCl\ |Pt$

(4) $Ni|Ni(NH_3)_4^{2+}(1.0\ mol/L),\ NH_3(1.0\ mol/L)\ ||\ Ni^{2+}(1.0\ mol/L)|Ni$

5. Write the shorthand notations for the voltaic cells using the following reactions.

(1) $H^+ + Zn \longrightarrow Zn^{2+} + H_2$

(2) $Fe^{3+} + Hg(l) + Cl^- \longrightarrow Hg_2Cl_2(s) + Fe^{2+}$

(3) $Pb^{2+} + CrO_4^{2-} \longrightarrow PbCrO_4(s)$

(4) $Ag^+ + 2NH_3 \longrightarrow Ag(NH_3)_2^+$

6. Under standard conditions, predict whether the following reactions at 298.15K will be spontaneous in the forward direction by using the potentials of half-reactions to calculate the emf associated with it.

(1) $Ag + Cu^{2+} \longrightarrow Cu + Ag^+$

(2) $Sn^{4+} + I^- \longrightarrow Sn^{2+} + I_2$

(3) $Fe^{2+} + H^+ + O_2 \longrightarrow Fe^{3+} + H_2O$

(4) $S + OH^- \longrightarrow S^{2-} + SO_3^{2-} + H_2O$

7. According to the following cell diagrams, write the corresponding half-cell reactions and overall cell reactions. Calculate $E_{cell}^{\ominus}$ and $\Delta_r G_m^{\ominus}$ for these cell reactions.

(1) $Ni|Ni^{2+}(1.00\ mol/L)\ ||\ Cu^{2+}(1.00\ mol/L)|Cu$

(2) $Fe|Fe^{2+}(1.00\ mol/L)\ ||\ Cl^-(1.00\ mol/L)|Cl_2(100\ kPa)|Pt$

Answer: (1) 0.5989 V, −115587 J/mol (2) 1.8053 V, −348423 J/mol

8. Calculate the values of the equilibrium constants at 298.15K for the following reactions.

(1) $Ag^+ + Fe^{2+} \rightleftharpoons Ag + Fe^{3+}$

(2) $5Br^- + BrO_3^- + 6H^+ \rightleftharpoons 3Br_2 + 3H_2O$

Answer: (1) 2.97 (2) 3.63×10^{76}

9. At 298.15K, knowing that K_{sp} for $Ag_2C_2O_4$ is 3.5×10^{-11}, and $E_{Ag^+/Ag}^{\ominus} =$

$+0.799$ V, calculate the $E^{\ominus}_{Ag_2C_2O_4/Ag}$ for the reaction: $Ag_2C_2O_4(s) + 2e \rightleftharpoons 2Ag + C_2O_4^{2-}$.

Answer: 0.490 V

10. Choose a proper oxidation agent from $Fe_2(SO_4)_3$ and $KMnO_4$ that will convert I^- to I_2 without converting Br^- to $Br_2(s)$ and Cl^- to Cl_2.

11. The following reaction is often used to prepare Cl_2 gas in laboratory

$$MnO_2(s) + 4HCl \longrightarrow MnCl_2 + Cl_2 + 2H_2O$$

(1) Under standard conditions, predict whether the reaction at 298.15 K will be spontaneous in the forward direction by using tabulated standard reduction potentials.

(2) Using 12.0 mol/L HCl solution, predict whether the reaction will be spontaneous in the forward direction? (Given that, $p_{Cl_2} = 100$ kPa, $c_{Mn^{2+}} = 1.00$ mol/L).

12. The galvanic cell shown as
Pt|H_2(100 kPa)|HA(0.500 mol/L) ∥ Cl^-(1.00 mol/L)|AgCl|Ag has a cell potential of 0.568 V at 298.15K. Calculate the K_a of the weak acid HA. ($E^{\ominus}_{AgCl/Ag} = +0.2223$ V).

13. Here are two standard reduction potentials

$$Cu^{2+} + e^- \rightleftharpoons Cu^+ \quad E^{\ominus} = 0.153 \text{ V}$$
$$Cu^{2+} + I^- + e^- \rightleftharpoons CuI(s) \quad E^{\ominus} = 0.860 \text{ V}$$

Calculate the K_{sp} of CuI at 298.15K.

Answer: 1.44×10^{-12}

14. Knowing that: $E^{\ominus}_A/V \quad Fe^{3+} \xrightarrow{0.771} Fe^{2+} \xrightarrow{-0.447} Fe$ and $E^{\ominus}_{Cu^{2+}/Cu} = +0.340$ V, explain why Fe can reduce Cu^{2+}, while Cu can reduce Fe^{3+}.

15. At 298.15K, consider the following reduction potentials

$$AgI(s) + e^- \rightleftharpoons Ag(s) + I^- \quad E^{\ominus}_{AgI/Ag} = -0.152 \text{ V}$$
$$Ag^+ + e^- \rightleftharpoons Ag(s) \quad E^{\ominus}_{Ag^+/Ag} = +0.799 \text{ V}$$

Calculate the K_{sp} of AgI.

Answer: 8.63×10^{-17}

16. At 298.15K, $Cu^{2+} + 2e^- \rightleftharpoons Cu \quad E^{\ominus} = 0.340$ V $\quad Cu^{2+} + e^- \rightleftharpoons Cu^+ \quad E^{\ominus} = 0.150$ V

(1) Can Cu^+ undergo disproportionation reaction? Explain why. (2) Calculate the equilibrium constant for the reaction $Cu^{2+} + Cu \rightleftharpoons 2Cu^+$. (3) If the $K_{sp} = 1.20 \times 10^{-6}$ for CuCl, calculate the equilibrium constant for the reaction $Cu^{2+} + Cu(s) + 2Cl^- \rightleftharpoons 2CuCl(s)$.

Answer: (2) 3.81×10^{-7} (3) 2.64×10^5

17. Based on the standard reduction potentials and on cell diagrams, explain the following phenomena.

(1) $E^{\ominus}_A/V \quad Fe^{3+} \xrightarrow{0.771} Fe^{2+} \xrightarrow{-0.447} Fe$ and $E^{\ominus}_{O_2/H_2O} = 1.299$ V, scrap iron is added to $FeSO_4$ solution in order to prevent that $FeSO_4$ is oxidized by O_2 in the air.

(2) $E^{\ominus}_A/V \quad Sn^{4+} \xrightarrow{0.15} Sn^{2+} \xrightarrow{-0.136} Sn$ and $E^{\ominus}_{O_2/H_2O} = 1.299$ V, Sn is added to $SnCl_2$ solution in order to prevent that $SnCl_2$ is oxidized by O_2 in the air.

Chapter 11 Chemistry of Coordination Compounds

Coordination compounds play an essential role in the methods of separation used in qualitative and quantitative chemical analysis, in various industrial processes and in life itself. In fact, knowledge of the properties of coordination compounds is necessary to the understanding of the chemistry of the metals. Coordination chemistry is the study of compounds formed between metal ions and other neutral or negatively charged molecules. The formation of coordination compound and some special feature of coordination compounds will be discussed in this chapter.

11.1 The History of Coordination Chemistry

Discovery of Coordination Compound

It is difficult for us to state exactly when the first metal complex was discovered. Perhaps the earliest one on record is Prussian blue, $KCN \cdot Fe(CN)_2 \cdot Fe(CN)_3$, which obtained by the artist color maker Diesbach, at the beginning of the 18th century. However, the date usually cited is that of the discovery of hexaamminecobalt(Ⅲ) chloride, $CoCl_3 \cdot 6NH_3$, which was most likely prepared in the late 1700s by Tassaert, a French chemist. He observed that NH_3 combined with Co^{3+} to yield a reddish brown product. His experimental observations could not be explained on the basis of the theory available at that time. Because it was necessary to understand how $CoCl_3$ and NH_3, each have a stable compound of saturate valence, could combine to make yet another very stable compound. This discovery marks the real beginning of coordination chemistry, but the answer was not to be found until about 100 years later.

Note that, during that time many such compounds were synthesized and their properties were studied by many chemists. For example, Cobalt ammoniate oxalate was prepared in 1822 by Gmelin, $CoCl_3 \cdot 5NH_3$ and other cobalt ammoniate were prepared by Claudet and Fremy individually. Much of the early work was done with ammonia, and the resulting compounds were known as metal ammines. It was also found that other amines and anions such as CN^-, NO_2^-, NCS^-, and Cl^- could form stable complex. Many compounds were prepared from these anions, and it was very interesting to note that each compound was named after the chemist who originally prepared it (see Table 11.1).

Table 11.1 Compounds named after chemist discoveries

Coordination Compound	Name	Coordination Compound	Name
$NH_4[Cr(NH_3)_2(NCS)_4]$	Reineck's salt	$[Pt(NH_3)_4][PtCl_4]$	Magnus's green salt
$K[Co(NH_3)_2(NO_2)_4]$	Erdman's salt	$K[Pt(C_2H_4)Cl_3]$	Zeise's salt

Later on, the name of these compounds was to name on the basis of color (Table 11.2), owing to many of the compounds are colored. Some of these names are still used,

Table 11.2 Compounds named according to their color

Coordination Compound	Name	Color	Coordination Compound	Name	Color
$[Co(NH_3)_5Cl]Cl_2$	Purpureocobaltic chloride	purple	cis-$[Co(NH_3)_4Cl_2]Cl$	Violeocobaltic chloride	violet
trans-$[Co(NH_3)_4Cl_2]Cl$	Praseocobaltic chloride	green	$[Co(NH_3)_5H_2O]Cl_3$	Roseocobaltic chloride	red

but it soon found not satisfactory, and it had to be abandoned. The nomenclature system now used will be introduced in section 11.3.

Other important early observations were that certain complex compounds might exist in two different forms having the same chemical composition, and differ in physical and chemical properties. For example, the α and β forms of $PtCl_2 \cdot 2NH_3$ are both a cream color, but they differ in solubility and in chemical reactivity.

It was necessary to account for all of these experimental facts with a suitable theory. However, all these early experimental observations could not be explained on the basis of the chemical theory available at that time. Several hypotheses and theories proposed were only to be discarded because they were inadequate to explain most subsequent experimental data. For example, the Blomstrand-Jorgensen chain theory, they tried to extend the established ideas of organic chemistry to account for the newer coordination compounds. The chain theory was used rather extensively and then proved to be wrong. The next, we shall discuss the coordination theory of Werner, which has provides a suitable explanation for most of the existence and behavior of metal coordination compounds.

The Modern Coordination Chemistry

The discovery and explanation of coordination compounds have been viewed against the larger picture of progress in understanding atomic structure, the periodic table and molecular bonding. Our present understanding of the true nature of coordination compound is due to the ingenious Swiss chemist, named Alfred Werner (1866~1919) and winner of a Nobel Prize in 1913 for his brilliant contribution to inorganic chemistry.

In 1893, at the age of only 26, Alfred Werner proposed what is now commonly called as Werner's coordination theory or the modern coordination chemistry. Working with the cobalt ammonia and other related series including Cr and Pt, he proposed that most elements exhibit two types of valence (combining ability), a primary valance and a secondary valence. The primary valence refers to what we now call the oxidation number of the metal, and the secondary valence corresponds to what we now call the coordination number. Every element tends to satisfy both its primary and secondary valence. The secondary valence is directed toward fixed positions in space, and this is also the basis for the stereochemistry of coordination compounds. For example, Werner explained that the compound originally written as $CoCl_3 \cdot 6NH_3$ is composed of the ion $[Co(NH_3)_6]^{3+}$, the primary valence or oxidation number of Co is 3. The three Cl^- ions saturate the primary valence or oxidation state of cobalt, the ions that neutralize the charge of the metal ion use the primary valence. With six NH_3 directly bonded to the Co^{3+}, the Co^{3+} has a secondary valence of 6. The ammonia molecules use the secondary valence and they are said to be coordinated to the metal. The compound, therefore, should be formulated as $[Co(NH_3)_6]Cl_3$, which states that both the primary and secondary valence are satisfied.

It is very interesting to note that Werner based the theory on a great variety of experimental evidence. Table 11.3 lists some cobalt compounds studied by Werner. According to his theory, these compounds consist of two kinds of Cl^- ions, those that can be precipitated immediately as AgCl by using $AgNO_3$ and those that not. Such as, one mole of $CoCl \cdot 5NH_3$ reacts with 2 mde of $AgNO_3$ to give a precipitate of two mole of AgCl. He knew that one Cl^- ion is strongly attached to the Co^{3+} ion and could not be precipitated by Ag^+ ion. The compound should, therefore, be formulated as $[Co(NH_3)_5Cl]Cl_2$. The number of free Cl⁻ ion determined this way agreed exactly with his theory.

Table 11.3 Some cobalt coordination compounds studied by Werner

Complex	Modern Formula	Number of Cl⁻ Ions	Number of Ions
$CoCl_3 \cdot 6NH_3$	$[Co(NH_3)_6]Cl_3$	3	4
$CoCl_3 \cdot 5NH_3$	$[Co(NH_3)_5Cl]Cl_2$	2	3
$CoCl_3 \cdot 4NH_3$	$[Co(NH_3)_4Cl_2]Cl$	1	2
$CoCl_3 \cdot 4NH_3$	$[Co(NH_3)_4Cl_2]Cl$	1	2

Over about twenty years, Werner performed many experiments on coordination compounds and all of them well agreed with his theory. However, Werner's theory was made a long time before it was possible to directly determine the structure of complex ions by X-ray crystallography. His theory has been a guiding principle in modern inorganic chemistry and in the concept of valence.

11.2 Formation of Coordination Compound

Definition of Coordination Compound

Transition-metal ions often function in chemical reactions as Lewis acid, accepting electron pairs from other molecules or ions, which function as Lewis bases. For example, Fe^{3+} and H_2O can bond to one another in a Lewis acid-base reaction.

Here, a lone pair of electrons on the oxygen atom of H_2O forms a coordinate covalent bond to Fe^{3+}. In water, the Fe^{3+} ion ultimately bonds to six H_2O molecules to give the $Fe(H_2O)_6^{3+}$ ion. Such an ion that consists of a metal atom or ion with Lewis bases attached to it through coordinate covalent bonds is called a complex ion. The neutral molecules or ions that act as Lewis bases in the complex ions are called ligands (from the Latin word ligare, meaning "to bind") or complexing agents. Every ligand has at least one atom with a lone pair of valence electrons. The ligands are said to be coordinated to the metal ion and bond to a central metal atom (or ion) by coordinate bonds. The atom of the ligand bound directly to the metal is called the donor atom. For example, nitrogen is the donor atom in the $[Cu(NH_3)_4]^{2+}$ complex ion. The number of donor atoms attached to a metal is called the coordination number of the metal ion, which is termed as the total number of points of attachment to the central element. For example, Ag^+ in $[Ag(NH_3)_2]^+$ ion has a coordination number of 2, in $[Cu(NH_3)_4]^{2+}$ copper has a coordination number of 4, and in $[Cr(NH_3)_4Cl_2]^+$ chromium has a coordination number of 6. The coordination number can vary from 2 to as many as 16, depending on the size, charge, and electron configuration of the metal ion. As shown in table 11.4, some metal ions exhibit constant coordination numbers.

Table 11.4 Coordination number of some metal ions

Coordination Number	Metal Ion	Complex Ion
2	Ag^+, Cu^+	$[Ag(NH_3)_2]^+$, $[Cu(CN)_2]^-$
4	Cu^{2+}, Zn^{2+}, Cd^{2+}, Hg^{2+}, Al^{3+}, Sn^{2+}, Pb^{2+}, Co^{2+}, Ni^{2+}, Pt^{2+}, Fe^{2+}, Fe^{3+}	$[HgI_4]^{2-}$, $[Zn(CN)_4]^{2-}$, $[Pt(NH_3)_2Cl_2]$, $[Ni(CN)_4]^{2-}$
6	Cr^{3+}, Al^{3+}, Pt^{4+}, Fe^{2+}, Fe^{3+}, Co^{2+}, Co^{3+}, Ni^{2+}, Pb^{4+}	$[Co(NH_3)_3(H_2O)Cl_2]$, $[Fe(CN)_6]^{3-}$, $[Ni(NH_3)_6]^{2+}$, $[Cr(NH_3)_4Cl_2]^+$

Many transition metal ions show more than one coordination number. But 6 is the most common coordination number, follows closely by 4, with a few metal ions showing a coordination number of 2 or 5. Based on the above discussed, the definition of the coordination compounds could be concluded.

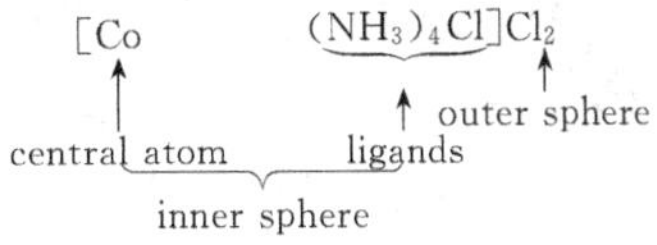

A coordination compound is a compound consisting either of complex ions, and counter ions, anions or cations to form a compound with no net charge or of a neutral complex species [Such as $Ni(CO)_5$]. In writing the chemical formula for a coordination compound, square brackets are used to set off the groups within the coordination sphere from other parts of the compound. $[Co(NH_3)_5Cl]Cl_2$ is a typical coordination compound. The square brackets indicate the composition of the complex ion, $[Co(NH_3)_5Cl]^+$ is called inner sphere. The two Cl^- counter ions are shown outside the brackets, and known as outer sphere. Note that in this compound one Cl^- act as a ligand along with the five NH_3 molecules (Figure 11.1).

Figure 11.1 The composition of $[Co(NH_3)_5Cl]Cl_2$

The net charge of a complex ion is the sum of the charges on the central metal atom and its surrounding ligands. In the $[CuCl_2]^-$ ion, for example, each Cl^- ion has an oxidation number of -1, so the oxidation number of Cu must be $+1$. If the ligands do not bear net charges, the oxidation number of the metal is equal to the charge of the complex ion. Thus in $[Cu(NH_3)_4]^{2+}$ each NH_3 is neutral, so the oxidation number of Cu is $+2$.

Classification of Ligand

Depending on the number of donor atoms present, ligands are classified as monodentate, bidentate, or polydentate.

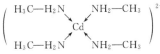

Figure 11.2 An example of monodentate ligands

1. Monodentate ligands

The ligands that we have discussed so far, bond to the metal atom through only one atom of the ligand, which is called a monodentate ligand or a unidentate ligand. The word monodentate comes from Greek monos and the Latin dentis. It means "one-toothed". A monodentate ligand has only one pair of electrons with which to attach to the metal center and occupy only one coordination site. Examples of monodentate ligands include NH_3, H_2O, CO and other neutral two-electron donors, such as, methylamine (Figure 11.2). As for ammonia, it bonds through the nitrogen atom and water through the oxygen atom. They are the unidentate ligands.

2. Polydentate Ligands

Some ligands, however, have more than one atom with a lone pair that can be used to bond to a metal ion. A bidentate ligand (two toothed ligand) is a ligand that bonds to a metal atom through two atoms of ligand. One such ligand is ethylenediamine, $H_2N-CH_2-CH_2-NH_2$ (commonly abbreviated "en"). Two nitrogen atoms at the ends of this molecule have also lone pairs of electrons, and both nitrogen atoms can coordinate with a metal atom (Figure 11.3).

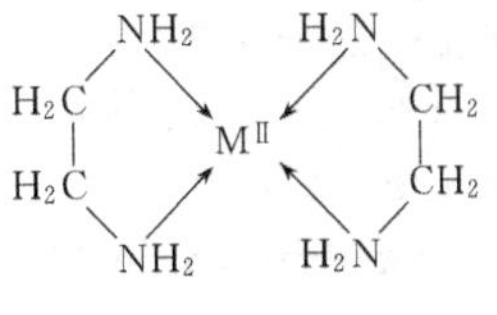

Some ligands can form more than two coordination bonds to a metal atom or ion and occupy more than one coordination site are called polydentate ligands (also termed multidentate ligands). Coordination compounds formed by bidentate ligands or polydentate ligands are called chelates, terms derived from the Greek chele, meaning "claw", because such ligands appear to grasp the metal ion of two or more donor

Figure 11.3 Structure of metal-ethylenediamine

atoms in the manner that resulting configuration rather resembling a crab clutching at its prey. Such ligands are called chelating agents.

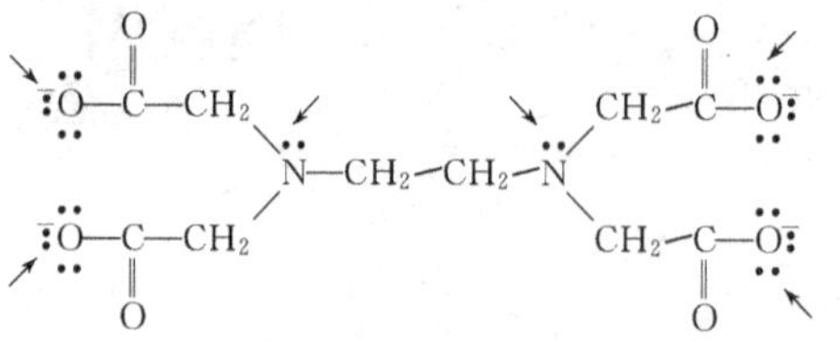

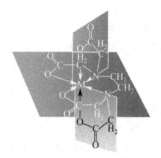

Figure 11.4 EDTA is a hexdentate ligand

In general, polydentate ligands form more stable complexes than do related monodentate ligands. Some polydentate ligands can form as many as six bonds to a metal. One example of such ligand is ethylenediaminotetraacetate ion $(-OOC-CH_2)_2N-CH_2-CH_2-N(CH_2-COO-)_2$, abbreviated EDTA, has six donor atoms, as shown in Figure. 11.4. This ligand virtual surrounds the metal ion (Figure 11.5 and Figure 11.6), coordinating to it through six atoms, two nitrogens and four oxygens (one from each carboxylate). So it is a hexdentate ligand. As might be expected from the large number of coordination sites, EDTA forms very stable complex ions with most metal ions. It is therefore used as a "scavenger" to remove toxic heavy metals such as lead from the blood and excreted from our body. EDTA is also used to clean up spills of radioactive metals and a reagent to analyze solutions for the metal ion content.

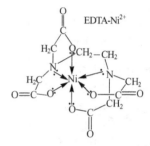

Figure 11.5 The coordination of EDTA with a 2+ transition metal ion

Figure. 11.6 The structure of [NiEDTA]$^-$ ion

11.3 Nomenclature of Coordination Compounds

Now that we have discussed the various types of ligands, the composition of coordination compound and the oxidation numbers of metals, our next step is to learn what to call these coordination compounds correctly. The rules for a systematic naming of coordination compounds are outlined below.

(1) In naming the entire coordination compound, the cation is named before the anion (just as for sodium chloride), no matter whether the cation or the anion is the complex species. For example, $K[PtCl_3NH_3]$ is called potassium trichloroammine-platinate(II).

(2) The name of the complex ion, whether anion, cation, or neutral species, consists of two parts written together as one word. Ligands are always named before the metal ion. For example, $Fe(CN)_4^{4-}$ is named hexacyanoferrate(III).

(3) In naming ligands, an "o" is added to the root name after anion. For a neutral ligand the name of the molecule is used, with the exception of H_2O, NH, CO, and NO, as illustrated in Table 11.5.

(4) The Greek prefixes mono-(1)(usually omitted), di-(2), tri-(3), tetra-(4), penta-(5), and hexa-(6) are used to denote the number of simple ligands. The prefixes bis-, tris-, tetrakis-, pentakis- and hexakis- are also used, especially for more complicated ligands or ones that already contain di-, tri-, and so on. For example, the name for

[Co(en)$_3$]Br$_3$ is tris (ethylenediamine) cobalt(Ⅲ) bromide.

Table 11.5 Names of Some Common Unidentate Ligands

Anions				Neutral Molecules	
F$^-$	fluoro	O^{2-}	oxo	H$_2$O	aqua
Cl$^-$	chloro	CO$_3^{2-}$	carbonato	NH$_3$	amine
Br$^-$	bromo	C$_2$O$_4^{2-}$	oxalato	CO	carbonyl
I$^-$	iodo	SO$_3^{2-}$	sulfito	NO	nitrosyl
CN$^-$	cyano	SO$_4^{2-}$	sulfato	CH$_3$NH$_2$	methylamine
OH$^-$	hydroxo	S$_2$O$_3^{2-}$	thio sulfato		
NO$_2^-$	nitro	NO$_3^-$	nitrato		
ONO$^-$	nitrito	SCN$^-$	thiocyanato		

(5) A Roman numeral in parentheses is used to indicate the oxidation state of the central metal ion, and it follows the name of the central metal (an oxidation state of zero is indicated by 0 on parentheses).

(6) When more than one type of ligands is present, ligands are named in alphabetical order. Prefixes do not affect the order. For example, NH$_3$ (ammine) would be come before H$_2$O (aqua). Thus, [Co(NH$_3$)$_5$H$_2$O]Cl$_3$ is named pentaammineaquocobalt(Ⅲ) chloride.

(7) If the complex ion is negative, the name of the metal always ends in the suffix-ate. For example, Scandium (Sc = scandate), Chromium (Cr = chromate), Zinc (Zn = zincate), Nickel (Ni = nickelate) and so on. For some metal the Latin name is used. Some of these names are shown in Table 11.6.

Table 11.6 Names of some negative complex ions

English Name	Latin Name	Anion Name	English Name	Latin Name	Anion Name
Iron	Ferrum	Fe=Ferrate	Silver	Argentum	Ag=Argentate
Copper	Cuprum	Cu=Cuprate	Lead	Plumbum	Pb=Plumbate
Gold	Aurum	Au=Aurate	Tin	Stannum	Sn=Stannate

We have seen many instances of these usage in the examples given above. Some examples of coordination compound are given in Table 11.7.

Table 11.7 Names of selected coordination compounds

Coordination Compound	Name	Coordination Compound	Name
[Co(NH$_3$)$_5$Cl]Cl$_2$	pentaamminechlorocobalt(Ⅲ)chloride	K$_4$[PtCl$_6$]	potassium hexachloroplatinate(Ⅱ)
[Co(NH$_3$)$_6$]Cl$_2$	hexaamminecobalt(Ⅱ)chloride	[Cr(NH$_3$)$_3$Cl$_3$]	triamminetrichlorochromium(Ⅲ)chloride
[Co(NH$_3$)$_5$Cl$_2$]Cl	pentaamminedichlorocobalt(Ⅲ)chloride	[Ag(CN)$_2$]$^-$	dicyanoargenate(Ⅰ)ion
[Co(H$_2$O)$_6$]Cl$_3$	hexaaquocobalt(Ⅲ)chloride		

11.4 Structure and Isomerism

Compounds with the same formula may have different properties. These compounds have the same composition, but a different arrangement of atoms, and which exhibit different properties are called isomers. Isomerism, the existence of isomers, is a characteristic feature of both inorganic and organic compounds. Although isomers are composed of the same collection of atoms, they differ in one or more chemical or physical properties such as solubility, color, or rate of reaction with some reagents. There are two main types of isomers in coordination compounds, structural isomers (which have different bonds) and stereoisomers (which have the same bonds but different spatial arrangement of the bonds). Each of these classes also has subclasses (see Figure 11.7).

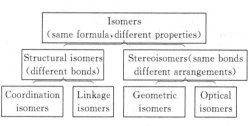

Figure 11.7 Forms of isomerism in coordination compounds

Structural (or constitutional) Isomers

A number of different types of structural isomerism are known in coordination chemistry. Figure 11.7 gives two examples, coordination isomerism and linkage isomerism.

The first kind of structural isomerism is coordination isomerism (also called coordination-sphere isomers). They differ in the ligands that are directly bonded to the metal, as opposed to being outside the complex ion in the solid lattice. For example, $CrCl_3 \cdot 6H_2O$ has three common forms.

$$[Cr(H_2O)_6]Cl_3 \text{ (a violet)}$$
$$[Cr(H_2O)_5Cl]Cl_2 \cdot H_2O \text{ (a green)}$$
$$[Cr(H_2O)_4Cl_2]Cl \cdot 2H_2O \text{ (also a green)}.$$

In the second and third compounds, the water has been displaced from the complex ion by chloride ions and occupies a site in the solid lattice.

The second type of structural isomerism is linkage isomerism, in which the composition of the complex ion is same. But the atom of a ligand that is coordination to the meal ion is different. In this case, a particular ligand is capable of coordinating to a metal in two different ways, such as thiocyanate, SCN^-, which can bond to the metal ions either with the electron pair on the nitrogen atom or with an electron pair on the sulfur atom.

Another ligand capable of coordinating through either of two donor atoms is NO_2^-, whose potential donor atoms are N and O. Because the nitrite ion, NO_2^-, can coordinate through either the nitrogen or an oxygen atom, as is shown in Figure 11.8. When it coordinates through the nitrogen atom (a), the NO_2^- ligand is called nitro. When it coordinates to the oxygen atom (b), it is called nitrite and is generally written ONO^-. The isomers differ in their chemical and physical properties. For example, the N-bonded isomer is yellow, whereas the O-bonded isomer is red. $[Co(NH_3)_5(ONO)]Cl_2$ and $[Co(NH_3)_5(NO_2)]Cl_2$ are linkage isomers.

Many other ligands are potentially capable of forming linkage isomers. Theoretically, all that is required is that two different atorms of the ligands contain an unshared electron pair.

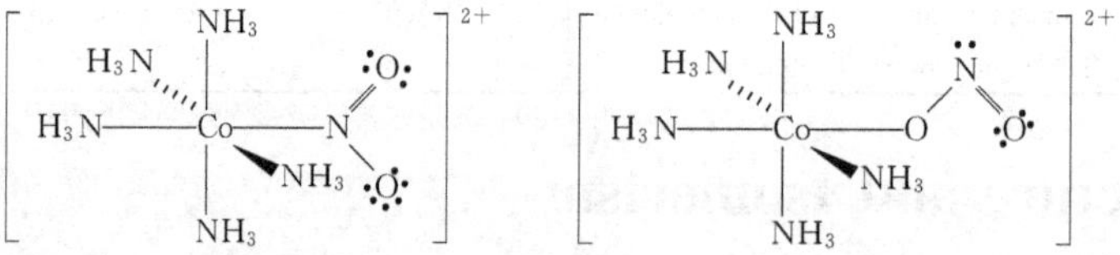

Figure 11.8 (a) N-bound; (b) O-bound isomers of $[Co(NH_3)_5NO_2]^{2+}$

Stereoisomers

Stereoisomerism is the most important form of isomerism. Stereoisomers have the same chemical bonds but exhibit different spatial arrangement. For example, consider in $[Pt(NH_3)Cl_2]$ the Cl^- ligands can be either adjacent to or opposite each other, as shown in Figure 11.9. This form of isomerism, in which the arrangement of the constituent atoms is different though the same bonds are present, is called geometric isomerism. Isomer (a), with like ligands in adjacent positions, is called the cis isomer. Isomer (b), with like ligands across from each other, is called the trans isomer.

In general, geometric isomers have different solubility, melting points, and boiling points. They may also have different chemical reactivities. For example, cis-$[Pt(NH_3)Cl_2]$ (called Cisplatin, yellow), Figure 11.9 (a), is effective in treatment of many cancers, whereas the trans-$[Pt(NH_3)Cl_2]$ (pale yellow) isomer is ineffective.

$$
\begin{array}{cc}
H_3N \diagdown \diagup Cl & Cl \diagdown \diagup NH_3 \\
Pt & Pt \\
H_3N \diagup \diagdown Cl & H_3N \diagup \diagdown Cl \\
(a) & (b)
\end{array}
$$

Figure 11.9 (a) cis; (b) trans geometric of square-planar $[Pt(NH_3)_2Cl_2]$

The cis and trans isomers of the tetraammine dichlorocobalt (Ⅲ) iron are shown in Figure 11.10. Note that these two isomers have different colors. Their salts also possess different solubilities in water.

(a) Violet   (b) Green

Figure 11.10 (a) cis; (b) trans geometric isomers of the octahedral $[Co(NH_3)_4Cl_2]^+$

Optical isomerism is the second type of stereoisomerism. Isomers of this type are mirror images that cannot be superimposed on each other, and they are called enantiomers. They bear the same resemblance to each other that our left hand bears to our right hand. If you look at your left hand in a mirror, the image is identical to your right hand. Note that, your two hands are not superimposed on one another. $[Co(en)_3]^{3+}$ ion is a good example of a complex that exhibits this type of isomerism. Figure 11.11 shows the two enantiomers of $[Co(en)_3]^{3+}$ and their mirror-image relationship to each other.

In general, the number of isomers of a complex can be determined by making a series of drawings of the structure with ligands in different locations. It is easy to over-estimate the number of geometric isomers. Sometimes different orientations of a single isomer are incorrectly thought to be different isomers. Therefore, one should keep in mind that if two structures can be rotated so that they are equivalent, they are not isomers of each other.

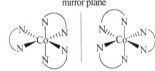

Figure 11.11 The two optical isomers of $[Co(en)_3]^{3+}$

11.5 Coordination Equilibrium in Solution

As a matter of fact, Metal ion in solution does not exist in isolation. But in combination with ligands (solvent molecules or simple ions) or chelating agents, giving rise to complex ion or coordination compounds. Metal ions exhibit marked preference for certain ligands. For example, Cu^{2+} ion coordinates NH_3 in preference to H_2O. In general, however, the factors that determine which ligand will coordinate best with a given metal ion are numerous and complicated, they are not completely understood. Later in this section some of these factors will be discussed.

The properties of a metal ion in solution are dependent on the nature of the ligands surrounding the metal. The stability of a complex is conveniently expressed by stability constants.

Why can most transition metal ions form very stable complex in solution? To answer this question, we must know the nature and the stability of the complex that metal ions can form with the solvent (H_2O) and potential ligands in solution.

Stability Constants

It needs to point out that the stability of a complex in solution refers to the degree of association between metal ion and ligands in the state of equilibrium. Qualitatively, the greater the association, the greater the stability of the compounds will be. The magnitude of the formation equilibrium constants for the association, quantitatively expresses the stability.

Now let us consider the formation of $[Cu(NH_3)_4]^{2+}$. When ammonia is added to a solution of copper(II) salt, there is a rapid reaction in which water coordinated to the metal ion is replaced by ammonia. The product of this reaction is normally represented as $[Cu(NH_3)_4]^{2+}$. The reaction equation is shown below

$$Cu^{2+} + 4NH_3 \rightleftharpoons [Cu(NH_3)_4]^{2+}$$

At a certain temperature, the equilibrium constant for this reaction is

$$K_s = \frac{[Cu(NH_3)_4^{2+}]}{[Cu^{2+}][NH_3]^4}$$

Where, K_s is called stability constants or complex ion formation constants.

Keep in mind that the larger the value of K_s, the more stable the coordination compound.

For it's reverse reaction, K_{dis} is used for the dissociation of $[Cu(NH_3)_4]^{2+}$ into Cu^{2+} and $4NH_3$. It is instability constant. K_{dis} is the reciprocal of K_s

$$K_s = \frac{1}{K_{dis}}$$

In fact, metal complexes are formed in solution by stepwise reaction, and equilibrium constants can be written for each step. The successive equilibrium constants for the formation of $[Cu(NH_3)_4]^{2+}$

$$Cu^{2+} + NH_3 \rightleftharpoons [Cu(NH_3)]^{2+} \quad (1), K_{s1} = \frac{[Cu(NH_3)^{2+}]}{[Cu^{2+}][NH_3]}$$

$$[Cu(NH_3)]^{2+} + NH_3 \rightleftharpoons [Cu(NH_3)_2]^{2+} \quad (2), K_{s2} = \frac{[Cu(NH_3)_2^{2+}]}{[Cu(NH_3)^{2+}][NH_3]}$$

$$[Cu(NH_3)_2]^{2+} + NH_3 \rightleftharpoons [Cu(NH_3)_3]^{2+} \quad (3), K_{s3} = \frac{[Cu(NH_3)_3^{2+}]}{[Cu(NH_3)_2^{2+}][NH_3]}$$

$$[Cu(NH_3)_3]^{2+} + NH_3 \rightleftharpoons [Cu(NH_3)_4]^{2+} \quad (4), K_{s4} = \frac{[Cu(NH_3)_4^{2+}]}{[Cu(NH_3)_3^{2+}][NH_3]}$$

The equilibrium constants K_{s1}, K_{s2}, K_{s3} and K_{s4} are stepwise stability constants (also called stepwise formation constants).

When reaction (1)+(2), it gives

$$Cu^{2+} + 2NH_3 \rightleftharpoons [Cu(NH_3)_2]^{2+}$$

The new equilibrium constants β_2 is

$$\beta_2 = \frac{[Cu(NH_3)_2^{2+}]}{[Cu^{2+}][NH_3]^2} = \frac{[Cu(NH_3)^{2+}]}{[Cu^{2+}][NH_3]} \times \frac{[Cu(NH_3)_2^{2+}]}{[Cu(NH_3)^{2+}][NH_3]}$$

Thus
$$\beta_2 = K_{s1} K_{s2}$$

Where β_2 is the second type of equilibrium constant, called an overall stability constant. Similarly we have

$$\beta_3 = K_{s1} K_{s2} K_{s3}$$
$$\beta_4 = K_{s1} K_{s2} K_{s3} K_{s4}$$

Respectively.

In general $\beta_n = K_{s1} K_{s2} \cdots K_{sn}$, $\beta_n = K_s$

K_s of some complex ions are given in Table 11.8.

Table 11.8 Selected K_s and $\lg K_s$ for complex ions in water (298K)

Complex Ion	K_s	$\lg K_s$	Complex Ion	K_s	$\lg K_s$
$[CdCl_4]^{2-}$	6.3×10^2	2.80	$[HgCl_4]^{2-}$	1.2×10^{15}	15.08
$[Co(NH_3)_6]^{2+}$	1.3×10^5	5.11	$[Cd(CN)_4]^{2-}$	6.0×10^{18}	18.78
$[Ag(NH_3)_2]^+$	1.1×10^7	7.04	$[AlF_6]^{3-}$	6.9×10^{19}	19.84
$[Cd(NH_3)_6]^{2+}$	1.3×10^7	7.11	$[Ag(CN)_2]^-$	1.3×10^{21}	21.11
$[Ni(NH_3)_6]^{2+}$	5.5×10^8	8.74	$[Cu(CN)_2]^-$	1.0×10^{24}	24.00
$[Zn(NH_3)_4]^{2+}$	2.9×10^9	9.46	$[HgI_4]^{2-}$	6.8×10^{29}	29.83
$[FeF_6]^{3-}$	1.1×10^{12}	12.04	$[Fe(CN)_6]^{4-}$	1.0×10^{35}	35.00
$[Cu(NH_3)_4]^{2+}$	2.1×10^{13}	13.32	$[Co(NH_3)_6]^{3+}$	2×10^{35}	35.30
$[Ag(S_2O_3)_2]^{3-}$	2.9×10^{13}	13.46	$[Fe(CN)_6]^{3-}$	1.0×10^{42}	42.00

Just as others equilibrium constant, the stability constant is very useful in that it summarizes a great deal in formation of complexes. It can be used to calculate species concentration in a complex equilibrium.

Sample 11.1 Calculate the Ag^+ ion concentration in a solution of $[Ag(NH_3)_2]^+$ prepared by adding 1.0×10^{-3} mol $AgNO_3$ to 1.0 L of 0.100 mol/L NH_3 solution.

Solution:
$$Ag^+ + 2NH_3 \rightleftharpoons Ag(NH_3)_2^+$$

$$K_s = \frac{[Ag(NH_3)_2^+]}{[Ag^+][NH_3]^2} = 1.1 \times 10^7 \text{ (from Table 11.8)}$$

Assume that most of the silver ion will form complex ions.

Let $[Ag^+] = x$;

Then
$[Ag(NH_3)_2^+] = 1.0 \times 10^{-3} - x \approx 1.0 \times 10^{-3}$, $[NH_3] = 0.100 - 2(1.0 \times 10^{-3}) + 2x \approx 0.098$

Inserting these values into the equilibrium-constant expression, we have

$$K_s = \frac{1.0 \times 10^{-3}}{x(0.098)^2} = 1.1 \times 10^7$$

Solving, $x = 9.5 \times 10^{-9}$; i.e. the equilibrium concentration of silver ion is 9.5×10^{-9} mol/L. From this we can see that a tiny fraction of the Ag^+ present in solution.

Sample 11.2 Calculate the formation constant for the reaction of a tri-positive metal ion with thiocyanate ion to form the mono complex if the total metal concentration in the solution is 2.0×10^{-3} mol/L, the total SCN^- concentration is 1.51×10^{-3} mol/L, the free SCN^- concentration is 1.0×10^{-5} mol/L.

Solution:
$$M^{3+} + SCN^- \rightleftharpoons [MSCN]^{2+}$$

The bound SCN^- concentration is $(1.51\times10^{-3}\ mol/L)-(1.0\times10^{-5}\ mol/L)=1.50\times10^{-3}\ mol/L$. That is also the concentration of $[MSCN]^{2+}$ ion and hence the bound concentration. The free M^{3+} concentration is therefore
$$(2.0\times10^{-3}\ mol/L)-(1.50\times10^{-3}\ mol/L)=0.50\times10^{-3}\ mol/L.$$
Thus $$K=\frac{[MSCN^{2+}]}{[M^{2+}][SCN^-]}=\frac{1.50\times10^{-3}}{(0.50\times10^{-3})(1.0\times10^{-3})}=3.0\times10^5$$

Complex Ions and Solubility

In complexation competition exists among H^+, metal ion, ligands, and precipitation of metal ions by various precipitants. It just as acid may be used to lower the concentration of anions in solution, so complexing agents may be used in some cases to lower the concentration of cations. For example, you have perhaps performed the experiment used to identify Ag^+ ion in solution before. Recall that if silver ion is present, the addition of Cl^- gives an immediate precipitate of AgCl. This precipitate dissolves in an excess of aqueous ammonia because the addition of NH_3 converts most of the silver in AgCl to the complex ion. But if an excess of nitrid acid is added to the clear solution, the white precipitate is formed again. This behavior is exactly due to the equilibria below.

$$Ag^+(aq)+Cl^-(aq)\rightleftharpoons AgCl(s) \qquad (1)$$
$$Ag^+(aq)+2NH_3(aq)\rightleftharpoons Ag(NH_3)_2^+(aq) \qquad (2)$$
$$H^++2NH_3\rightleftharpoons NH_4^+ \qquad (3)$$

A white precipitate forms as shown in (1), because AgCl is not soluble in water. It does, however, dissolve in excess NH_3 because of the formation of stable complex ion of Equation (2). The addition of excess HNO_3 to the clear solution causes equilibrium (2) to shift to the left, and the white precipitate of AgCl reappears. The reappearance is due to a lowering of the concentration of NH_3 its reaction with H^+ to form NH_4^+, as shown in equation (3).

Sample 11.3 How much AgBr could dissolve in 1.0L of 0.40 mol/L NH_3? Assume that $Ag(NH_3)_2^+$ is the only complex formed. K_{sp} of AgBr is 5.0×10^{-13}.

Solution:
$$AgBr(s)\rightleftharpoons Ag^+(aq)+Br^-(aq)$$
$$Ag^+(aq)+2NH_3(aq)\rightleftharpoons Ag(NH_3)_2^+(aq)$$

Adding the solubility and complex-ion equilibria, it obtain the overall equation
$$AgBr(s)+2NH_3(aq)\rightleftharpoons Ag(NH_3)_2^++Br^-(aq) \qquad K$$
$$K=\frac{[Ag(NH_3)_2^+][Br^-]}{[NH_3]^2}=\frac{[Ag(NH_3)_2^+]}{[Ag^+][NH_3]^2}[Ag^+][Br^-]=K_sK_{sp}$$

Let $x=$solubility. Then $x=[Br^-]=[Ag^+]+[Ag(NH_3)_2^+]$
$$K_s=\frac{[Ag(NH_3)_2^+]}{[Ag^+][NH_3]^2}=1.1\times10^7 \quad \text{(from Table 11.8)}$$

Since the overwhelming majority of silver is in the form of the complex, $x=[Br^-]=[Ag(NH_3)_2^+]$, and we have
$$K_s\cdot K_{sp}=\frac{[Ag(NH_3)_2^+][Br^-]}{[NH_3]^2}=\frac{x^2}{(0.40-2x)^2}=(5.0\times10^{-13})\times(1.1\times10^7)$$

Solving $\qquad x=9.36\times10^{-4}\ mol/L$

The solubility of AgBr(s) in 0.40mol/L NH_3 is 9.36×10^{-4} mol/L.

Factors that Affect Complex Stability

The term "stable complex" can be defined in terms of the equilibrium constant for the formation of complex. Many factors have effect on the formation and stability of complex, they are outlined as follow.

(1) Metals effects

a. The smaller the ionic radius, the more stable the complex.

b. The larger the oxidation state, the more stable the compound.

c. Crystal field effects. Natural order of stability.

$Ca^{2+} < Sc^{2+} < Ti^{2+} < V^{2+} < Cr^{2+} < Mn^{2+} < Fe^{2+} < Co^{2+} < Ni^{2+} < Cu^{2+} < Zn^{2+}$

(2) Ligand effects

a. The greater the base strengths (affinities for H^+) of a ligand, the greater are the tendency of the ligand to form stable complexes. From this point of view, F^- should form more stable complexes than Cl^-, Br^-, or I^-, and NH_3 should be a better ligand than H_2O.

b. Chelate effect The stability of a metal chelate is greater than that of an analogous nonchelated metal complex, such as, $[Co(en)_3]^{3+} > [Co(NH_3)_6]^{3+}$. The more extensive the chelation, the more stable the system. We recall the very stable complexes of hexadentate EDTA. Chelating ligands in general form more stable complexes than their monodentate analogus.

c. Chelate ring size The most stable metal chelates contain saturated ligands that form five-membered chelate rings or unsaturated ligands that form six-membered rings.

d. Steric strain Because of steric factors, large ligands form less stable metal complexes than do the smaller ligands. For example, $H_2N-CH_2-CH_2-NH_2$ forms more stable complexes than $(CH_3)_2N-CH_2-CH_2-N(CH_3)_2$.

11.6 Valence Bond Theory of Coordination Compounds

In earlier chapters, we have seen that the modern valence bond theory, which includes the hybrid atomic orbital model, is a very useful model for describing the bonding in molecules. It has also been extensively used in coordination compounds. The central feature of the model is the formation of hybrid atomic that are used for sharing electron pairs to from s bonds between atoms. The same model can be used to account for the bonding in complex ions. When a ligand containing unshared electron pair overlaps an unoccupied orbital on the metal ion, a coordinate covalent bond is formed. The hybrid orbitals used by the metal ion depend on the number and arrangement of the ligands.

Octahedral Complexes

We can easily assume that a complex ion with a coordination number of 6 has an octahedral arrangement of ligands because of obeying the VSEPR model. In order to minimize the repelling forces between them, they should be as far away as possible. Hybrid orbitals with an octahedral arrangement can be formed with two d orbitals, one 4s orbital, and three 4p orbitals. For example, $[Cr(H_2O)_6]^{3+}$ ion has the coordination number of 6 and thus is an octahedral complex.

The valence electron configuration of Cr^{3+} is $3d^3$, its orbital diagram is

Cr^{3+} : ↑ ↑ ↑ _ _ _ _ _ _ _ _ _ _
      3d   4s   4p   4d

The d orbitals used to form the Hybrid orbitals should be 3d. According to previous knowledge, we will call the hybrid orbitals containing 3d orbitals d^2sp^3 to indicate that the d orbitals have a principal quantum number of 1 less than that of the 4s and 4p orbitals. The d^2sp^3 is called inner orbitals. The six orbitals containing electron pairs from ligands overlap these unoccupied d^2sp^3 orbitals on the metal ion to from six coordinate covalent bonds. The orbital diagram for the Cr^{3+} in the complex is shown as bellow.

                        inner orbitals
$[Cr(H_2O)_6]^{3+}$ ↑ ↑ ↑   ↑↓ ↑↓ ↑↓ ↑↓ ↑↓ ↑↓   _ _ _ _ _
                    3d 3d   4s 4p 4p 4p   4d
                    d^2sp^3 hybridized orbitals

Note that there are still three unpaired electrons in 3d orbitals on the metal ion, which accounts for the paramagnetism of this complex ion. The bonding in other octahedral complexes of Cr(Ⅲ) is as the same.

The bonding in complexes of Fe(Ⅱ) is more diverse. Most of its complexes are octahedral and paramagnetic. The magnitude of this paramagnetism indicates four unpaired electrons in the complex ions. A good example is the hexaaquarion(Ⅱ) ion, $[Fe(H_2O)_6]^{2+}$.

We now consider the bonding in $[Fe(H_2O)_6]^{2+}$. The valence electron configuration of Fe^{2+} is $3d^6$. The orbital diagram of Fe^{2+} is

Fe^{2+} ↑↓ ↑ ↑ ↑ ↑ _ _ _ _ _ _ _ _ _
         3d   4s   4p   4d

Because the 3d orbitals have electrons in the five equivalent orbitals, and they cannot be used for bonding to ligand orbitals unless some of these electrons are moved. Suppose we use two of the four empty 4d orbitals instead. We will call the hybrid orbitals formed from them sp^3d^2 to emphasize that the d orbitals have the same principal quantum number as that of the s and p orbitals. The hybrid orbitals of sp^3d^2 are called outer orbitals. Then, we have the orbital diagram of $[Fe(H_2O)_6]^{2+}$ ion.

                              outer orbitals
                           4s 4p 4p 4p 4d 4d
Fe^{2+} ↑↓ ↑ ↑ ↑ ↑   ↑↓ ↑↓ ↑↓ ↑↓ ↑↓ ↑↓   _ _ _
             3d   six lone pair electrons donated by $6H_2O$ 4d
                       sp^3d^2 hybridized orbitals

There are still four unpaired electrons in $[Fe(H_2O)_6]^{2+}$, which explain very well the paramagnetism of this complex ion, in agreement with the experiment results.

However, the hexacyanoferrate(Ⅱ) ion, $[Fe(CN)_6]^{4-}$, is an example of a diamagnetic Fe(Ⅱ) complex ion. Suppose that in forming a complex ion of $[Fe(CN)_6]^{4-}$, two of the 3d electrons in Fe^{2+} pair up so that two 3d orbitals are unoccupied and can be used to form d^2sp^3 hybrid orbitals(inner orbitals). The valence electron configuration of this excited state of the Fe^{2+} ion is

Fe^{2+} (excited)   ↑↓ ↑↓ ↑↓ _ _ _ _ _ _   _ _ _ _ _
                    3d   3d 4s   4p   4d

The 3d orbitals are lower in energy than that of the 4d orbitals, so it will be preferred for bonding if available. We could consider this type of bonding to form only if the bonding were sufficiently strong to provide the energy needed for this electron pairing. $[Fe(CN)_6]^{4-}$ is exactly the case, because CN^- ion always shows a strong tendency to bond to metal ions, in contrast to the weaker bonding of H_2O. The orbital diagram for Fe^{2+} ion in this complex should be

$[Fe(CN)_6]^{4-}$ ↑↓ ↑↓ ↑↓ ↑↓ ↑↓ ↑↓ ↑↓ ↑↓ ↑↓ — — — — —
            3d   3d 3d 4s 4p 4p 4p   4d
                    d^2sp^3 hybridized orbitals

Apparently, this diagram shows us that it is a diamagnetic complex ion.

When examine the first series of transition-metal ions having configurations d^4, d^5, d^6, or d^7, it will be found two bonding possibilities, outer orbital sp^3d^2 hybridization and inner orbital d^2sp^3 hybridization. The number of unpaired electrons in the latter type of complex is lower than that with sp^3d^2 bonding, because some electrons are paired to give unoccupied 3d orbitals. A low-spin complex ion is a complex ion in which there is maximum pairing of electrons in the orbitals of the center metal ion. Whereas, a high-spin complex ion is a complex ion in which there is minimum pairing of electrons in the orbitals of the center metal ion. Strongly bonding ligands (according to the spectrochemical series) form low-spin complex, because low-spin complexes require energy for pairing of electrons. Bonding ligands form high-spin complexes weakly.

For the metal ions with the valence electron configurations d^8 and d^9, only one spins type of octahedral complex possible. Similarly, the d^1, d^2 and d^3 electron configurations should give only one type octahedral complex.

Sample 11.4 Why all octahedral complexes of nickel(II) are high-spin?
Solution:
The valence electronic configuration of Ni^{2+} is as follows.
Ni^{2+} ↑↓ ↑↓ ↑↓ ↑↑ __ __ __ __ __ __ __
       3d   4s   4p   4d

Since only one 3d orbital can be made available by pairing of the electrons, it cannot be inner orbital d^2sp^3 hybridization. The only octahedral hybridization possible is sp^3d^2, using 4d orbitals, and so all octahedral complexes of Ni(II) must be high-spin.

Sample 11.5 Predict the number of unpaired electrons in $[Cr(NH_3)_6]^{3+}$. Why is it only one spin type of octahedral complex possible?
Solution:
The orbital diagram for the Cr^{3+} ion in this complex is as follows.
                         inner orbitals
$[Cr(NH_3)_6]^{3+}$ ↑ ↑ ↑ ↑↓ ↑↓ ↑↓ ↑↓ ↑↓ ↑↓ — — — — —
                     3d 3d 4s 4p 4p 4p   4d
                     d^2sp^3 hybridized orbitals

The Cr(III) ion has three 3d electrons, all unpaired. The three electrons leave two orbitals vacant for inner orbital bonding, and in this case, outer orbital bonding is not necessary. Thus only one spin type of octahedral complex is possible.

Tetrahedral and Square Planar Complexes

The hybrid orbitals predicted for the metal ion in four-coordinate complexes depend on whether the structure is tetrahedral or square planar. For a tetrahedral arrangement of ligands, an sp^3 hybrid set is needed. In the tetrahedral $[Ni(NH_3)_4]^{2-}$ ion, for example, the Ni^{2+} is described as sp^3 hybridized. A square planner arrangement of ligands requires a set of four dsp^2 hybrid orbitals on the metal ion. For example, $Ni(CN)_4^{2-}$, the Ni^{2+} is described as dsp^2 hybridized. This complex is diamagnetic, which gives evidence for a square planar geometry using dsp^2 orbitals.

Linear Complex

Two hybrid orbitals are required in a linear complex ion and it has 180° from each other. This arrangement is given by a sp hybrid set. Thus, in the linear $Ag(NH_3)_2^+$ ion, the Ag^+ is described as sp hybridized.

A summary of hybrid orbitals and geometry in complex ions are given in Table 11.9.

Table 11.9 The hybrid orbitals and geometry of some complex ions

Hybrid Orbitals	Coordination Numbers	Geometry	Example
sp	2	linear	$[Ag(NH_3)_2]^+$ $[Ag(CN)_2]^-$ $[Cu(NH_3)_2]^+$
sp^2	3	Trigonal planar	$[Cu(CN)_3]^{2-}$ $[HgI_3]^-$ $[CuCl_3]^{2-}$
sp^3	4	tetrahedral	$[Zn(NH_3)_4]^{2+}$ $[Ni(NH_3)_4]^{2+}$ $[Cd(NH_3)_4]^{2+}$ $[Co(SCN)_4]^{2-}$ $[FeCl_4]^{2-}$
dsp^2	4	Square planar	$[Ni(CN)_4]^{2-}$ $[PtCl_4]^{2-}$ $[Cu(H_2O)_4]^{2+}$
sp^3d^2 d^2sp^3	6 6	Octahedral	$[FeF_6]^{3-}$ $[Fe(H_2O)_6]^{2+}$ $[Fe(CN)_6]^{3-}$ $[Cr(NH_3)_6]^{3+}$

11.7 The Crystal Field Theory

Scientists have recognized that the magnetic properties and colors of transition-metal complexes are related to the presence of d electrons in metal orbitals. However, the Valence Bond Theory and hybrid atomic orbital model cannot explain the color of complexes, and it is difficult to extend quantitatively. Consequently, another theory has emerged as prevailing theory of transition-metal complexes. This is crystal field theory (CFT), it is a model of the electronic of transition metal complexes that considers how the energies of the d orbitals of a center metal ion are affected by the electric field of the ligands, which accounts for many of the observed properties of coordination compounds. We will begin our discussion of crystal field theory with complex having octahedral geometry. Then we will see how it is applied to tetrahedral and square-planar complexes.

CFT for Octahedral Complex

One has already noted that the ability of a transition-metal ion to attract ligands such as H_2O around itself can be viewed as a Lewis acid-base interaction. The base, that is, the ligand can be considered to donate a lone pair of electrons into a suitable empty orbital on the metal. However, it can assume that much of the attractive interaction between the metal ion and the surrounding ligands is due to the electrostatic forces between the positive charge on the metal and negative charge on the ligands. If the ligand is ionic, as in the case of F^- or CN^-, the electrostatic interaction occurs between the positive charge on the metal center and the negative charge on the ligand. When the ligands are neutral, such as in the case of water or ammonia, the negative ends of the polar molecules contain an unshared electron pair are all directed toward the center metal atom. In such case, the attractive interaction is of the ion-dipole type. In either case the result is the same, the ligands are attracted toward the metal center. The assembly of metal ion and ligands is lower in energy than that of the fully separated charges (Figure 11.12).

As shown in Figure 11.12, in the crystal-field model the bonding between metal ion and donor atoms (from the ligand) is considered to be largely electrostatic. The energy of the metal ion plus coordinated ligands is lower than that of the fully separated metal ion plus ligands because of the electrostatic attraction. At the same time, the energies of the metal d electrons are increased by the repulsive interaction between these electrons and the electrons of the ligands. These repulsive interactions give rise to the splitting of the metal d-orbital energies. This interaction is called the crystal field. The crystal field causes

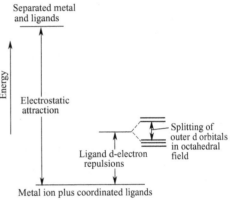

Figure 11.12 The energies of the metal d electrons in the crystal-field model

the energies of the d electrons on the metal to increase, as shown in Figure 11.12. The d orbitals of the metal ion, however, do not all behave in the same way under the influence of this crystal fields.

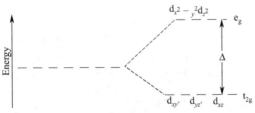

Figure 11.13 Energies of the d orbitals in an octahedral crystal field

In a six-coordinate octahedral complex, electrons in the $d_{x^2-y^2}$ and d_{z^2} orbitals (called e_g set) experience stronger repulsion than those in the d_{xy}, d_{yz} and d_{xz} orbitals (called the t_{2g} set). As a result, an energy separation, or splitting, occurs between the three lower-energy d orbitals and two higher-energy ones. This splitting of the d orbitals energies by the crystal field is depicted in Figure 11.13. Note that the difference in energy between the two sets of d orbitals on a metal ion that arises from the interaction of the orbital with the electric field of the ligands is called the crystalfield splitting energy, symbolized by Δ.

The greatest achievement of CFT is its success in interpreting the observed colors of transition-metal complexes. The energy gap, Δ, is of the same order of magnitude as the energy of a photon of visible light. Therefore, it is possible for a transition-metal complex to absorb visible light, which excites an electron from the lower-energy d orbitals into the higher-energy ones (called the d-d electronic transition). This causes the complex to appear colored. The $[Ti(H_2O)_6]^{3+}$ ion provides a good example because Ti(Ⅲ) has only one 3d electron. Experimental results show that $[Ti(H_2O)_6]^{3+}$ has a single absorption peak in the visible region of the spectrum. The maximum absorption is at 510 nm. Light of this wavelength causes the d electron to move from the lower-energy set of d orbitals into the higher-energy set, i. e. $t_{2g} \rightarrow e_g$. The absorption of 510 nm radiation that produces this transition causes substances containing $[Ti(H_2O)_6]^{3+}$ ion to appear purple. Notice that the absorption spectra of complexes containing more than one d electron are more complicated, because a greater number of electronic transitions are possible. In general, the color of a complex depends on the particular metal, its oxidation state, and the ligands bound to the metal. The presence of a partially filled d subshell on the metal is usually necessary for a complex to exhibit color.

The magnitude of the energy gap, Δ, and the color of a complex consequently depend on both the metal and the surrounding ligands. For example, $[Fe(H_2O)_6]^{3+}$ is light yellow, $[Cr(NH_3)_6]^{3+}$ is yellow, and $[Cr(H_2O)_6]^{3+}$ is violet. Ligands can be arranged in order of their abilities to increase the energy gap, Δ. As follows

$$\xrightarrow{\text{increasing } \Delta}$$

$I^- < Br^- < Cl^- < F^- < OH^- < ox, O^{2-} < H_2O < NH_3 < en < bpy, phen < NO_2^- \text{ (N-bonded)} < CN^- < CO$

This list is called the spectrochemical series, which is a list of ligands arranged in increasing order of their abilities to split the d orbital energy levels.

It is possible to arrange the metals according to a spectrochemical series as well. The approximate order is

$Mn^{2+} < Ni^{2+} < Co^{2+} < Fe^{2+} < Fe^{3+} < Co^{3+} < Mn^{3+} < Mo^{3+} < Rh^{3+} < Ru^{3+} < Pd^{4+} < Ir^{3+} < Pt^{4+}$

Normally, oxidation state of the metal could influence the magnitude of Δ. The higher the charge of the metal, the greater Δ. Considering nature of the metal ion, Δ increases on going from 5d>4d>3d. The magnitude of Δ has a direct effect on the color and magnetic of complex ions.

Ligands that lie on the low (or the left) end of the above spectrochemical series are called weak-field ligands, those on the high (or the right) end are termed strong-field ligands. Figure 11.14 shows schematically what happens to the crystal field splitting when ligand is varied in a series of Cr(Ⅲ) complexes. As the field exerted by the six surrounding

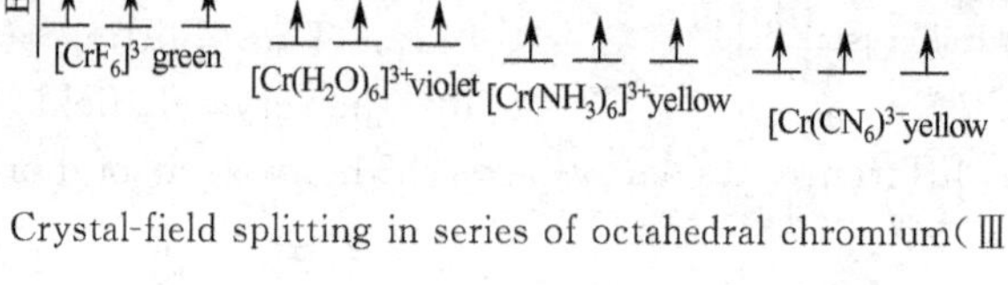

Figure11.14 Crystal-field splitting in series of octahedral chromium(Ⅲ) complexes

ligands increases, the splitting of the d orbitals increases. Because the absorption spectrum is related to this energy separation, and complexes vary in color.

Sample 11.6 A solution of $[Ni(H_2O)_6]^{2+}$ is green, but a solution of $[Ni(CN)_6]^{2-}$ is colorless, give an explanation for these observations.

Solution:

The value of Δ for the H_2O complex is in the visible region. For the cyano complex is in the ultraviolet region, because CN^- ion is a much stronger ligand than H_2O.

The crystal-field model also helps us understand the magnetic properties and some important chemical properties of transition-metal ions. From our earlier discussion of electronic structure in atoms (see section 2.8), we know that electrons will always occupy the lowest-energy vacant (t_{2g}) orbitals first and will occupy a set of degenerate orbitals one at a time with their spins parallel (According to Hund's rule). Thus, if we have one, two, or three electrons to add to the d orbitals in an octahedral complex ion, the electrons will first go into the lower-energy set of orbitals, with their spins parallel. When a fourth electron is added, one problem arises. If the electron is added to the lower-energy orbital, an energy gain of magnitude Δ is realized, as compared with placing the electron in the higher-energy orbital. However, we know that the electron must now be paired up with the electron already occupying the orbital. To do this, energy is needed for putting it in another orbital with parallel spin. The energy is called the spin-pairing energy. The spin-pairing energy arises from the greater electrostatic repulsion of two electrons that share an orbital compared with two that are in different orbitals.

The ligands surround the metal ion, the charge on the metal ion often play major roles in determining which of the two electronic arrangements arises. Strong-field ligands create a splitting of d-orbital energies that is large enough to overcome the spin-pairing energy. The d electrons then preferentially pair up in the lower-energy orbitals, producing a low-spin complex. When the ligands exert a weak crystal field, the splitting of the d-orbitals is small. The electrons then occupy the higher-energy d-orbitals in preference to pairing up in the lower-energy set, producing a high-spin complex. For example, in an octahedral complex of Co^{3+} (a metal ion with six 3d electrons), there are two possible ways to place the electrons in the splitting 3d orbitals. If the splitting produced by the ligands is very large (such as CN^-), a situation called the strong-field case, also know as low-spin case, the electrons will pair in the lower-energy t_{2g} orbitals. This may give a diamagnetic complex. On the other hand, if the splitting is small (the weak-field case, also called the high-spin case), the electrons will occupy all five orbitals before pairing. A paramagnetic complex occurs with four unpaired electrons on the metal ion. Note that, the ligands that surround the metal ion and the charge on the metal ion often play major roles in determining which of the two electronic arrangements arises.

Sample 11.7 Describe the distribution of d electrons in the complex $[CoF_6]^{4-}$ and $[Co(CN)_6]^{4-}$ ions. How many unpaired electrons are there in each ion?

Solution:

Consider the $[CoF_6]^{4-}$ and $[Co(CN)_6]^{4-}$ ions. The valence electron configuration of Co^{2+} is $3d^7$. In both cases, the ligands have a 1-charge. However, the F^- ion, on the low end of the spectrochemical series, is a weak-field ligand and so $[CoF_6]^{4-}$ is a high-spin complex. The CN^- ion, on the high end of the spectrochemical series, is a strong-field ligand. It produces a larger energy gap than does the F^- ion and so $[Co(CN)_6]^{4-}$ is a low-spin complex. The high-spin and low-spin distributions in the d orbitals are

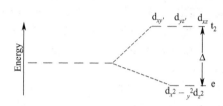

Thus $[CoF_6]^{3-}$ has three unpaired electrons, $[Co(CN)_6]^{4-}$ has only one unpaired electron.

CFT for Tetrahedral and Square-Planar Complexes

So far we have considered the crystal-field model only for the complexes of octahedral geometry. The crystal-field model is also applicable to tetrahedral and square-planar complexes. However, the ordering of the d-orbital energy in these complexes is different from that in octahedral complexes.

When there are only four ligands about the metal, the geometry is tetrahedral, except for the special case of metal ions with a d^8 electron configuration. If metal ion bonds with tetrahedrally arranged ligands, the d orbitals of the ion split to give two d orbitals at lower energy and three at higher energy. Four equivalent ligands can interact with a central metal ion most effectively by approaching along the vertices of a tetrahedron. That is, three of the metal d orbitals are raised in energy, and the other two are lowered, as shown in Figure 11.15. There are only four ligands instead of six, as in the octahedral case,

Figure 11.15 Energies of the d orbitals in a tetrahedral crystal field

and the crystal-field splitting is much smaller for tetrahedral complexes. Calculations show that for the same metal ion and ligand set, the crystal-field splitting for a tetrahedral complex is only four ninths as large as for the octahedral complexes. Thus, all tetrahedral complexes are high spin. The crystal field is never large enough to overcome the spin-pairing energies, and the crystal-field splitting of the metal d orbitals in tetrahedral complexes differs from that in octahedral complexes. In fact, the splitting pattern for a tetrahedral ion is just reverse of that for octahedral complexes.

Square-planar complexes, in which four ligands are arranged about the metal ion in a plane, represent a common geometric form. We can envision the square-planar complex as formed by removing two ligands from along the vertical z axis of the octahedral complex. When this happens, the four ligands in the planar are drawn in more tightly. The changes that occur in the energy levels of the d orbitals are shown in Figure11.16. The splitting pattern for square-planar complexes is the most complicated.

Normally, square-planar complexes are characteristic of metal ions with a d^8 electron configuration. They are nearly always low spin, that is, the eight d electrons are spin-paired to form a diamagnetic complex. Such an electronic arrangement is particularly common among the ions of heavier metals, such as Pd^{2+}, Pt^{2+} and Au^{3+}. However, the relative placement of the d_{z^2}

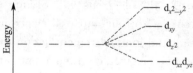

Figure 11.16 Energies of the d orbitals in a square-planar crystal field

and the d_{yz} and d_{xz} orbitals cannot be determined simply by inspection, and it must be calculated.

As stated above, we can now conclude that crystal field theory is a model of the electronic structure of transition-metal complexes that considers how the energies of the d orbitals of a metal ion are affected by the electric field of the ligands. In its simplest from, the theory assumes that the ligands can be approximately regarded as negative point charges, and the five d orbitals of the metal ion no longer have exactly the same energy. The splitting of d orbital energies and its consequences are at the heart of crystal field theory. We have also seen that the crystal-field model provides a basis for explaining many features of transition-metal complexes. In fact, it can be used to explain many observations in addition to those we have discussed.

Note that, molecular orbital theory can also be used to describe the bonding in complexes. However, the application of molecular orbital theory to coordination compounds is beyond the scope of our discussion. The crystal-field model, although not entirely accurate in all details, provides an adequate and useful description.

11.8 The Biological Effects of Coordination Compounds

Most of the first-row transition metals are essential for human health especially. So far, nine of the 24 elements known to be necessary for human life are metals. These nine, V, Cr, Mn, Fe, Co, Ni, Cu, Zn, and Mo, owe their roles in living systems mainly to their ability to form stable complexes with a variety of donor groups present in biological systems. Metallic elements are essential components of many enzymes operating within our bodies. Carbonic anhydrase, which contains Zn^{2+}, is responsible for rapidly interconverting dissolved CO_2 and bicarbonate ion, HCO_3^-. The zinc in carbonic anhydrase is coordinated by three nitrogen-containing groups and a water molecule. The enzyme's action depends on the fact that the coordinated water molecule is more acdic than bulk solvent molecules.

Another good example is the iron, which plays an important role in almost all living cells. Iron-deficiency anemia is a common problem in humans. Although oxygens are the oxidizing agent for the processes of the oxidation of carbohydrates, proteins, and fats, from which the principal source of energy come in mammals, it does not react directly with the nutrient molecules. Instead, the electrons from the breakdown of these nutrients are passed along a complex chain of molecules, called the respiratory chain, eventually reaching the O_2 molecule. Respiratory chains are ion-containing species called cytochromes.

In addition to participating role in the transfer of electrons from nutrients to oxygen, iron also plays a principal role in the transport and storage of oxygen in all mammalian blood and tissues. In general, oxygen is stored using a molecule called myoglobin, which consists of a heme complex and a protein in a structure very similar to that of the cytochromes. In myoglobin, the Fe^{2+} ion is coordinated to four nitrogen atoms of the porphyrin ring and to one nitrogen atom of the protein chain (Figure 11.17). Since Fe^{2+} is normally six-coordinate, this leaves one position open for bonding to an O_2 molecule.

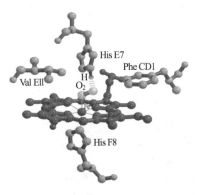

Figure 11.17 The structure of heme

The understanding of the biological role of iron helps us to explain the toxicities of substances such as the cyanide ion and carbon monoxide. Both CN^- and CO are excellent ligands toward iron So they can strongly interfere with normal workings of the iron complexes in our body. For example, CO binds to the iron atom in hemoglobin some 200 times more strongly than O_2. This is the reason why CO is a deadly toxic substance. Because it is a very stable complex, carboxyhemoglobin, prevents the normal uptake of O_2 thus depriving the body of needed oxygen. The mechanism for the toxicity of CN^- ion is somewhat different from carbon monoxide. Cyanide coordinates strongly to affect cytochrome oxidase, an iron-containing cytochrome enzyme that catalyzes the oxidation-reduction reactions of certain cytochromes. The coordinated cyanide prevents the normal electron transfer process in living cell, and cause a rapid death. Because of this behavior, cyanide is known as a respiratory inhibitor.

In fact, most trace metal ions exist in the bloodstream as complex with amino acids or small peptides. Many biologically important molecules, such as the porphyrins, are complexes of chelating agents. A related group of plant pigments known as chlorophylls are important in photosynthesis, the process by which plants use solar energy to convert CO_2 and H_2O into carbohydrates.

Key Words

coordination chemistry [kəuˌɔːdiˈneiʃən ˈkemistri] 配位化学
coordination compound [kəuˌɔːdiˈneiʃən ˈkɔmpaund] 配位化合物
coordination number [kəuˌɔːdiˈneiʃən ˈnʌmbə] 配位数
coordination equilibrium [kəuˌɔːdiˈneiʃən ˌiːkwiˈlibriəm] 配位平衡
coordinate covalent bond [kəuˈɔːdinit kəuˈveilənt bɔnd] 配位键
complex ion [ˈkɔmpleks ˈaiən] 配离子
ligand [ˈligənd, ˈlaigənd] 配体
primary valence [ˈpraiməri ˈveiləns] 主价
secondary valence [ˈsekəndəri ˈveiləns] 副价
complexing agent [kəmˈpleksiŋ ˈeidʒənt] 配位剂
inner sphere [ˈinə sfiə] 内界
outer sphere [ˈautə sfiə] 外界
monodentate ligand [ˈmɔnəuˈdenteit ˈligənd] 单齿配体
unidentate [ˈjuːnidenteit] 单齿配体, 一合配体
polydentate ligand [ˌpɔliˈdenteit ˈligənd] 多齿配体
chelate [ˈkiːleit] 螯合的
claw [klɔː] 爪, 脚爪
stability [stəˈbiliti] 稳定性
stabilize [ˈsteibilaiz] 稳定
stable [ˈsteibl] 稳定的
stability constant [stəˈbiliti ˈkɔnstənt] 稳定常数
stepwise stability constants [stepwaiz stəˈbiliti ˈkɔnstənt] 逐级稳定常数
instability constants [ˌinstəˈbiliti ˈkɔnstənt] 不稳定常数
scavenger [ˈskævindʒə] 清除剂、清道夫
nomenclature [nəuˈmenklətʃə] 命名法, 术语
systematic naming [ˌsistiˈmætik ˈneimiŋ] 系统命名法
numerical prefixes [njuː(ː)ˈmerikəl ˈpriːfiks] 数字词头
isomerism [aiˈsɔmərizm] 异构性
isomer [ˈaisəumə] 异构体
structural isomer [ˈstrʌktʃərəl ˈaisəumə] 结构异构
stereoisomer [ˌstiəriəu ˈaisəmə] 立体异构
coordination isomerism [kəuˌɔːdiˈneiʃən aiˈsɔmərizm] 配位异构
linkage isomerism [ˈliŋkidʒ aiˈsɔmərizm] 键合异构
geometric isomerism [dʒiəˈmetrik aiˈsɔmərizm] 几何异构
optical isomerism [ˈɔptikəl aiˈsɔmərizm] 光学异构
enantiomer [iˈnæntiəumə] 对映异构体
superimpose [ˈsjuːpərimˈpəuz] 双重叠加, 重合
dextrorotatory [ˈdekstrəuˈrəutətəri] 右旋的
levorotatory [liːvəuˈrəutətəri] 左旋的
optically active [ˈɔptikəli ˈæktiv] 光学活性的

Chapter 11 CHEMISTRY of Coordination Compounds

racemic [rə'si:mik] 外消旋的
inner orbital ['inə 'ɔ:bitl] 内轨的
outer orbital ['autə 'ɔ:bitl] 外轨的
paramagnetism [pærə'mægnitizəm] 顺磁性
a low-spin complex ion 低自旋配离子
a high-spin complex ion 高自旋配离子
crystal field theory 晶体场理论
separated charges 电荷分离
crystal-field splitting energy 晶体场理论分

裂能
spectrochemical series [,spektrə'kemistri 'siəri:z] 光谱系列
weak-field ligand 弱场配体
strong-field ligand 强场配体
spin-pairing energy 自旋成对能
cytochrome ['saitəukrəum] 细胞色素
myoglobin [,maiə'gləubin] 肌球素,肌红素

Exercises

1. Write the systematic names of the following coordination compounds.
   (1) $[Co(NH_3)_5Cl]Cl_2$ (4) $K_3[Fe(CN)_6]$
   (2) $[Co(NH_3)_6]Cl_2$ (5) $[Cr(en)_3]Cl_2$
   (3) $K_2[PtCl_4]$ (6) $[Co(NH_3)_2](NO_2)_3$

   Answer: (1) pentaaminechlorocobalt(Ⅲ) chloride.
   (2) hexaaminecobalt(Ⅱ) chloride
   (3) potassium tetrachloroplatinum(Ⅱ)
   (4) Potassium hexacyanoferrate(Ⅲ) or Potassium ferricyanide
   (5) tris (ethylenediamine) chromium(Ⅲ) chloride
   (6) diaminecobalt(Ⅱ) nitrate

2. Write the formulas and specify the coordination number of the following coordination compounds.
   (1) dichlorobis (ethylenediamine) platinum(Ⅳ) nitrate.
   (2) pentaammineiodochromium(Ⅲ) iodide.
   (3) amminetrichloroplatinate(Ⅱ) ion.
   (4) pentaaminechlorocobalt(Ⅲ) chloride.

   Answer: (1) $[Pt(en)_2Cl_2](NO_3)_2$, 6; (2) $[Cr(NH_3)_5I]I_2$, 6;
   (3) $[PtNH_3Cl_3]^-$, 4; (4) $[Co(NH_3)_5Cl]Cl_2$, 6

3. Specify the oxidation number of the central metal atom in each of the following compounds:
   (1) $[Co(NH_3)_6]Cl_2$ (4) $[Cr(NH_3)_6](NO_3)_3$
   (2) $[Ru(NH_3)_6H_2O]Cl_2$ (5) $K_3[Fe(CN)_6]$
   (3) $K_4[PtCl_6]$ (6) $[Fe(CO)_5]$

   Answer: (1) Co+2, (2) Ru+2, (3) Pt+2, (4) Cr+3, (5) Fe+3, (6) Fe0

4. What is the coordination number of
   (1) molybedenum in $[Mo(CN)_8]^{4-}$?
   (2) copper in $[Cu(en)_2]^{2+}$?

   Answer: (1) 8, (2) 4

5. What is the distinction between a covalent bond and a coordination covalent bond? How many covalent bonds and how many coordinate covalent bonds are there in the NH_4^+ ion? Is it possible to distinguish between the coordinate covalent bond(s) and the other covalent bond(s) in this ion? Explain.

6. How many geometric isomers are there for $[Cr(H_2O)_2Br_4]^-$?

7. A freshly prepared aqueous of $Pd(NH_3)_2Cl_2$ does not conduct electricity. Is this

compound to be regarded as a strong or weak electrolyte? Explain in terms of its structure.

8. In the reaction $[CoCl_2(NH_3)_4]^+ + Cl^- \longrightarrow [CoCl_3(NH_3)_3] + NH_3$, only one isomer of the complex product is obtained. Is the initial complex *cis* or *trans*?

Answer: trans

9. Which of the following ligands are capable of linkage isomerism? Explain your answer.

$$SCN^-, N_3^-, NO_2^-, NH_2CH_2CH_2NH_2, OCN^-, I^-$$

Answer: SCN^-, NO_2^- and OCN^- are capable of linkage isomerism.

10. Calculate the concentration of $[Cu^{2+}]$ in 1.0L of a solution made by dissolving 0.10mol Cu^{2+} in 1.00 mol/L aqueous ammonia. $K_s = 1.0 \times 10^{12}$

Answer: 8.0×10^{-13} mol/L

11. Manganese(II) forms a complex with bromide ion. Its paramagnetism indicates five unpaired electrons. What are its probable formula and geometry?

Answer: $[MnBr_4]^{2-}$, the hybrid orbitals are sp^3, tetrahedral

12. On the basis of valence bond theory, predict whether square planar complexes of palladium (2+) are high spin or low spin.

Answer: low spin

13. In terms of valence bond method, cite an example in which the term inner orbital does not imply low spin. In terms of crystal field theory, cite an example in which the term strong field does not imply low spin.

Answer: Cromium(III)

14. For an ion located in a square planar field of negative charge situated in the x and y directions, which d orbital(s) will have the highest energy? The lower?

Answer: The orbital with highest energy is $d_{x^2-y^2}$; the lowest are the degenerate pair d_{xz} and d_{yz}.

15. Both $[Fe(CN)_6]^{4-}$ and $[Fe(H_2O)_6]^{2+}$ appear colorless in dilute solutions. The former ion is low-spin and the latter is high-spin.

(1) How many unpaired electrons are in each of these ions?

(2) Why should both ions be colorless, in view of the apparent significant difference in their Δ values?

Chapter 12 Nonmetals and Semimetals

12.1 General Concepts

Nonmetals and Semimetals totaly include 24 elements. They are found on the right hand side of the Periodic Table, mainly distribute from the group IV A to the group VIII A except for hydrogen and Boron. They have valence electron configuration $ns^{1-2}\,np^{1-6}$. Contrast to metals, non-metals are poor thermal conductors, good electrical insulators and are neither malleable nor ductile. Semimetals have mostly nonmetallic properties. One of their distinguishing characteristics is that their conductivity increases as their temperature increases. This is the opposite of what happens in metals. The semimetals include elements such as silicon (Si) and germanium (Ge).

12.2 Hydrogen

Hydrogen is the most abundant element in the universe, about 70 percent of the universe is composed of hydrogen, but most of them are found associated with oxygen. So water is the most abundant hydrogen compound with 11 percent hydrogen by mass.

Hydrogen was found by the English chemist Henry Cavendish. There are three isotopes of hydrogen, protium (^{1}H), deuterium (^{2}H) and tritium (^{3}H). ^{1}H is the most common isotope of hydrogen with a nucleus consisting of a single proton. It comprises 99.9844 percent of naturally occurring hydrogen. The ^{2}H isotope comprises 0.0156 percent of naturally occurring hydrogen with a proton and a neutron. The third isotope, ^{3}H, is radioactive and only trace quantities exist naturally.

H_2 is a colorless, odorless and tasteless gas at room temperature which is composed of diatomic molecules. H_2 is nonpolar, and it has only two electrons, so attractive forces between molecules are extremely weak. As a result, the melting point (-259℃) and boiling point (-253℃) of H_2 are very low. H_2 is flammable. Mixtures of air containing hydrogen are explosive.

The H-H bond enthalpy (436 kJ/mol) is higher than a single bond. H_2 has a strong bond, most reactions of H_2 are slow at room temperature. However, the molecule is readily activated by heat, irradiation, and catalysis. The activation process generally produces hydrogen atoms, which are very reactive. Once H_2 is activated, it reacts rapidly and exothermically with a wide variety of substances. Hydrogen forms strong covalent bonds with many elements.

Hydrogen is an effective reducing agent for many metal oxides. For example, when H_2 is passed over heated CuO, copper is produced.

$$CuO(s) + H_2(g) = Cu(s) + H_2O(g)$$

In the laboratory, hydrogen is usually obtained by the reaction between an active metal and a dilute strong acid, such as zinc and HCl.

$$Zn + 2HCl = ZnCl_2 + H_2(g)$$

Hydrogen has a very low solubility in water, so it can be collected by displacement of water.

Large quantities of hydrogen are produced by reacting carbon or methane (CH_4) with steam at 1100℃.

$$CH_4(g) + 2H_2O(g) \Longrightarrow CO_2(g) + 4H_2(g)$$
$$C(s) + H_2O(g) \Longrightarrow H_2(g) + CO(g)$$

Hydrogen also can be obtained by electrolysis of brine (NaCl) solutions.

$$2NaCl(aq) + 2H_2O(l) \Longrightarrow H_2(g) + Cl_2(g) + 2NaOH(aq)$$

Hydrogen is a commercially important substance. In the United States, about 2×10^8 kg hydrogen is produced annually. Over two thirds of the hydrogen produced is used to synthesize ammonia by the Haber process.

$$N_2(g) + 3H_2(g) \Longrightarrow 2NH_3(g)$$

Hydrogen is also used to manufacture methanol, CH_3OH, by the catalytic reaction of CO and H_2 at high pressure and temperature.

$$CO(g) + 2H_2(g) \Longrightarrow CH_3OH(g)$$

12.3 Boron

Boron is the only nonmetal element in the group ⅢA, its outer-shell electron configuration is $2s^2 2p^1$. Boron has an extended network structure. The melting point of boron is 2300℃. Boron has 216 allotropic forms, and a-rhombic boron is the most universal.

The Hydride of Boron

The hydride of boron is very important. It is a family compounds, usually they are generally called as boranes. The simplest borane is BH_3 with six valence electrons. It is unstable, and it can react with itself to form diborane, B_2H_6, that accords with the octet rule. The diborane can be obtained by the reactions as following.

$$2BCl_3 + 6H_2 \Longrightarrow B_2H_6 + 6HCl$$
$$2BF_3 + LiAlH_4 \Longrightarrow B_2H_6 + 3LiF + 3AlF_3$$

Diborane is so reactive that it can be flaming spontaneously in air. This reaction is extremely exothermic.

$$B_2H_6 + 3O_2 \Longrightarrow B_2O_3 + 3H_2O$$

Other boranes also undergoes a very exothermic reaction with O_2. According to this reaction, Boranes have been explored as solid fuels for rockets.

Moreover, diborane reacts with water to give out hydrogen. This reaction is also exothermic.

$$B_2H_6 + 6H_2O \Longrightarrow 2H_3BO_3 + 6H_2$$

Boranes usually are used as reducing agents and Lewis acids.

The Oxide and Oxyacid of Boron

Boric oxide, B_2O_3, is the only important oxide of boron. It is the anhydride of boric acid (H_3BO_3).

$$B_2O_3 + 3H_2O \Longrightarrow 2H_3BO_3 \text{ (endothermic reaction)}$$

Boric acid, H_3BO_3, is a very weak acid. The solutions of boric acid even can wash eyes. Upon heating, boric acid loses water to form tetraboric acid, $H_2B_4O_7$.

$$4H_3BO_3(s) \Longrightarrow H_2B_4O_7(s) + 5H_2O(g)$$

The hydrated sodium salt of tetraboric acid, $Na_2B_4O_7 \cdot 10H_2O$, is commonly called borax. It is widely used in cleaning products.

12.4 The Group ⅣA Elements, the Carbon Group Elements

The group ⅣA consists of carbon (C), silicon (Si), germanium (Ge), tin (Sn), and lead (Pb). The general trend from nonmetallic to metallic is very evident in group ⅣA. Carbon is a nonmetal. Silicon and germanium both are semimetals. Tin and lead both are metals.

The general properties of group ⅣA elements are shown in Table 12.1. The group ⅣA elements possess the outer-shell electron configuration ns^2np^2. The each element in this group has four electrons in its outer energy level. The tendency to lose electrons increases as the size of the atom increases, as it does with increasing atomic number. Usually, they form $+2$ or $+4$ oxidation except carbon can form -4 oxidation.

Table 12.1 The general properties of group ⅣA elements

Property	C	Si	Ge	Sn	Pb
Atomic number	6	14	32	50	82
Valence shell electron configuration	$2s^22p^2$	$3s^23p^2$	$4s^24p^2$	$5s^25p^2$	$6s^26p^2$
Main oxidation number	$+4$	$+4$	$+4$	$+4,+2$	$+2,(+4)$
Covalent radius/pm	77	117	122	140	147
Ionic radius(M^{4+})/pm	16	42	53	71	84
Ionic radius(M^{2+})/pm	—	—	—	93	120
Ionization energy/(kJ/mol)	1086	786	762	709	716
Electronegativity	2.55	1.90	2.01	1.80	1.87

Carbon

Carbon is the sixth most abundant element in the universe. Most of carbon exists in combined form. It is also commonly obtained from coal deposits. Three naturally occurring allotropes of carbon are known to exist, amorphous, graphite and diamond. Amorphous carbon is formed when a material containing carbon is burned without enough oxygen for it to burn completely. Graphite, one of the softest materials known, is a form of carbon that is primarily used as a lubricant. It is also actually used in pencils. In addition, graphite, in a form known as coke, is used in large amounts in the production of steel. Diamond is one of the hardest substances known. Although graphite and diamond differ only in their crystal structure, they possess very different physical properties.

Some allotropes of carbon were researched. A fourth allotrope of carbon, known as white carbon, was produced in 1969. It is a transparent material that can split a single beam of light into two beams, a property known as birefringence. Very little is known about this form of carbon. Large molecules consisting only of carbon, known as buckminster-fullerenes, or buckyballs. They have recently been discovered and are currently the subject of much scientific interest. A single buckyball consists of 60 or 70 carbon atoms (C_{60} or C_{70}) linked together in a structure that looks like a soccer ball. They can trap other atoms within their framework. They appear to be capable of withstanding great pressures and have magnetic and superconductive properties.

Carbon forms two oxides, carbon monoxide (CO) and carbon dioxide (CO_2). Carbon monoxide is a colorless, odorless and toxic gas. It can bind to hemoglobin and interfere with O_2 transport. Carbon monoxide is mainly from the incompletely burning of carbon. It is a air pollutant.

$$2C + O_2 = 2CO$$

Carbon monoxide is used as a fuel for its combustibility.

$$2CO + O_2 = 2CO_2$$

It is also used in metallurgical procedure for its reductive property.

$$4CO + Fe_3O_4 = 3Fe + 4CO_2$$

In addition, carbon monoxide is used in organic synthesis as an intermediate.

Carbon dioxide is also a colorless and odorless gas. It is not toxic, but overmuch carbon dioxide can produce the so-called greenhouse effect although it is a component of Earth's atmosphere.

Solid CO_2 is called Dry Ice, it sublimes at atmospheric pressure at $-78\ ℃$. According to this property, carbon dioxide can be used as a refrigerant and the production of carbonated beverage. Carbon dioxide is also an intermediate of chemical industry, such as the manufacture of $Na_2CO_3 \cdot H_2O$ and $NaHCO_3$.

Carbon dioxide dissolves in water to form carbonic acid, H_2CO_3.

$$CO_2 + H_2O = H_2CO_3$$

Carbonic acid is unstable and can give out CO_2.

$$H_2CO_3 = CO_2(g) + H_2O$$

Carbonic acid is a weak diprotic acid, it can form two types of salts, hydrogen carbonates (bicarbonates) and carbonates.

There are nearly ten million known carbon compounds. An entire branch of chemistry, known as organic chemistry, is devoted to their study. Many carbon compounds are essential for life as we know it. Some of the most common carbon compounds are carbon dioxide (CO_2), carbon monoxide (CO), carbon disulfide (CS_2) with similar bonding to CO_2, chloroform ($CHCl_3$), carbon tetrachloride (CCl_4), methane (CH_4), ethylene (C_2H_4), acetylene (C_2H_2), benzene (C_6H_6), ethyl alcohol (C_2H_5OH) and acetic acid (CH_3COOH).

Silicon

Silicon is a hard, lustrous gray solid. It widely exists in silicic acid and silicates. It is the second most abundant element in the earth's crust. It has numerous known isotopes, with mass numbers ranging from 22 to 44. ^{28}Si is the most abundant isotope.

Silicon dioxide (SiO_2), also known as silica, is an important oxide of silicon. It is most commonly found in nature as sand or quartz, as well as in the cell walls of diatoms. Silica is the most abundant mineral in the earth's crust. Silicon dioxide is a principal component of most types of glass and substances such as concrete.

Silicon dioxide is attacked by hydrofluoric acid to produce hexafluorosilicic acid. So hydrofluoric acid is used to remove or pattern silicon dioxide in the semiconductor industry.

$$SiO_2 + 6HF = H_2SiF_6 + 2H_2O$$

Silicon dioxide dissolves in hot concentrated alkali.

$$SiO_2 + 2NaOH = Na_2SiO_3 + H_2O$$

Silicon dioxide can react with basic metal oxides [e.g. sodium oxide, potassium oxide, lead(II) oxide, zinc oxide] or mixtures of oxides forming silicates. Glasses as the Si—O—Si bonds in silica are broken successively. As an example the reaction of sodium oxide, SiO_2 can produce sodium orthosilicate, sodium silicate and glasses, depending on the proportions of reactants.

$$2Na_2O + SiO_2 = Na_4SiO_4$$

$$Na_2O + SiO_2 = Na_2SiO_3$$
$$(0.25-0.8)Na_2O + SiO_2 \longrightarrow glasses$$

Silicon is commercially prepared by the reaction of high-purity silica with wood, charcoal, and coal, in an electric arc furnace using carbon electrodes. At temperatures over 1900℃, the carbon reduces the silica to silicon according to the chemical equation.

$$SiO_2 + C = Si + CO_2$$
$$SiO_2 + 2C \longrightarrow Si + 2CO$$

As the second most abundant element in the earth's crust, silicon is vital to the construction of industry as a principal constituent of natural stone, glass, concrete and cement. Silicon's greatest impact on the modern world's economy and lifestyle has resulted from silicon wafers used as substrates in the manufacture of discrete electronic devices.

Germanium

Germanium is a hard, grayish-white solid. It has a metallic luster and the same crystal structure as diamond. It is a semiconductor for electronic devices.

Tin

Tin is a soft, silvery metal. Tin is obtained chiefly from the mineral cassiterite, where it occurs as an oxide. Tin can be highly polished and is used as a protective coat for other metals in order to prevent corrosion or other chemical action. This metal directly combines with chlorine and oxygen, and it displaces hydrogen from dilute acids. Tin is malleable at ordinary temperatures, but it is brittle when it is cooled. Tin has two allotropes at normal pressure and temperature, gray tin and white tin. A third allotrope, called brittle tin, exists at temperatures above 161℃. Tin bonds readily to iron, and they has been used for coating lead or zinc and steel to prevent corrosion. Tin-plated steel containers are widely used for food preservation, and this forms a large part of the market for metallic tin. It can form compounds with $+2$ and $+4$ oxidation states, for example, $SnCl_2$ and $SnCl_4$.

Lead

Lead is a soft, malleable, heavy metal. Lead has a silvery-white color when freshly cut, but it tarnishes to a dull grayish color when it is exposed to air. Lead is used in building construction, lead-acid batteries, bullets and shot, weights, and it is a part of solder, pewter, and fusible alloys. Lead has the highest atomic number of all stable elements. Although the next element bismuth has a half-life so long (longer than the estimated age of the universe) it can be considered stable. Like mercury, another heavy metal, lead is a potent neurotoxin that accumulates in soft tissues and bone over time.

Various oxidized forms of lead are easily reduced to the metal. An example is heating PbO with mild organic reducing agents such as glucose. A mixture of the oxide and the sulfide heated together without any reducing agent will also form the metal.

$$2PbO + PbS = 3Pb + SO_2$$

Metallic lead is attacked only superficially by air, forming a thin layer of oxide that protects it from further oxidation. The metal is not attacked by sulfuric or hydrochloric acids. It does, however, dissolve in nitric acid with the evolution of nitric oxide gas to form dissolved $Pb(NO_3)_2$.

$$3Pb + 8H^+ + 8NO_3^- = 3Pb^{2+} + 6NO_3^- + 2NO + 4H_2O$$

PbO is representative of lead's II oxidation state. It is soluble in nitric and acetic acids, from which solutions it is possible to precipitate halide, sulfate, chromate, carbonate ($PbCO_3$), and basic carbonate [$Pb_3(OH)_2(CO_3)_2$] salts of lead. The sulfide can also be precipitated from acetate solutions. These salts are all poorly soluble in water. Among the halides, the iodide is less soluble than the bromide, which is less soluble than the chloride in turn. The II oxide is also soluble in alkali hydroxide solutions to form the corresponding plumbite salt.

$$PbO + 2OH^- + H_2O = Pb(OH)_4^{2-}$$

Chlorination of plumbite solutions causes the formation of lead's IV oxidation state.

$$Pb(OH)_4^{2-} + Cl_2 = PbO_2 + 2Cl^- + 2H_2O$$

PbO_2 is representative of the IV state, and is a powerful oxidizing agent. The chloride of this oxidation state is formed only with difficulty and decomposes readily into the II chloride and chlorine gas. The bromide and iodide of IV lead are not known to exist. Lead dioxide dissolves in alkali hydroxide solutions to form the corresponding plumbates.

$$PbO_2 + 2OH^- + 2H_2O = Pb(OH)_6^{2-}$$

Lead also has an oxide that is a hybrid between the II and IV oxidation states. Red lead (also called minium) is Pb_3O_4.

Lead(II) forms a series of complexes with chloride, the formation of which alters the corrosion chemistry of the lead. This will tend to limit the solubility of lead in saline media.

$$Pb^{2+} + Cl^- \longrightarrow [PbCl]^+$$
$$[PbCl]^+ + Cl^- \longrightarrow PbCl_2$$
$$PbCl_2 + Cl^- \longrightarrow [PbCl_3]^-$$
$$[PbCl_3]^- + Cl^- \longrightarrow [PbCl_4]^{2-}$$

Lead is a major constituent of the lead-acid battery. It is used as a coloring element in ceramic glazes, notably in the colors red and yellow. Lead is used to form glazing bars for stained glass or other multi-lit windows. The practice has become less common, not for danger but for stylistic reasons.

12.5 The Group VA Elements, the Nitrogen Group Elements

The group VA elements possess the outer-shell electron configuration ns^2np^3. The oxidation number may range from -3 to $+5$. This group consists of nitrogen (N), phosphorus (P), arsenic (As), antimony (Sb) and bismuth (Bi). Nitrogen and phosphorus are nonmetals. Arsenic is semimetal. Antimony and bismuth both are metals. The general properties of group VA elements are listed in Table 12.2.

Table 12.2 The general properties of the group VA elements

Property	N	P	Ar	Sb	Bi
Atomic number	7	15	33	51	83
Valence shell electron configuration	$2s^22p^3$	$3s^23p^3$	$4s^24p^3$	$5s^25p^3$	$6s^26p^3$
Main oxidation number	$-3,+1,+2,$ $+3,+4,+5$	$-3,+3,+5$	$-3,+3,+5$	$-3,+3,+5$	$+3,+5$
Covalent radius/pm	75	110	121	143	152
Ionic radius(M^{3-})/pm	171	212	222	245	—
Ionic radius(M^{3+})/pm	—	—	69	92	108
Ionic radius(M^{5+})/pm	11	34	47	62	74
Ionization energy/(kJ/mol)	1314	1000	941	869	812
Electronegativity	3.04	2.19	2.18	2.05	2.02

Nitrogen

The element nitrogen was discovered by Daniel Rutherford in 1772. Nitrogen occurs in all living organisms, it is a constitute element of amino acids, proteins, and nucleic acids. Nitrogen resides in the chemical structure of almost all neurotransmitters, and it is a defining component of alkaloids, biological molecules produced by many organisms.

Nitrogen is a nonmetal, with an electronegativity of 3.0. It has five electrons in its outer shell and is therefore trivalent in most compounds. The triple bond in molecular nitrogen (N_2) is very stable. It is difficulty for N_2 to react directly with other elements. N_2 is a colorless, odorless, tasteless and mostly inert diatomic gas at standard conditions. It constitutes 78% by volume of Earth's atmosphere.

At atmospheric pressure molecular nitrogen liquifies at $-195.8°C$ and freezes at $-210.0°C$ into the beta (β) hexagonal close-packed crystal allotropic form. Below $-237.6°C$ nitrogen assumes the alpha (α) cubic crystal allotropic form. Liquid nitrogen, a fluid resembling water, but with 80.8% of the density it is a common cryogen.

Unstable allotropes of nitrogen consisting of more than two nitrogen atoms have been produced in the laboratory, like N_3 and N_4. There are two stable isotopes of nitrogen: ^{14}N and ^{15}N. By far the most common is ^{14}N (99.634%), Nitrogen is generally unreactive at standard temperature and pressure. N_2 reacts spontaneously with few reagents, being resilient to acids and bases as well as oxidants and most reductants. When nitrogen reacts spontaneously with a reagent, the net transformation is often called nitrogen fixation.

Lithium burns in an atmosphere of N_2 to give lithium nitride.

$$6Li + N_2 = 2Li_3N$$

Magnesium also burns in nitrogen, forming magnesium nitride.

$$3Mg + N_2 = Mg_3N_2$$

Many industrially important compounds, such as ammonia, nitric acid, organic nitrates, and cyanides, contain nitrogen. The very strong bond in elemental nitrogen dominates nitrogen chemistry. It causes difficulty for both organisms and industry in converting the N_2 into useful compounds, and releases large amounts of energy when these compounds burn or decay back into nitrogen gas. Nitrogen gas has a wide variety of applications including serving as the replacement for air where oxidation is undesirable.

N_2 forms a variety of adducts with transition metals. The first example of a dinitrogen complex is $[Ru(NH_3)_5(N_2)]^{2+}$. Such compounds are now numerous, other examples include $IrCl(N_2)(PPh_3)_2$, $W(N_2)_2(Ph_2CH_2CH_2PPh_2)_2$. These complexes illustrate how N_2 might bind to the metal(s) in nitrogenase and the catalyst for the Haber-Bosch Process.

The starting point for industrial production of nitrogen compounds is the Haber-Bosch process, in which nitrogen is fixed by reacting N_2 and H_2 over a ferric oxide (Fe_3O_4) catalyst at about 500°C and 200 atmospheres pressure. Biological nitrogen fixation in free-living cyanobacteria and in the root nodules of plants also produces ammonia from molecular nitrogen. The reaction, which is the source of the bulk of nitrogen in the biosphere, is catalyzed by the nitrogenase enzyme complex which contains Fe and Mo atoms, using energy derived from hydrolysis of adenosine triphosphate (ATP) into adenosine diphosphate and inorganic phosphate (-20.5 kJ/mol).

The main hydride of nitrogen is ammonia (NH_3). Ammonia is weak basic. In solution, ammonia forms the ammonium ion (NH_4^+).

Other classes of nitrogen anions (negatively charged ions) are the poisonous azides

(N_3^-), which are linear and isoelectronic to carbon dioxide, but bind to important iron-containing enzymes in the body in a manner more resembling cyanide. Another molecule of the same structure is the colorless and relatively inert anesthetic gas dinitrogen monoxide N_2O, it is also known as laughing gas. This is one of a variety of oxides, the most prominent of which are nitrogen monoxide (NO) (known more commonly as nitric oxide in biology), a natural free radical molecule used by the body as a signal for short-term control of smooth muscle in the circulation. Another notable nitrogen oxide compound (a family often abbreviated NO_x) is the reddish and poisonous nitrogen dioxide NO_2, which also contains an unpaired electron and is an important component of smog. Nitrogen molecules containing unpaired electrons show an understandable tendency to dimerize (thus pairing the electrons), and they are generally highly reactive.

The more standard oxides, dinitrogen trioxide N_2O_3 and dinitrogen pentoxide N_2O_5, are actually fairly unstable and explosive a tendency which is driven by the stability of N_2 as a product. The corresponding acids are nitrous HNO_2 and nitric acid HNO_3, with the corresponding salts called nitrites and nitrates. Dinitrogen tetroxide N_2O_4 (DTO) is one of the most important oxidisers of rocket fuels, used to oxidise hydrazine in the Titan rocket and in the recent NASA MESSENGER probe to Mercury. DTO is an intermediate in the manufacture of nitric acid HNO_3, one of the few acids is stronger than hydronium and a fairly strong oxidizing agent.

Phosphorus

Phosphorus is a multivalent nonmetal of the nitrogen group, phosphorus is commonly found in inorganic phosphate rocks. Phosphorus is never found as a free element in nature on Earth for its high reactivity,. Phosphorus is a component of DNA and RNA. It is an essential element for all living cells. The most important commercial use of phosphorus-based chemicals is the production of fertilizers. Phosphorus compounds are also widely used in explosives, nerve agents, friction matches, fireworks, pesticides, toothpaste and detergents.

The two most common allotropes are white phosphorus and red phosphorus. A third form, scarlet phosphorus, is obtained by allowing a solution of white phosphorus in carbon disulfide to evaporate in sunlight. A fourth allotrope, black phosphorus, is obtained by heating white phosphorus under very high pressures (12,000 atmospheres). In appearance, properties and structure of it is very like graphite, being black and flaky, a conductor of electricity. It has puckered sheets of linked atoms. Another allotrope is diphosphorus which is highly reactive. White phosphorus (P_4) exists as individual molecules made up of four atoms in a tetrahedral arrangement, resulting in very high ring strain and instability. It contains 6 single bonds.

White phosphorus is a white, waxy transparent solid. This allotrope is thermodynamically unstable at normal condition and will gradually change to red phosphorus. This transformation, which is accelerated by light and heat, makes white phosphorus almost always contain some red phosphorus and appear yellow. For this reason, it is also called yellow phosphorus. It glows greenish in the dark (when exposed to oxygen), is highly flammable and pyrophoric (self-igniting) upon contact with air as well as toxic (causing severe liver damage on ingestion). The incendiary bomb Napalm relies, among others, on this principle to spontaneously ignite. The odor of combustion of this form has a characteristic garlic smell, and samples are commonly coated with white "(di) phosphorus pentoxide", which consists of P_4O_{10} tetrahedral with oxygen inserted between the phosphorus atoms and at their vertices. White phosphorus is insoluble in water but soluble in carbon disulfide.

White phosphorus (P_4) exists as individual molecules made up of four atoms in a tetrahedral arrangement, resulting in very high ring strain and instability. It contains 6 single bonds.

Arsenic

Arsenic is commonly found as arsenide and arsenate compounds. The most common oxidation states for arsenic are -3, $+3$, and $+5$. Arsenic forms an unstable, gaseous hydride, arsine (AsH_3). Arsenic has two oxides, As_2O_3 and As_2O_5, they are soluble in water and then form acidic solutions. Arsenic (V) acid, like phosphorous acid, is a weak acid. Arsenic and its compounds are used as pesticides, herbicides, and various alloys.

Antimony

Antimony in its elemental form is a silvery white, brittle, fusible and crystalline solid. It exhibits poor electrical and heat conductivity properties. As a metalloid, antimony resembles a metal in its appearance and in many of its physical properties, but it does not chemically react as a metal. Antimony is increasingly being used in the semiconductor industry in the production of diodes, infrared detectors, and Hall-effect devices. As an alloy, this metalloid greatly increases lead's hardness and mechanical strength. The most important use of antimony is as a hardener in lead for storage batteries. Antimony is used in flame-proofing, paints, ceramics, enamels, a wide variety of alloys, electronics, and rubber.

Bismuth

Bismuth compounds are used in cosmetics and medical procedures. As the toxicity of lead has become more apparent in recent years, alloy uses for bismuth metal as a replacement for lead become an increasing part of bismuth's commercial importance.

Bismuth is a brittle metal with a white, silver-pink hue. it often occurs in its native form with an iridescent oxide tarnish showing many refractive colors from yellow to blue. When it is combusted with oxygen, bismuth burns with a blue flame and its oxide forms yellow fumes. Its toxicity is much lower than that of its neighbors in the periodic table such as lead, tin, tellurium, antimony, and polonium.

Though virtually unseen in nature, high-purity bismuth can form distinctive hopper crystals. These colorful laboratory creations are typically sold to collectors. Bismuth is relatively nontoxic and has a low melting point just above 271℃, so crystals may be grown using a household stove, although the resulting crystals will tend to be lower quality than lab-grown crystals. Bismuth oxychloride is sometimes used in cosmetics. Bismuth subnitrate and bismuth subcarbonate are used in medicine. Bismuth subsalicylate [the active ingredient in Pepto-Bismol and (modern) Kaopectate] is used as an antidiarrheal and to treat some other gastro-intestinal diseases. Also, the product bibrocathol is an organic molecule containing bismuth and is used to treat eye infections. Bismuth subgallate (the active ingredient in Devrom) is used as an internal deodorant to treat malodor from flatulence (or gas) and faeces.

12.6 The Group ⅥA Elements, the Oxygen Group Elements

The group VIA of the periodic table includes five chemical elements as oxygen (O), sulfur (S), selenium (Se), tellurium (Te), and polonium (Po). Oxygen and sulfur are nonmetals. Selenium and tellurium are semimetal. Polonium is metal. Because polonium has no stable isotopes and is very rare, the introduction about it will be little. The general properties of group ⅥA Elements are shown in Table 12.3.

Table 12.3 The general properties of group ⅥA Elements

Property	O	S	Se	Te
Atomic number	8	16	34	52
Valence shell electron configuration	$2s^2 2p^4$	$3s^2 3p^4$	$4s^2 4p^4$	$5s^2 5p^4$
Main oxidation number	$-1, -2$	$-2, +2, +4, +6$	$-2, +2, +4, +6$	$-2, +2, +4, +6$
Covalent radius/pm	66	104	117	137
Ionic radius(M^{2-})/pm	132	184	191	211
Electronegativity	3.5	2.5	2.4	2.1

The oxygen group elements are very important. Oxygen makes up nearly 21% of the earth's atmosphere and nearly half of the Earth's crust by mass. Oxygen accounts for two thirds of the mass of the human body and nine tenths of the mass of water. Sulfur is also an abundant element, mined for the extensive production of sulfuric acid.

Oxygen

Oxygen is the element with atomic number 8. It is a highly reactive nonmetallic period 2 element that readily forms compounds with almost all other elements. Oxygen is the third most abundant element in the universe by mass after hydrogen and helium, and it is the most abundant element by mass in the Earth's crust. Oxygen constitutes 88.8% of the mass of water and 20.9% of the volume of air. All major classes of structural molecules in living organisms, such as proteins, carbohydrates, and fats, contain oxygen, as do the major inorganic compounds that comprise animal shells, teeth, and bone.

At standard temperature and pressure, two atoms of the element bind to form dioxygen, O_2. O_2 is a colorless, odorless and tasteless diatomic gas. It has a bond length of 121pm and a bond energy of 498kJ/mol. Oxygen is soluble in water and the solubility of oxygen in water varies with temperature. Oxygen condenses at 90.20K, and freezes at 54.36K. Both liquid and solid O_2 are clear substances with a light sky-blue color.

In nature, free oxygen is produced by the light-driven splitting of water during oxygenic photosynthesis. Green algae and bacteria in marine environments provide about 70% of the free oxygen and the rest is produced by plants. A simplified overall formula for photosynthesis is

$$6CO_2 + 6H_2O + \text{photons} \longrightarrow C_6H_{12}O_6 \text{(glucose)} + 6O_2$$

Oxygen is produced industrially by fractional distillation of liquefied air, electrolysis of water and other means. In the laboratory, oxygen can also be produced by heating potassium chlorate.

$$2KClO_3 = 2KCl + 3O_2 \uparrow \text{ (MnO}_2 \text{ as a catalyst)}$$

The oxygen has three common oxidation states, -2, -1 and 0. The oxidation state of oxygen is -2 in almost all known compounds of oxygen. The oxidation state of -1 is found in a few compounds such as peroxides. The oxidation state of 0 is found in dioxygen. Some compounds containing oxygen in other oxidation states are very uncommon, $-1/2$ (superoxides), $-1/3$ (ozonides), $+1$ (dioxygen difluoride), and $+2$ (oxygen difluoride).

Water (H_2O) is the most familiar oxygen compound. Oxides, such as zinc oxide, ZnO, form when oxygen combines with other elements. Due to its electronegativity, oxygen forms chemical bonds with almost all other elements to give corresponding oxides. However, different elements have different conditions to form oxides. For example, some elements, such as the rusting of iron, readily form oxides at standard conditions for temperature and pressure. The surface of metals like aluminium and titanium are oxidized in the presence of air and become coated with a thin film of oxide that passivates the metal and slows further corrosion. Some of the transition metal oxides are found in nature as nonstoichiometric compounds, with a slightly less metal than the chemical formula would show. For example, the natural occurring FeO (wüstite) is actually written as $Fe_{1-x}O$, where x is usually around 0.05.

Oxygen as a compound is present in the atmosphere in trace quantities in the form of carbon dioxide (CO_2). The earth's crustal rock is composed in large part of oxides of silicon (silica SiO_2, found in granite and sand), aluminium (aluminium oxide Al_2O_3, in bauxite and corundum), iron [iron(Ⅲ) oxide Fe_2O_3, in hematite and rust] and other metals.

The rest of the Earth's crust is also made of oxygen compounds, in particular calcium carbonate (in limestone) and silicates (in feldspars). Water-soluble silicates in the form of Na_4SiO_4, Na_2SiO_3, and $Na_2Si_2O_5$ are used as detergents and adhesives.

Oxygen also acts as a ligand for transition metals, forming metal O_2 bonds with the iridium atom in Vaska's complex, with the platinum in PtF_6 and the iron center of the heme group of hemoglobin.

Oxygen is used as the production of steel, plastics and textiles, rocket propellant, oxygen therapy, and life support in aircraft, submarines, spaceflight and diving. Oxygen is also used medically for patients who require mechanical ventilation, often at concentrations above the 21% found in ambient air. Low pressure pure O_2 is used in space suits.

Most commercially produced O_2 is used to smelt iron into steel. Smelting of iron ore into steel consumes 55% of commercially produced oxygen. In this process, O_2 is injected through a high-pressure lance into molten iron, which removes sulfur impurities and excess carbon as the respective oxides, SO_2 and CO_2. The reactions are exothermic, so the temperature increases to 1700℃.

Another 25% of commercially produced oxygen is used by the chemical industry. Ethylene is reacted with O_2 to create ethylene oxide, which is converted into ethylene glycol in turn. The primary feeder material is used to manufacture a host of products, including antifreeze and polyester polymers (the precursors of many plastics and fabrics).

Most of the remaining 20% of commercially produced oxygen is used in medical applications, metal cutting and welding, as an oxidizer in rocket fuel and water treatment. Oxygen is used in oxyacetylene welding burning acetylene with O_2 to produce a very hot flame. In this process, metal up to 60 cm thick is first heated with a small oxy-acetylene flame and then quickly cut by a large stream of O_2. Rocket propulsion requires a fuel and an oxidizer. Larger rockets use liquid oxygen as their oxidizer, which is mixed and ignited with the fuel for propulsion.

Trioxygen (O_3) is usually known as ozone and is a very reactive allotrope of oxygen that is damaging to lung tissue. It is a rare gas on Earth found mostly in the stratosphere. Ozone is produced in the upper atmosphere when O_2 combines with atomic oxygen made by the splitting of O_2 by ultraviolet (UV) radiation. Since ozone absorbs strongly in the UV region of the spectrum, it functions as a protective radiation shield for the planet. Near the earth's surface, however, it is a pollutant formed as a by-product of automobile exhaust.

Ozone helps protect the biosphere from ultraviolet radiation with the high-altitude ozone layer, but is a pollutant near the surface where it is a by-product of smog.

The metastable molecule tetraoxygen (O_4) was discovered in 2001, and was assumed to exist in one of the six phases of solid oxygen. It was proven in 2006 that that phase, created by pressurizing O_2 to 20 GPa, is in fact a rhombohedral O_8 cluster. This cluster has the potential to be a much more powerful oxidizer than either O_2 or O_3 and may therefore be used in rocket fuel. A metallic phase was discovered in 1990 when solid oxygen is subjected to a pressure of above 96 Gpa and it was shown in 1998 that at very low temperature, this phase becomes superconducting.

Sulfur

Usually, sulfur is a yellow crystalline solid. In nature, it can be found as the pure element, sulfide and sulfate minerals. Sulfur also exists in many types of meteorites. Even it is also found in two amino acids, cysteine and methionine as an essential element for life.

Sulfur has 18 isotopes, but only four of them are stable, $^{32}S(95.02\%)$, $^{33}S(0.75\%)$, $^{34}S(4.21\%)$, and $^{36}S(0.02\%)$. Sulfur forms more than 30 solid allotropes, which is more than any other element. At room temperature, the thermodynamically stable form is the rhombic sulfur, S_8.

The commercial use of sulfur is primarily in fertilizers, but it is also widely used in gunpowder, matches, insecticides and fungicides.

Hydrogen sulfide (H_2S) is the most important sulfide. It has the characteristic smell of rotten eggs. Dissolved in water, hydrogen sulfide is acidic and will react with metals to form a series of metal sulfides. H_2S is normally prepared by the reaction as following.

$$FeS + HCl = H_2S\uparrow + FeCl_2$$

Sulfur can form compounds with many elements. A complex family of compounds usually derived from S^{2-}, common naturally occurring sulfur compounds include the sulfide minerals, such as pyrite (FeS), cinnabar (HgS), galena (PbS), sphalerite (ZnS) and stibnite (SbS).

Sulfur has two important oxides, the dioxide (SO_2) and the trioxide (SO_3). Sulfur dioxide (SO_2) is a colorless gas with a pungent odor, and it is poisonous. Sulfur burns in air to give SO_2. SO_2 is very soluble in water. The solution formed by SO_2 is acidic, it is called as sulfurous acid (H_2SO_3).

$$SO_2(g) + H_2O(l) = H_2SO_3(aq)$$

Sulfurous acid is a diprotic acid, it can form two types salts, sulfites (SO_3^{2-}) and bisulfites (HSO_3^-). Sulfurous acid and the corresponding sulfites are fairly strong reducing agents. Sulfites are heavily used to bleach paper. They are also used as preservatives in dried fruit.

Sulfur trioxide (SO_3) is a corrosive gas with choking odor. SO_3 can produce from SO_2 and O_2. But due to the activation energy barrier, the reaction needs a catalyst to accelerate.

$$SO_2 + O_2 = SO_3$$

Sulfur trioxide reacts violently with water to produce sulfuric acid (H_2SO_4). For this reason, sulfur trioxide is of great commercial importance.

$$SO_3(g) + H_2O(l) = H_2SO_4(aq)$$

Sulfuric acid is classified as a strong acid. It is also a diprotic acid. Consequently, it also forms two series of compounds, sulfates and bisulfates. They have many uses. For example, magnesium sulfate is better known as Epsom salts, it can be used as a laxative, a bath additive, a magnesium supplement for plants, and a desiccant.

Sulfuric acid also reacts with SO_3 in equimolar ratios to form pyrosulfuric acid ($H_2S_2O_7$), pyrosulfuric acid is added to water to form H_2SO_4 again.

$$SO_3(g) + H_2SO_4(l) = H_2S_2O_7(l)$$
$$H_2S_2O_7(l) = SO_3(g) + H_2SO_4(l)$$

Commercial sulfuric acid is 98 percent H_2SO_4. It is a dense and oily liquid with the boiling point of 340℃. It is a powerful dehydrating agent and a moderately good oxidizing agent. For example, hot concentrated sulfuric acid can dissolve inactive copper metal.

$$Cu(s) + 2H_2SO_4(c) = CuSO_4(aq) + 2H_2O(l) + SO_2(g)$$

Elemental sulfur is mainly converted to sulfuric acid (H_2SO_4). It is very important to the world's economies. The production and consumption of sulfuric acid is considered a standard measure of a nation's industrial development. The principal use for H_2SO_4 is the extraction of phosphate ores for the production of fertilizer manufacturing. Other applications of sulfuric acid include wastewater processing, oil refining, and mineral extraction.

12.7 The Group ⅦA Elements, the Halogens

The halogens include a series of nonmetal elements from group ⅦA of the periodic table, comprising fluorine (F), chlorine (Cl), bromine (Br), iodine (I) and astatine (At). All isotopes of astatine are radioactive and unstable, so the study of it is very little.

Owing to their outer electron configuration of ns^2np^5, the halogens have high reactivity, so they are found in the environment only in compounds and as ions. Halide ions and oxoanions such as iodate (IO_3^-) can be found in many minerals and in seawater. Halogenated organic compounds can also be found as natural products in living organisms.

Properties

In their elemental forms, the halogens exist as diatomic molecules, but these only have a fleeting existence in nature, and they are much more common in the laboratory and in industry. At room temperature and pressure, fluorine is yellowish brown gas, chlorine is a pale green gas, bromine is a red volatile liquid, iodine and astatine are solids. Therefore, group ⅦA is the only periodic table group exhibiting all three states of matter at room temperature.

The halogens show a number of trends when moving down the group for instance, decreasing electronegativity and reactivity, and increasing melting and boiling point (see Table 12.4).

Table 12.4 The general properties of group ⅦA Elements

Halogen	Standard Atomic Weight	Melting Point/K	Boiling Point/K	Electronegativity(Pauling)
Fluorine	18.998	53.53	85.03	3.98
Chlorine	35.453	171.6	239.11	3.16
Bromine	79.904	265.8	332.0	2.96
Iodine	126.904	386.85	457.4	2.66
Astatine	(210)	—	—	2.2

Due to the atoms being one electron short of a full outer shell of eight electrons, the simple substance of halogens is usually used as oxidizing agent. The oxidizability of them is weakening from F_2 to I_2, so the replacement reaction can happen among them.

$$Cl_2 + 2Br^- = 2Cl^- + Br_2$$

$$Br_2 + 2I^- = 2Br^- + I_2$$
$$Cl_2 + 2I^- = 2Cl^- + I_2$$

Solubility in Water

Fluorine reacts vigorously with water to produce oxygen (O_2) and hydrogen fluoride (HF).

$$2F_2 + 2H_2O = O_2 + 4HF$$

Chlorine also can react with water to form hydrochloric acid (HCl) and hypochlorous acid, a solution that can be used as a disinfectant or bleach.

$$Cl_2 + H_2O = HCl + HClO$$

Bromine slowly reacts to form hydrogen bromide (HBr) and hypobromous acid (HBrO).

$$Br_2 + H_2O = HBr + HBrO$$

However, iodine is minimally soluble in water (0.03 g/100 g water at 20℃) and does not react with it.

Preparation

The hydrogen fluoride is added to potassium fluoride (KF) to make potassium bifluoride (KHF_2). Electrolysis of potassium bifluoride produces fluorine gas at the anode.

$$HF + KF = KHF_2$$
$$2KHF_2 = 2KF + H_2\uparrow + F_2\uparrow$$

Industrially, elemental chlorine is usually produced by the electrolysis of sodium chloride dissolved in water. Along with chlorine, this electrolytic process yields hydrogen gas and sodium hydroxide, according to the following chemical equation.

$$2NaCl + 2H_2O = Cl_2\uparrow + H_2\uparrow + 2NaOH$$

Because of its commercial availability and long shelf-life, bromine is not typically prepared. Small amounts of bromine can however be generated through the reaction of solid sodium bromide with concentrated sulfuric acid.

$$2NaBr + 3H_2SO_4 = Br_2\uparrow + 2NaHSO_4 + SO_2\uparrow + 2H_2O$$

Uses

Chlorine is a powerful oxidant, and it is used in bleaching and disinfectants. As a common disinfectant, chlorine compounds are used in swimming pools to keep them clean and sanitary. In the upper atmosphere, chlorine-containing molecules have been implicated in the destruction of the ozone layer.

The main applications for bromine are in fire retardants and fine chemicals.

Iodine and its compounds are primarily used in medicine, photography and in dyes.

Interhalogen Compounds

The halogens react with each other to form interhalogen compounds. Diatomic interhalogen compounds such as BrF, ICl, and FCl) bear resemblance to the pure halogens in some respects. The properties and behaviour of a diatomic interhalogen compound tend to be intermediate among those of its parent halogens. Some properties, however, are found in neither parent halogen. For example, Cl_2 and I_2 are soluble in CCl_4, but ICl is not soluble since it is a polar molecule due to the relatively large electronegativity difference between I and Cl.

Hydrogen Halides

The halogens all form binary compounds with hydrogen, the hydrogen halides (HF, HCl, HBr, HI, and HAt), a series of particularly strong acids. When in aqueous solution, the hydrogen halides are known as hydrohalic acids. HAt, should also qualify, but it is not typically included in discussions of hydrohalic acid due to astatine's extreme instability toward alpha decay.

Industrial fluorine production starts with fluorspar, which is heated with sulfuric acid to produce anhydrous hydrogen fluoride.

$$CaF_2 + H_2SO_4 = 2HF\uparrow + CaSO_4$$

Hydrogen chloride also can be obtained by the reaction of available salt with concentrated sulfuric acid.

$$NaCl + H_2SO_4 \stackrel{\triangle}{=} HCl\uparrow + NaHSO_4$$

Hydrogen bromide and hydrogen iodide can not be prepared by their salts and concentrated sulfuric acid, because Br^- and I^- can be further oxidized. Usually, phosphoric acid, H_3PO_4, is used.

$$NaBr + H_3PO_4 = HBr\uparrow + NaH_2PO_4$$
$$NaI + H_3PO_4 = HI\uparrow + NaH_2PO_4$$

The Oxyacids of Halogens

Except fluorine (oxidation number -1 and 0), other halogens exists in all odd numbered oxidation states from -1 to $+7$, as well as the elemental state of zero. So they form different oxyacids and their compounds (see Table 12.5).

Table 12.5 The comparison of different oxyacids

Oxidation Number	Cl	Br	I	The Name of Acid
-1	HCl	HBr	HI	hydrohalic acid
$+1$	HClO	HBrO	HIO	hypohalous acid
$+3$	HClO$_2$	—	—	halous acid
$+5$	HClO$_3$	HBrO$_3$	HIO$_3$	halic acid
$+7$	HClO$_4$	HBrO$_4$	HIO$_4$	perhalic acid

Progressing through the states, hydrochloric acid can be oxidized using manganese dioxide, or hydrogen chloride gas oxidized catalytically by air to form elemental chlorine gas. The solubility of chlorine in water is increased if the water contains dissolved alkali hydroxide. This is due to disproportionation.

$$Cl_2 + 2OH^- = Cl^- + ClO^- + H_2O$$

In hot concentrated alkali solution disproportionation continues

$$2ClO^- = Cl^- + ClO_2^-$$
$$ClO^- + ClO_2^- = Cl^- + ClO_3^-$$

Sodium chlorate and potassium chlorate can be crystallized from solutions formed by the above reactions. If their crystals are heated, they undergo the final disproportionation step.

$$4ClO_3^- = Cl^- + 3ClO_4^-$$

This same progression from chloride to perchlorate can be accomplished by electrolysis. The anode reaction progression is as follows.

Iodine is oxidized to iodate by nitric acid.

$$I_2 + 10HNO_3 = 2HIO_3 + 10NO_2\uparrow + 4H_2O$$

or by chlorates
$$I_2 + 2ClO_3^- = 2IO_3^- + Cl_2\uparrow$$

Iodine is converted in a two stage reaction to iodide and iodate in solutions of alkali hydroxides (such as sodium hydroxide).

$$I_2 + 2OH^- = I^- + IO^- + H_2O$$
$$3IO^- = 2I^- + IO_3^-$$

12.8　The Group ⅧA Elements, the Noble Gases

The group ⅧA elements are generally called the noble gases. They are in the rightmost group (group 0 or group 18 depending on which nomenclature is being used) of the Periodic Table of the Elements. The group ⅧA elements are all gases at room temperature. The noble gases have a completed octet of valence shell electrons except helium (He) with a filled 1s shell. So they are stable and chemically unreactive.

The group ⅧA elements include helium (He), neon (Ne), argon (Ar), krypton (Kr), xenon (Xe) and radon (Rn). They exist in Earth's atmosphere except for radon. Argon is relatively abundant of them.

All Noble Gases are widely used in industrial applications and in scientific research. With the exception of radon they are all used as luminescent gases, e.g. in lasers or in neon lighting. They are also widely used as protective gases, e.g. in semiconductor manufacturing. All the gases, except helium and radon, can be obtained from the fractional distillation of liquid air. Helium (originating from radioactive decay) is not held in the atmosphere by the earth's gravity. However, it is found sealed or trapped deep inside the earth, commonly along with gas and petroleum, from where it is extracted. Helium is the most well-known and most widely used noble gas, e.g. to fill balloons and dirigibles. And usually liquid helium is used as a coolant. Radon is very difficult to obtain as a pure gas, since it is not a stable element. Even so, in the past it has found uses as a radioactive tracer gas for medical applications.

The Compounds of Noble Gases

The noble gases are comparatively stable. For a long time it was thought that the noble gases were unable to perform any chemical reactions and form compounds. In the 1960s, Neil Bartlett prepared the first noble gases' compound. Bartlett made xenon react with PtF_6, a strong oxidant, to obtain a yellow solid, xenon hexafluoro-platinate ($XePtF_6$) successfully. Other noble gas compounds were synthesized after this. However, they haven't found any broad applications and are mainly of academic interest.

Xenon fluorides are all exergonic, and quite stable at normal temperatures. They do react readily with water, even pulling water from air, so they must be kept in anhydrous conditions.

$$XeF_6 + H_2O = XeOF_4 + 2HF$$

Key Words

nonmetal ['nɔn'metl]　非金属（元素）
semimetal ['semi'metl]　半金属（元素），过渡金属（元素）
protium ['prəutiəm]　氕（氢的质量最轻的同位素）
deuterium [dju:'tiəriəm]　氘
tritium ['tritiəm]　氚（氢的放射性同位素）
boron ['bɔ:rən]　硼
hydrogen ['haidrədʒən]　氢
carbon ['kɑ:bən]　碳

silicon ['silikən]　硅，硅元素
germanium [dʒə:'meiniəm]　锗
tin [tin]　锡，马口铁，罐
lead [li:d]　铅
nitrogen ['naitrədʒən]　氮
phosphorus ['fɔsfərəs]　磷
arsenic ['ɑ:sənik]　砷，砒霜
antimony ['æntiməni]　锑
bismuth ['bizməθ]　铋
oxygen ['ɔksidʒən]　氧

sulfur ['sʌlfə] 硫磺，硫黄
selenium [si'li:niəm, -njəm] 硒
tellurium [te'ljuəriəm] 碲
polonium [pə'ləuniəm] 钋
halogen ['hælədʒən] 卤素
fluorine ['flu(:)əri:n] 氟
chlorine ['klɔ:ri:n] 氯
bromine ['brəumi:n] 溴

iodine ['aiədi:n] 碘，碘酒
astatine ['æstəti:n] 砹
helium ['hi:ljəm, -liəm] 氦
neon ['ni:ən] 氖
argon ['ɑ:gɔn] 氩
krypton ['kriptɔn] 氪
xenon ['zenɔn] 氙
radon ['reidɔn] 氡

Exercises

1. What is the relationship of O_2 and O_3?

Answer: O_2 and O_3 is allotrope

2. How to produce large quantities of hydrogen?

Answer: $CH_4(g) + 2H_2O(g) = CO_2(g) + 4H_2(g)$ or $C(s) + H_2O(g) = H_2(g) + CO(g)$

3. Write the oxidation number of nitrogen in the following compounds:
NO, NO_2, N_2O_3, N_2O_5, HNO_3, HNO_2, NH_4Cl

Answer: NO: +2; NO_2: +4; N_2O_3: +3; N_2O_5: +5; HNO_3: +5; HNO_2: +3 NH_4Cl: −3

4. Give two reasons why F_2 is the most reactive of the halogens.

Answer: ① The unexpectedly low value for the electron affinity of fluorine.
② The unexpectedly small bond energy of the F_2 molecule.

5. Why incomplete combustion of carbon can cause poisoning?

Answer: $2C + O_2 = 2CO$, CO is a toxic gas.

6. Write the balanced chemical equation for the reaction that occur between Br^- (aq) and Cl_2

Answer: $2Br^-(aq) + Cl_2(aq) \longrightarrow Br_2(l) + 2Cl^-(aq)$

7. How to produce hydrogen halides?

Answer: $2CaF_2 + H_2SO_4 = 2HF\uparrow + CaSO_4$; $NaCl + H_2SO_4 = HCl\uparrow + NaHSO_4$
$NaBr + H_3PO_4 = HBr\uparrow + NaH_2PO_4$; $NaI + H_3PO_4 = HI\uparrow + NaH_2PO_4$

8. Please discuss the acidity, oxidizability and the thermal stability about the oxyacids of chlorine.

Answer: Acidity: $HClO < HClO_2 < HClO_3 < HClO_4$
Oxidizability: $HClO > HClO_2 > HClO_3 > HClO_4$
Thermal stability: $HClO < HClO_2 < HClO_3 < HClO_4$

Chapter 13 Metals

13.1 The Group IA Elements

The elements of group IA contain lithium (Li), sodium (Na), potassium (K), rubidium (Rb), caesium (Cs), and francium (Fr). They are often called alkali metals, after the alkaline properties of their hydroxides such as NaOH. As with all metals, the alkali metals are malleable, ductile, and are good conductors of heat and electricity. The alkali metals are softer, and they have lower melting, boiling point and lower densities than nearly any other metals. The alkali metals provide the best examples of regular trends with no significant exceptions. All the elements are extremely reactive elements, acting as powerful reducing agents. These metals have only one electron in their outer shell. The relatively weak effective nuclear charge makes it easier to remove the single electron to form positive ions. Thus, alkali metals do not occur freely in nature. They are stored under oil to prevent air oxidation.

Lithium

Lithium is a soft alkali metal with a silver-white color. Under standard conditions, it is the lightest metal and the least dense solid element. Like all alkali metals, lithium is highly reactive. When cut open, lithium exhibits a metallic luster, but returns to a dull silvery grey color quickly because of oxidation in air. For this reason, lithium metal is typically stored under the cover of oil.

In group I A, lithium exhibits some atypical properties. The physical and chemical properties of lithium and its compounds show more similarity with those of magnesium and its compounds than those of other group I A element. This is an example of diagonal relationship.

Like all alkali metals, lithium reacts easily in water, though it is far less dangerous than other alkali metals in this regard. The lithium-water reaction at normal temperatures is brisk but not violent.

$$2Li+2H_2O=2LiOH+H_2$$

In moist air, lithium will react with nitrogen, oxygen, and water vapor. Consequently, the lithium surface becomes coated with a mixture of lithium hydroxide (LiOH), lithium carbonate (Li_2CO_3), and lithium nitride (Li_3N). Lithium is the only alkaline metal that reacts with nitrogen at ambient temperature.

$$Li_2O+H_2O=2LiOH$$
$$6Li+N_2=2Li_3N$$
$$2LiOH+CO_2=Li_2CO_3+H_2O$$

Alkali metals react with oxygen gas to form different types of oxides, simple oxide, peroxide and superoxide. Lithium is the only alkali metal that forms a simple oxide (Li_2O), sodium forms peroxide (Na_2O_2), potassium and the other form superoxide (MO_2).

$$4Li+O_2=2Li_2O$$

Lithium oxide is white ionic solid, which dissolves in water to form alkaline solution of LiOH. It is a basic oxide and forms salts with acidic oxides and acids as usual.

The solubility and thermal stability of lithium compounds are often closer to those of magnesium than to those of other group I A elements. Lithium carbonate, fluoride, hydroxide, and phosphate are much less soluble in water than corresponding salts of the other alkali metals. For example, Alkali metal halides are all soluble in water except LiF. LiCl, LiBr and LiI are much more soluble in polar organic solvent, such as ethanol and acetone, than the halides of Na and K. Alkali metal hydroxides are soluble in water except LiOH. Lithium carbonate and lithium hydroxide are decomposed by heat to form the oxide, whereas the carbonate and hydroxide of other alkali metals are thermally stable.

Lithium is used to make lightweight batteries such as those used in cameras, pacemakers, and computers. Lithium is a desirable battery electrode because of its lightness. It produces high voltage when combined with proper oxidizing agent. Lithium is also used in alloys with other light metals, it imparts high-temperature strength to aluminum and ductility to magnesium when added in small quantities.

Sodium

Sodium is the most abundant alkali metals. The natural abundances in the Earth's crust is about 2.6% by mass. Sodium is also present in great quantities in the Earth's oceans as sodium chloride (common salt). Sodium is a soft, silvery white, highly reactive element, which quickly oxidizes in air and is violently reactive with water, so it must be stored in an inert medium, such as kerosene. At room temperature, sodium metal is so soft that it can be easily cut with a knife.

Compared with other alkali metals, sodium is generally less reactive than potassium and more reactive than lithium. Sodium reacts exothermically with water more violently than lithium, producing very caustic sodium hydroxide and highly flammable hydrogen gas. These are extreme hazards.

$$2Na + 2H_2O = 2NaOH + H_2 \uparrow$$

When burned in air, sodium forms sodium peroxide Na_2O_2 as the main product, or with limited oxygen, the oxide Na_2O. If burned in oxygen under pressure, sodium superoxide NaO_2 will be produced.

$$4Na + O_2 = 2Na_2O$$
$$2Na + O_2 = Na_2O_2$$
$$Na + O_2 = NaO_2$$

Sodium peroxide is hydrolyzed by water to form sodium hydroxide plus hydrogen peroxide.

$$Na_2O_2 + 2H_2O = 2NaOH + H_2O_2$$

Solution of sodium peroxide is unstable and gradually loses oxygen due to the decomposition of H_2O_2.

$$2Na_2O_2 + 2H_2O = 4NaOH + O_2 \uparrow$$

The decomposition is rapid in hot solution.

In chemistry preparations, sodium peroxide is used as an oxidizing agent. It is also used as an oxygen source by reacting it with carbon dioxide to produce oxygen and sodium carbonate. It is thus particularly useful in scuba gear, submarines

$$2Na_2O_2 + 2CO_2 = 2Na_2CO_3 + O_2 \uparrow$$

Sodium can reduce hydrogen gas to form the ionic hydride (salt-like hydride).

$$2Na + H_2 = 2NaH$$

NaH is an industrial base and reducing agents.

Sodium hydroxide is an important industrial chemical, being easily formed by the reaction of oxide with water and soluble in water behaving as strong base. Commercially, it is produced by the electrolysis of an aqueous solution of sodium chloride.

$$2NaCl + 2H_2O = H_2 + 2NaOH + Cl_2$$

Sodium hydroxide is hygroscopic solids which attack the skin, hence the old name, caustic soda. It is widely used as source of the hydroxide ion OH^- both in laboratory and on a large scale.

$$NaOH = Na^+ + OH^-$$

It is a strong alkaline, and it reacts readily with acids to form water and their corresponding salt.

$$NaOH + HCl = NaCl + H_2O$$

It reacts with carbon dioxide in the air to give sodium carbonate.

$$2NaOH + CO_2 = Na_2CO_3 + H_2O$$

With excess carbon dioxide, i.e. if the gas is passed through a solution of the hydroxide, a hydrogencarbonate is formed.

$$NaOH + CO_2 = NaHCO_3$$

Sodium carbonate is a white solid, which is thermally stable. It is soluble in water, and gives alkaline solutions due to hydrolysis.

$$CO_3^{2-} + H_2O = HCO_3^- + OH^-$$

Sodium carbonate is used in the manufacture of glass, soaps and caustic soda.

Sodium bicarbonate is a white solid. Sodium bicarbonate is an amphoteric compound. Aqueous solutions are mildly alkaline.

$$HCO_3^- + H_2O = H_2CO_3 + OH^-$$

NaHCO$_3$ may be obtained by the reaction of excess carbon dioxide with an aqueous solution of sodium hydroxide. The initial reaction produces sodium carbonate.

$$CO_2 + 2NaOH = Na_2CO_3 + H_2O$$

Further addition of carbon dioxide produces sodium bicarbonate, which at sufficiently high concentration will precipitate out of solution.

$$Na_2CO_3 + CO_2 + H_2O = 2NaHCO_3$$

Sodium reacts vigorously with halogens to produce ionic halides, the most important of which is NaCl. Sodium chloride, salt, is the primary sodium compound. It is currently mass-produced by evaporation of seawater or brine from other sources, such as brine wells and salt lakes, and by mining rock salt, called halite. Sodium chloride is a source of many chemicals, including sodium metal, chlorine gas, hydrochloride and sodium hydroxide.

Owing to its high reactivity, sodium is found in nature only as a compound and never as the free element. Sodium is found in many different minerals, of which the most common is sodium chloride, which occurs in vast quantities dissolved in seawater, as well as in solid deposits. Sodium is now produced commercially through the electrolysis of liquid sodium chloride. The electrolysis of molten NaCl yields sodium metal and Cl$_2$ gas, which is industrially carried out in the Downs cell in which the NaCl is mixed with calcium chloride. The melting point of NaCl is 801 ℃, too high a temperature to carry out this electrolysis economically, adding calcium chloride to the mixture lowering the melting point below 700 ℃. As calcium is less electropositive than sodium, no calcium will be formed.

$$2NaCl(l) \xrightarrow{electrolysis} 2Na(l) + Cl_2(g)$$

The most important use of sodium is as a reducing agent, for example, in obtaining metals such as titanium, zirconium.
$$4Na + TiCl_4 = 4NaCl + Ti$$
Another use of sodium metal is as heat-transfer medium in nuclear reactors because of the low melting point, high boiling point and low vapor pressure.

Potassium

Potassium is the second most abundant alkali metals. The natural abundance in the Earth's crust is 2.4% by mass. Potassium is silvery when first cut but it oxidizes rapidly in air and tarnishes within minutes, so it is generally stored under oil or grease. In many respects, potassium and sodium are chemically similar.

When burning in air, K forms the superoxide KO_2 as the main product.
$$K + O_2 = KO_2$$
An important use of KO_2 is used as a portable source of oxygen and a carbon dioxide absorber. It is widely used in portable respiration systems, submarines and spacecraft as it takes less volume than $O_2(g)$.
$$4KO_2 + 2CO_2 = 2K_2CO_3 + 3O_2 \uparrow$$
Like the other alkali metals, potassium reacts violently with water producing hydrogen. The reaction is notably more violent than that of lithium or sodium with water, and generates sufficient heat to ignite the evolved hydrogen.
$$2K + 2H_2O = H_2 \uparrow + 2KOH$$
Like sodium hydroxide, potassium hydroxide is soluble in water behaving as strong base. It is produced by the electrolysis of potassium chloride solutions, analogous to the method of manufacturing sodium hydroxide.
$$2KCl + 2H_2O = 2KOH + H_2 \uparrow + Cl_2 \uparrow$$
It is hygroscopic solids which attack the skin, hence the old name, caustic potash. It is also widely used as source of the hydroxide ion OH^- both in laboratory and on a large scale.
$$KOH = K^+ + OH^-$$
It reacts strongly with carbon dioxide to produce potassium carbonate, and is used to remove traces of CO_2 from air.

Potassium chloride is the important compound of potassium, which is obtained from naturally occurring brines. The majority of the potassium chloride produced is used for making fertilizer, because potassium is a major essential element for plant growth. Potassium chloride is also used as raw material in the manufacture of KOH, KNO_3, and other industrially important compounds of potassium.

Potassium carbonate is a white salt, soluble in water, which forms a strongly alkaline solution. It is prepared commercially by carbonated using carbon dioxide of potassium hydroxide, the product of the electrolysis of potassium chloride.
$$2KOH + CO_2 = K_2CO_3 + H_2O$$
Potassium carbonate is used in the production of soap and glass.

Potassium bicarbonate, also known as potassium hydrogen carbonate or potassium acid carbonate, is a colorless, odorless, slightly basic and salty substance. It is manufactured by reacting potassium carbonate with carbon dioxide and water.
$$K_2CO_3 + CO_2 + H_2O = 2KHCO_3$$
Most potassium occurs in the Earth's crust as minerals, such as feldspars and clays.

Potassium cannot be obtained by the electrolysis of KCl, because potassium is soluble in the melt. It is produced by the reduction of molten KCl by liquid sodium.
$$KCl(l) + Na(l) = K(g) + NaCl(l)$$

Rubidium

Rubidium is a soft, silvery-white metallic element of the alkali metal group. It is one of the most electropositive and most alkaline elements. Rubidium is very soft and highly reactive, with properties similar to other elements in group I A. Rubidium can be liquid at ambient temperature, but only on a hot day given that its melting point is about 40 ℃. It ignites spontaneously in air and reacts violently with water and even with ice at -100 ℃.

Rubidium and its salts have few commercial uses. The metal is used in the manufacture of photocells and in the removal of residual gases from vacuum tubes. Rubidium salts are used in glasses, ceramics and in fireworks to give them a purple color. Potential uses are in ion engines for space vehicles, as working fluid in vapor turbines, and as getter in vacuum tubes.

Cesium

Cesium is a soft, silvery-gold alkali metal with a melting point of 28 ℃, which makes it one of the metals that are liquid at or near room temperature. It is the most electropositive and most alkaline element. Cesium metal oxidized rapidly when exposed to the air and can form the dangerous superoxide on its surface. Cesium reacts explosively in cold water and also reacts with ice at -116 ℃. Cesium hydroxide (CsOH) is a very strong base and will rapidly etch the surface of glass.

Francium

Francium is the second rarest naturally occurring element, after astatine. It is the least electronegative of all the known elements. Francium is a highly radioactive metal that decays into astatine, radium, and radon. Its most stable isotope, francium-223, has a half-life of only 22 minutes.

13.2　The Group II A Elements

The Group II A consists of beryllium (Be), magnesium (Mg), calcium (Ca), strontium (Sr), barium (Ba) and radium (Ra). The elements are often called alkaline earths metals, which are named after their oxides, beryllia, magnesia, lime, strontia and baryta. These oxides are basic (alkaline) when combined with water. "Earth" is an old term applied by early chemists to nonmetallic substances that are insoluble in water and resistant to heating, properties shared by these oxides. Alkaline earths metals are silver colored, generally harder than alkali metals and have higher melting and boiling points, as well as greater ionization energy due to stronger effective nuclear charge. Alkaline earths metals have the valence shell electron configuration ns^2, the energetically preferred state of achieving a filled electron shell is to lose two electrons to form doubly charged positive ions M^{2+}. All elements, with the exception of beryllium, have very similar chemical properties. Like the alkali metals, alkaline earths metals are strong reducing agents.

Beryllium

Beryllium is a relatively rare element in both the Earth and the universe. It is a steel grey, strong, light-weight yet brittle alkaline earth metal.

Beryllium is distinctly different from the other elements in group ⅡA in its chemical properties. It participates primarily in covalent rather than ionic bonding. In some ways the compounds of beryllium share similar characteristics with those of Al, suggesting a diagonal relationship.

When added to water, beryllium is totally unreactive, and doesn't even react with steam.

BeO does not react with water, whereas the oxide of other alkaline earths metals form hydroxide. BeO is amphoteric, which means that it has the properties of both an acid and a base, whereas, all oxides of other alkaline earth metals are basic. BeO dissolve in acid to form Be^{2+}, and in strong base solution to form BeO_2^{2-}.

$$BeO + 2HCl = BeCl_2 + H_2O$$
$$BeO + NaOH = NaHBeO_2$$

Beryllium hydroxide is also amphoteric. The two reactions below show this

$$Be(OH)_2 + H_2SO_4 = BeSO_4 + 2H_2O$$
$$Be(OH)_2 + 2NaOH = Na_2Be(OH)_4$$

This distinguishes beryllium hydroxide from the other hydroxides of group ⅡA which are not amphoteric. This amphoterism is also shown by aluminum hydroxide in group ⅢA.

Beryllium forms halides by direct combination. As a consequence of the high ionization energy of beryllium, its halides are essentially covalent, whereas the halides of other members of group ⅡA are ionic. In the gaseous state, beryllium chloride consists of linear molecules. In the solid state it exists as a structure somewhere between an infinite assembly of chains and an ionic lattice (Figure 13.1). The halides are soluble in organic solvents.

Beryllium is found in some 30 mineral species, the most important of which are bertrandite, beryl chrysoberyl and phenacite. Aquamarine and emerald are precious forms of beryl. Beryl and bertrandite are the most important commercial sources of beryllium and its compounds. Most of the metal is now prepared by reducing beryllium fluoride with magnesium metal. First, the mineral is converted to BeF_2, which is then reduced with magnesium to beryllium. The reaction is carried out at about 1000 ℃.

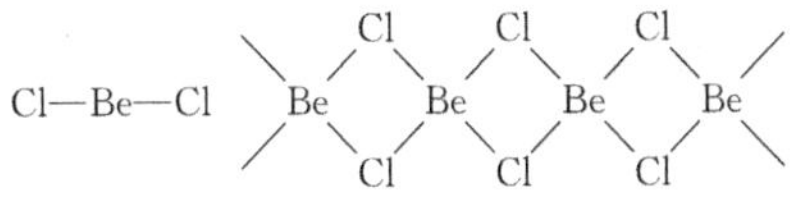

Figure 13.1 Structure of gaseous and solid beryllium chloride

$$BeF_2 + Mg = Be + MgF_2$$

As the lightest of the alkaline earth metals, beryllium is widely used in alloy with copper, nickel and other metals. For example, when added in small amount to copper, beryllium increase the strength of metal dramatically and improves the corrosion resistance while preserving high conductivity and other desirable property.

Magnesium

Magnesium rank 6th most abundant elements in the Earth's crust, with a natural abundance of 2.76% by mass. Several magnesium compounds occur naturally, either in mineral form or in brine. These include the carbonate, chloride, hydroxide, and sulfate. Other magnesium compounds can be prepared from these.

The chemical properties of magnesium are dominated by the strong reducing power.

Magnesium form oxides when it is exposed to air at room temperature, although unlike the alkaline metals, storage in an oxygen-free environment is unnecessary. Because magnesium is protected by a thin layer of oxide which is fairly impermeable and hard to remove.

$$2Mg+O_2=2MgO$$

Mg does not react with water, but it reacts with steam to form $Mg(OH)_2$ and H_2.

$$Mg+H_2O=Mg(OH)_2+H_2\uparrow$$

Magnesium also reacts exothermically with most acids, such as hydrochloric acid (HCl), producing the chloride of the metal and releases hydrogen gas.

$$Mg+2H^+=Mg^{2+}+H_2\uparrow$$
$$Mg_3N_2+6H_2O=Mg(OH)_2+2NH_3$$

The hot metals are also sufficiently strong reducing agents to reduce nitrogen gas and form nitrides.

$$3Mg+N_2=Mg_3N_2$$

Magnesium can also reduce and burn in carbon dioxide.

$$2Mg+CO_2=2MgO+C$$

Magnesium reacts with hydrogen to form ionic hydrides.

$$Mg+H_2=MgH_2$$

The hydrides react with water to produce hydrogen gas and can be a source of fuel.

$$MgH_2+2H_2O=Mg(OH)_2+2H_2$$

Magnesium oxide is obtained by the calcination of $MgCO_3$. It is a basic oxide and dissolves in acid to form Mg^{2+} solution.

$$MgO+2H^+=Mg^{2+}+H_2O$$

Magnesium oxide is not very soluble in water, but its aqueous suspection behaves as a moderately strong base.

Magnesium hydroxide can be precipitated by the metathesis reaction among magnesium salts and sodium, potassium, or ammonium hydroxide.

$$Mg^{2+}+2OH^-=Mg(OH)_2$$

Magnesium hydroxide is not very soluble in water. While the solubility of magnesium hydroxide is low, all of the magnesium hydroxide which dissolves in the water does dissociate. Since the dissociation of this small amount of dissolved magnesium hydroxide is complete, magnesium hydroxide is considered a strong base.

Magnesium is the third most abundant structural metal in the earth's crust, only exceeded by aluminum and iron. It is found in many rocky minerals, like dolomite, magnetite, olivine and serpentine. It's also found in seawater, underground brines and salty layers. Magnesium is produced by the electrolysis of molten $MgCl_2$, yielding Mg metal and Cl_2 gas.

$$MgCl_2(l) \xrightarrow{electrolysis} Mg(l)+Cl_2(g)$$

With a density of only two thirds of the aluminum's, magnesium has countless applications in cases where weight reducing is important, i.e. in aeroplane and missile construction. It also has many useful chemical and metallurgic properties, which make it appropriate for many other non-structural applications.

Calcium

Calcium rank 5th most abundant elements in the Earth's crust, with a natural abundance of 4.66% by mass. Calcium does not occur as the metal itself in nature. Instead it is found in various minerals including limestone, gypsum and fluorite.

Calcium rapidly forms a grey-white oxide and nitride coating when exposed to air. It is somewhat difficult to ignite, unlike magnesium, but when lit, the metal burns in air with a brilliant high-intensity red light.

$$2Ca + O_2 = 2CaO$$

Calcium metal reacts with water, evolving hydrogen gas at a rate rapid enough to be noticeable, but not fast enough at room temperature to generate much heat. The slowness of the reaction results from the metal being partly protected by sparingly soluble white calcium hydroxide.

$$Ca + 2H_2O = Ca(OH)_2 + H_2 \uparrow$$

In aqueous solutions of acids where the salt is water soluble, calcium reacts vigorously.

$$Ca + 2H^+ = Ca^{2+} + H_2$$

Calcium can reduce hydrogen gas when heated, forming the hydride.

$$Ca + H_2 = CaH_2$$

The hydride is stable in dry air, but it reacts with water to produce hydrogen gas and can be a source of fuel.

$$CaH_2 + H_2O = Ca(OH)_2 + H_2 \uparrow$$

Calcium oxide (CaO), commonly known as burnt lime, lime or quicklime, is a cheapest and most widely used bases. It is usually made by the thermal decomposition of materials such as limestone in a lime kiln.

$$CaCO_3 \stackrel{\triangle}{=\!=} CaO + CO_2 \uparrow$$

Calcium oxide reacts with water to form hydroxide (slaked lime).

$$CaO + H_2O = Ca(OH)_2$$

Quicklime is used to treat acidic soils of lawns, gardens, and farmland. It is also used to treat excess acidity in lakes.

$$CaO + 2H^+ = Ca^{2+} + H_2O$$

Quicklime is also used in air pollution control. When coal is burned in an electric power plant, sulfur in the coal is converted to SO_2, a toxic gas that forms acid rain. When powdered limestone is mixed with powered coal before combustion, the limestone decomposes to CaO, which reacts with SO_2 gas to form calcium sulfite.

$$CaO + SO_2 = CaSO_3$$

Calcium hydroxide, known as slaked lime, is an important compound of calcium. It is formed by the action of water on CaO. It is sparingly soluble in water. A suspension of fine calcium hydroxide particles in water is called milk of lime. The solution is called lime water and is a medium strength base.

$Ca(OH)_2$ is used to test for the acidic gas carbon dioxide by being bubbled through a solution. It turns cloudy where CO_2 is present.

$$Ca(OH)_2 + CO_2 = CaCO_3 + H_2O$$

$Ca(OH)_2$ is commercially used as a cheap base, used in all applications where high water solubility is not essential.

Calcium carbonate ($CaCO_3$) is one of the common compounds of calcium. Limestone is a naturally occurring form of calcium carbonate.

Although calcium carbonate is not very soluble in water, it is basic. Thus it dissolves readily in acidic solution. It reacts with strong acids and releases carbon dioxide.

$$CaCO_3 + 2HCl = CaCl_2 + CO_2 \uparrow + H_2O$$

It reacts with water saturated with carbon dioxide to form the soluble calcium bicarbonate.

$$CaCO_3 + CO_2 + H_2O = Ca(HCO_3)_2$$

This reaction is important in the erosion of carbonate rocks, forming caverns, and leads to hard water in many regions.

The main use of calcium carbonate is in the construction industry. Limestone is used as building stones. It is also used to manufacture other building materials, for example, quicklime (CaO) and slaked lime [$Ca(OH)_2$]. Quicklime and slaked lime are the cheapest and most widely used bases. In additional, pure $CaCO_3$ is extensively used as filler, providing bulk materials such as paint, plastics, printing inks, and rubber. It is also used in toothpastes, food, cosmetics, and antacids.

Calcium is not naturally found in its elemental state. Commercially it can be made by the electrolysis of molten calcium chloride, $CaCl_2$.

$$CaCl_2 = Ca + Cl_2 \uparrow$$

Strontium

Strontium is a soft silver-white or yellowish metallic element that is highly reactive chemically. Strontium reacts with water on contact to produce strontium hydroxide and hydrogen gas. It burns in air to produce both strontium oxide and strontium nitride, but since it does not react with nitrogen below 380 ℃ it will only form the oxide spontaneously at room temperature. It should be kept under kerosene to prevent oxidation.

The primary use for strontium compounds is in glass for color television cathode ray tubes to prevent X-ray emission.

Due to its extreme reactivity with oxygen and water, this element occurs naturally only in compounds as the form of the sulfate mineral celestite ($SrSO_4$) and the carbonate strontianite ($SrCO_3$). The metal can be prepared by electrolysis of melted strontium chloride mixed with potassium chloride.

$$SrCl_2 = Sr + Cl_2 \uparrow$$

Barium

Barium is a soft silvery metallic alkaline earth metal. It is never found in nature in its pure form due to its reactivity with air. Barium is chemically similar to calcium and strontium, but more reactive. This metal oxidizes very easily when exposed to air. It is highly reactive with water or alcohol, while producing hydrogen gas. Burning in air or oxygen produces not only barium oxide (BaO) but also the peroxide.

All water or acid soluble barium compounds are extremely poisonous. Barium sulfate ($BaSO_4$) is highly insoluble in water, and is eliminated completely from the digestive tract. So barium sulfate can be taken orally and used as a radiocontrast agent for X-ray imaging of the digestive system ("barium meals" and "barium enemas").

Barium is surprisingly abundant in the Earth's crust. The most common naturally occurring minerals are the very insoluble barium sulfate, $BaSO_4$ (barite), and barium carbonate, $BaCO_3$ (witherite). Barium is extracted from the mineral barite which is crystallized barium sulfate. Because barite is so insoluble, the ore is heated with carbon to reduce it to barium sulfide.

$$BaSO_4 + 2C = BaS + 2CO_2$$

The barium sulfide is then hydrolyzed or reacted with acids to form other barium compounds such as the chloride, nitrate. Barium is commercially produced through the electrolysis of molten barium chloride ($BaCl_2$).

$$BaCl_2 = Ba + Cl_2$$

Radium

Radium is a radioactive alkaline earth element. Its appearance is almost pure white, but it readily oxidizes on exposure to air, turning black. It is extremely radioactive. Radium has 25 different known isotopes. ^{226}Ra is the most common and most stable isotope, with a half-life of 1602 years.

This metal is found in tiny quantities in the uranium ore pitchblende, and various other uranium minerals. Radium, chemically similar to barium, reacts violently with water to form radium hydroxide and is slightly more volatile than barium.

13.3 The Group ⅢA Elements

Metal aluminum, gallium, indium, and thallium are among the group ⅢA Elements. These elements have electron configuration $(ns)^2(np)^1$ and the chemistry is dominated by the +3 oxidation state.

Aluminum

Aluminum is in fact the most abundant metal, the third most abundant element in nature. It is commercially the most important metal. Pure aluminum is malleable, ductile, silvery-colored, and light. When exposed to air, it has a dull silvery appearance because of a thin layer of oxidation that forms quickly.

Like most of other main group metals, aluminum is an active metal. Because it is easily oxidized to +3 ion, it is a strong reducing agent.

Aluminum is easily oxidized by oxygen.

$$4Al + 3O_2 = 2Al_2O_3$$

The oxide forms thin impervious film which does not react with water or oxygen, and it preserves the aluminum metal underneath it from chemical attack.

Aluminum is amphoteric. It reacts with dilute HCl and H_2SO_4 to form $AlCl_3$ and $Al_2(SO_4)_3$, respectively.

$$2Al + 6HCl = 2AlCl_3 + 3H_2 \uparrow$$
$$2Al + 3H_2SO_4 = Al_2(SO_4)_3 + 3H_2 \uparrow$$

But the reaction with both dilute and concentrated nitric acid produces Al_2O_3, an impervious oxide layer, which forms a protective coating and prevents further oxidation by the acid.

$$2Al + 2HNO_3 = Al_2O_3 + 2NO \uparrow + H_2O$$

It reacts with a strong base to form hydrous aluminum ion and hydrogen gas.

$$2Al + 6H_2O + 2OH^- = 2[Al(OH)_4]^- + 3H_2 \uparrow$$

Aluminum oxide or alumina Al_2O_3 exists in two polymorphic forms, α-alumina or corundum and γ-alumina.

γ-Alumina is a white solid and insoluble in water with a very high melting point. It is amphoteric, it reacts violently with strong bases to form aluminates, and with strong acids to form salt-like substances. It has a large surface area, and it is a good absorbent, so it is widely used in dehydration, decolorisation and chromatography.

α-Alumina or corundum is the most common form of crystalline aluminum oxide. Corundum is insoluble in acids and alkalis, and it can only be brought into solution by first fusing it with sodium or potassium hydroxide when an aluminate is formed.

Hardness of corundum is only exceeded by diamond, boron nitride, and carborundum. Corundum may be colored by traces of impurity. Rubies are given their characteristic deep red color and their laser qualities by traces of the metallic element chromium. Sapphires come in different colors given by various other impurities, such as iron and titanium.

Aluminum hydroxide is also amphoteric, and it will dissolve in acid and alkaline solution. When it dissolves in acid, $Al(H_2O)_6^{3+}$ or its hydrolysis products. It also dissolves in strong alkali, forming $Al(OH)_4^-$.

$$Al(OH)_3 + OH^- = [Al(OH)_4]^-$$
$$Al(OH)_3 + 3H_2O + 3H^+ = [Al(H_2O)_6]^{3+}$$

Aluminum can form halides with all halogens. Aluminum fluoride (AlF_3) has considerable ionic character, it has high melting point, When it is molten it is a conductor of electric current. In contrast, the other aluminum halides are covalent and dimeric Al_2X_6 in vapour state.

Aluminum chloride is the most important aluminum halide. It can be prepared not only by the direct combination of the elements but also by the passage of dry hydrogen chloride over heated aluminum.

$$2Al + 3Cl_2 = Al_2Cl_6$$
$$2Al + 6HCl = Al_2Cl_6 + 3H_2 \uparrow$$

The dimeric structure of Al_2Cl_6 consists of two chlorine atom bonded exclusively to each aluminum atom and two chlorine atoms bridging the two aluminum atoms (see Figure 13.2). Bonding in this molecule can be described by sp^3 hybridization of two aluminum atoms. Each bridging chlorine atoms appear to bond to two aluminum atom in two ways. The bond to one aluminum atom is a conventional covalent bond, each atom contributes one electron to the bond. The bond to a second aluminum atom is a coordinate covalent bond, where the chlorine atom provides the pairs of electrons for the bond.

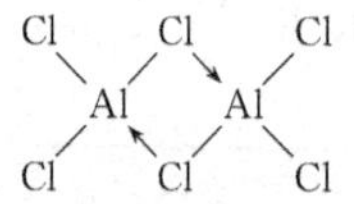

Figure 13.2 The structure of Al_2Cl_6

Aluminum halides are electron-deficient compound. They easily accept a pair of electrons, so aluminum halides are very reactive Lewis acid. They find widespread application in organic chemistry as a catalyst, such as, Friedel-Crafts acylations and alkylations reactions.

Aluminum is an abundant element in Earth's crust, but it is very rare in its free form. It is found in soils as aluminum silicates and in bauxite. Pure aluminum was produce by electrolysis of molten Al_2O_3-cryolite (Na_3AlF_6) mixture. Bauxite, which contains hydrated oxides such as $Al_2O_3 \cdot H_2O$, was used as the source of aluminum. Bauxite contains Fe_2O_3 as an impurity that must be removed first. Bauxite is dissolved in hot sodium hydroxide solution, and insoluble (mostly Fe_2O_3) are filtered off.

$$Al_2O_3 + 2OH^- + 3H_2O \longrightarrow 2[Al(OH)_4]^-$$

The solution containing $[Al(OH)_4]^-$ is diluted with water and slightly acidified to precipitate the hydroxide, $Al(OH)_3$.

$$[Al(OH)_4]^- + H_3O^+ \longrightarrow Al(OH)_3 + 2H_2O$$

$Al(OH)_3$ is then calcined to give aluminum oxide.

$$2Al(OH)_3 \longrightarrow Al_2O_3 + 3H_2O$$

Because of the high melting point of Al_2O_3 (>2000 ℃), electrolysis is not carried out on pure Al_2O_3. Instead, Al_2O_3 is dissolved in molten cryolite, Na_3AlF_6, which melts at a significantly lower temperature (<1000 ℃).

Pure aluminum is not very strong, but its alloys with copper, magnesium, and silicon are very strong and light, and they are suitable for the construction of aircraft parts. Because of its excellent electrical conductivity and low density, aluminum is widely used in the electrical industry. Aluminum is also used to make beverage cans, cookware, and aluminum foil.

Gallium

Gallium is a soft silvery metallic poor metal. Gallium is a brittle solid at low temperatures. The melting point temperature of 29.76 ℃ allows the metal to be melted in one's hand, thus, gallium can be used in high-temperature thermometers. Gallium is found and extracted as a trace component in bauxite, coal, diaspore, germanite, and sphalerite. Gallium does not exist in free form in nature, but it is easily obtained by smelting. Ultra-pure gallium has a beautiful, silvery appearance and the solid metal exhibits a conchoidal fracture similar to glass. The metal expands 3.1 percent on solidifying. Therefore, it should not be stored in glass or metal containers, because they may break as the metal solidifies. The compounds gallium nitride and gallium arsenide, used as a semiconductor, most notably in light-emitting diodes (LEDs).

Indium

Indium is a very soft, silvery-white metal. The most common isotope is slightly radioactive. It very slowly decays by beta emission to tin. This radioactivity is not considered hazardous, mainly because its half-life is 4.41×10^{14} years, four orders of magnitude larger than the age of the universe and nearly 50000 times longer than that of natural thorium. This metal is chemically similar to aluminum or gallium but more closely resembles zinc. Indium is most frequently associated with zinc materials, and it is produced mainly from residues generated during zinc ore processing. Its current primary application is to form transparent electrodes from indium tin oxide in liquid crystal displays. It's also used in making bearing alloys, germanium transistors, rectifiers, thermistors, and photoconductors. It can be plated onto metal and evaporated onto glass, forming a mirror as good as that made with silver but with more resistance to atmospheric corrosion.

Thallium

Thallium is a very soft and malleable metal, which can be cut with a knife. Twenty five isotopic forms of thallium, with atomic masses ranging from 184 to 210 are recognized. Natural thallium is a mixture of two isotopes. When freshly exposed to air, thallium exhibits a metallic luster, but soon develops a bluish-gray tinge, resembling lead in appearance.

Oddly, thallium acts much like an alkali metal in its physical and chemical properties. Thallium tends to appear in the +1 (thallous) oxidation state as well as the +3 (thallic) oxidation state. Thallous oxide (Tl_2O) and the metal react violently with water to form a hydroxide TlOH that completely dissociates into Tl^+ and hydroxide OH^- ions as if it were an alkali metal hydroxide. Its halides are quite salt-like, dissociating into ions into solutions that, like solutions of salts of the alkali metals, conduct electricity.

Thallium occurs in crooksite, lorandite, and hutchinsonite. It is also present in pyrites, and it is recovered from the roasting of this ore in connection with the production of sulfuric acid. It is also obtained from the smelting of lead and zinc ores.

Approximately 60%~70% of thallium production is used in the electronics industry, and the rest is used in the pharmaceutical industry and glass manufacturing. It is also used in infrared detectors. Thallium compound thallium sulfate is highly toxic and has been widely employed in rat poisons and insecticides. It is odorless and tasteless, giving no warning of its presence, however its use has been cut back or eliminated in many countries.

13.4 A Survey of Transition Metals

Properties of the Transition Metals

Transition elements (also called transition metals) lie in the middle part of the periodic table from group ⅢB to group ⅡB in periods 4, 5, and 6. These three rows of transition elements are usually called the first, second, and third transition series. All of them except palladium have either one or two s electrons in the outermost shell. In a given transition series, the elements differ from each other in the number of d electrons in the $(n-1)$ shell or f electrons in the $(n-2)$ shell.

The transition elements have some properties in common.

Except mercury, the transition elements are all metals, and are solids at room temperature. Most of them have high melting, boiling points and high densities. They are also hard and strong. Going across the period from the fourth to sixth row, melting points, boiling points and densities all reach maximums near the center of the transition series. As we all know, the more considerable covalent bonding takes place, the more strongly atoms are bonded to each other, so the atoms with half-filled valence orbits are hard, strong, and high-melting.

Most transition elements exhibit multiple oxidation states. For example, the compounds of Mn have several oxidation states, which are $+2$, $+3$, $+4$, $+5$, $+6$ and $+7$ oxidation states. Among them, the $+2$ (Mn^{2+}) and $+7$ (MnO_4^-) oxidation states are commonly found in solution, while the $+4$ oxidation state (as in MnO_2) are found in solid, and the rest are less found.

Table 13.1 lists the common nonzero oxidation states of the first transition-series elements.

Table 13.1 Properties of the First Transition-Series Elements

Group	ⅢB	ⅣB	ⅤB	ⅥB	ⅦB	ⅧB	ⅧB	ⅧB	ⅠB	ⅡB
Element	Sc	Ti	V	Cr	Mn	Fe	Co	Ni	Cu	Zn
Electron configuration	$3d^1 4s^2$	$3d^2 4s^2$	$3d^3 4s^2$	$3d^5 4s^1$	$3d^5 4s^2$	$3d^6 4s^2$	$3d^7 4s^2$	$3d^8 4s^2$	$3d^{10} 4s^1$	$3d^{10} 4s^2$
Atomic radii /pm	162	147	134	128	127	126	125	124	128	134
First ionization energy/(kJ/mol)	631	658	650	653	717	759	758	737	745	906
Melting point /℃	1541	1660	1917	1857	1244	1537	1494	1455	1084	420
Boiling point /℃	2831	3287	3380	2672	1962	2750	2870	2732	2567	907
Density /(g/cm³)	3.0	4.5	6.1	7.9	7.2	7.9	8.7	8.9	8.9	7.1
Oxidation states	+3	+2,+3, +4	+2,+3, +4,+5	+2,+3, +4,+5,	+2,+3, +4,+5, +6,+7	+2,+3, +6	+2,+3	+2,+3, +4	+1,+2	+2

From the table, we can summarize that almost every element have the $+2$ oxidation state, which is due to the loss of their two outer 4s electrons. With 3d electrons successively are lost, some elements also exhibit the oxidation states above $+2$. From Sc to Mn, the highest oxidation number is equal to the group number, that is, to the total number of 3d and 4s electrons in the atom. And it is observed only when the elements are combined with the most electronegative elements, such as O and F. We can also see the maximum oxidation state decreases after Mn. For the fifth-period and sixth-period transition elements, the multiplicity of oxidation states is also common.

Many of transition compounds in some oxidation states are colored. For example, MnO_4^{2-} ion is green, and MnO_4^- ion is purple.

Most transition metals are able to form stable complexes. For example, Cu^{2+} exists in water as complex ion, $Cu(H_2O)_4^{2+}$. When excess CN^- ions are added into the solution, $Cu(CN)_4^{2-}$ is given. Many of them have characteristic color and special chemical properties. So they are often used in chemical analysis and industrial processes.

Transition metals are often good catalysts. For example, vanadium oxide is often used in for the contact process, and platinum is used to speed up the manufacture of nitric acid. Because of their numerous oxidation states, transition metals can form new compounds during a reaction, which may provide an alternative route with a lower overall activation energy.

Transition metals are often paramagnetic, because their atoms or ions possess unpaired electrons.

All above properties of the transition metals are related to atomic structure and position in the periodic table.

Atomic Radii

The radii of transition metals show very similar variations across each series. As shown in Figure 13.3 they exhibit the same pattern of variation in the three series, first decrease and then increase across a period. In each series, the minimum atomic radius is near the center. And in each groups, the second-and third- transition metals are about the same size. For example, in group V B, Ta has virtually the same radius as Nb. This effect is caused by the lanthanide series, which are the elements between La and Hf. With electrons filling in 4f orbital, the effective nuclear charge steadily increases, while 5d and 6s electron can not be shielded by 4f electron completely, producing a contraction in size, called the lanthanide contraction. It results that chemical properties of the second-and third-transition series in a given group are much like. It can be used to explain why some transition metals, such as Zr and Hf, always occur together in nature.

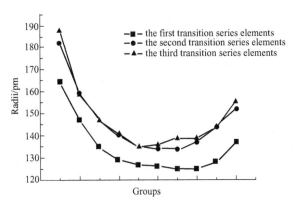

Figure 13.3   Radii of the three transition series

Electron Configurations

Table 13.1 lists the electron configurations of the first transition-series elements. As one of striking characteristic, the transition metals within in a given period are much similar in chemical properties. This phenomenon is related to their especial electron configuration: the last electrons are added to their inner shells. However, when these transition metals lose electrons, their outer s electrons will be lost firstly, then their $(n-1)$ shell d electrons. For example, when Fe is oxidized to Fe^{2+}, the electron configuration will changes from $3d^6 4s^2$ to $3d^6$. If it continues being oxidized, one electron will be lost from 3d orbit, forming Fe^{3+}.

13.5 Chemistry of Some Transition Metals

In this section we will consider mainly the chemical properties of the first series transition metals, and we will find they have a lot of characteristic that make them different from the main-group elements.

Chromium

Chromium belongs to the group VI B. Because this metal is shiny and has very little tendency to corrode, it can be used as the protective coating.

The most common oxidation states of chromium are +2, +3, +4 and +6. And solutions of chromium compounds or complex ions are richly colored.

Cr(II) is instable and shows relatively strong reducibility. It can be obtained by reducing Cr(III) and Cr(VI) with some reducers.

(1) Chromium(III)

Cr(III) is the most stable oxidation state of chromium. Its hydroxide and oxide can dissolves in both acid and base, because they are amphoteric substances.

$$Cr(OH)_3 + 3H^+ \rightleftharpoons Cr^{3+} + 3H_2O$$
$$Cr(OH)_3 + OH^- \rightleftharpoons CrO_2^- + 2H_2O$$

In basic solution, chromic(III) salt also shows relatively strong reducibility. For example, it can be oxidized by H_2O_2 to chromate.

$$2CrO_2^- + 3H_2O_2 + 2OH^- = 2CrO_4^{2-} + 4H_2O$$

(2) Chromium(VI)

In basic solution the bright yellow chromate ion, CrO_4^{2-}, is the most stable. If acid is added to an basic solution of chromate ion, the yellow chromate ion color changes to deep orange red. This color is due to dichromate ion, $Cr_2O_7^{2-}$. There exists an equilibrium between the dichromate and chromate ions when the pH changes.

$$2CrO_4^{2-} + 2H^+ = 2Cr_2O_7^{2-} + H_2O$$

Chromate and bichromate are both strong oxidants. They can be reduced to Cr(III) with green or blue color.

$$Cr_2O_7^{2-} + 3H_2S + 8H^+ = 2Cr^{3+} + 3S\downarrow + 7H_2O$$

When Ag^+, Pb^{2+} or Ba^{2+} ion is added to the solution of chromate or bichromate ion, the precipitate will be obtained.

$$2Pb^{2+} + Cr_2O_7^{2-} + H_2O = 2H^+ + 2PbCrO_4\downarrow \text{ (yellow)}$$
$$4Ag^+ + Cr_2O_7^{2-} + H_2O = 2H^+ + 2Ag_2CrO_4\downarrow \text{ (red)}$$

In acidic solution, $Cr_2O_7^{2-}$ can react with H_2O_2 and form chromium peroxide, $CrO(O_2)_2$, which looks blue in ethylene. This reaction can identify chromium.

$$Cr_2O_7^{2-} + 4H_2O_2 + 2H^+ = 2CrO(O_2)_2 + 5H_2O$$

Manganese

Manganese belongs to the group ⅦB. Manganese can exist in all oxidation states from $+2$ to $+7$, while $+2$ and $+7$ are the most common.

(1) Manganese(Ⅱ)

Manganese(Ⅱ) salts are characteristically very pale pink or colorless. It is stable in acidic solution. In HNO_3 solution, Mn^{2+} can be oxidized by $NaBiO_3$, and produce purple amaranthine $KMnO_4$, which is often be used to identify Mn^{2+}.

$$5NaBiO_3 + 2Mn^{2+} + 14H^+ = 2MnO_4^- + 5Bi^{3+} + 5Na^+ + 7H_2O$$

(2) Manganese(Ⅳ)

Mn(Ⅳ) behaves as both an oxidizing agent and a reducing agent.

In acidic solutions, its oxidizability is relatively strong. It can oxidize concentrated hydrochloric acid to chlorine (Cl_2). This reaction provides a convenient method for preparing a little of chlorine in the laboratory.

$$MnO_2 + 4HCl = MnCl_2 + Cl_2 + 2H_2O$$

In basic solution, it acts as a reducing agent. Many strong oxidizing agents can oxidize it to green MnO_4^{2-}.

$$3MnO_2 + 6KOH + KClO_3 = 3K_2MnO_4 + 3H_2O + KCl$$

(3) Manganese(Ⅶ)

Manganese(Ⅶ) chemistry is dominated by the intensely purple permanganate ion (MnO_4^-). MnO_4^- ion behaves as a strong oxidizing agent. But its reduction product changes with the medium. For example, if it reacts with SO_3^{2-} solution in acidic, neutral and basic conditions, it will be reduced to Mn^{2+}, MnO_2 and MnO_4^{2-}.

$$2MnO_4^- + 5SO_3^{2-} + 6H^+ = 2Mn^{2+} (\text{colorless}) + 5SO_4^{2-} + 3H_2O$$
$$2MnO_4^- + 3SO_3^{2-} + H_2O = 2MnO_2 \downarrow (\text{brown}) + 3SO_4^{2-} + 2OH^-$$
$$2MnO_4^- + SO_3^{2-} + 2OH^- = 2MnO_4^{2-} (\text{green}) + SO_4^{2-} + H_2O$$

Iron

Iron belongs to group Ⅷ. It has only two important oxidation states, $+2$ and $+3$.

Iron reacts with nonoxidizing acids such as chloride acid or dilute sulfuric acid to form Fe^{2+}. However, in the presence of air, Fe^{2+} tends to be oxidized to Fe^{3+}.

$$4Fe^{2+} + O_2 + 4H^+ = 4Fe^{3+} + 2H_2O$$

But when iron reacts with an oxidizing acid such as dilute nitric acid, Fe^{3+} is formed directly.

$$Fe + NO_3^- + 4H^+ = Fe^{3+} + NO \uparrow + 2H_2O$$

The outer electronic configuration of Fe^{3+} ($3d^5$) is stable, so Fe(Ⅲ) is more stable than Fe(Ⅱ).

(1) Hydroxides

Iron(Ⅱ) hydroxide and iron(Ⅲ) hydroxide are all slightly soluble. The former is white and shows relatively strong reducibility, while the latter is red brown. $Fe(OH)_2$ can be oxidized by air to produce $Fe(OH)_3$.

$$4Fe(OH)_2 + O_2 + 2H_2O = 4Fe(OH)_3$$

Iron(Ⅲ) hydroxide is amphoteric, while alkalinity is relatively strong. So it can dissolve in strongly acidic solution but not in basic solution.

(2) Salts

The ferric(Ⅱ) salts shows quite strong reducibility. It can be oxidized by many oxidize agents, such as Cl_2, MnO_4^-, etc.

$$2Fe^{2+} + Cl_2 = 2Fe^{3+} + 2Cl^-$$

The ferric(Ⅲ) salts have faintish oxidizability, and can be reduced to Fe(Ⅱ) only by stronger reducer, for example, KI.

(3) Coordination compounds

Iron has relatively strong coordination ability. Most iron complex has a coordination number of 6, such as $Fe[(CN)_6]^{4-}$, $Fe[(CN)_6]^{3-}$, $Fe[(NCS)_6]^{3-}$, etc.

$K_3Fe[(CN)_6]$ [potassium hexacyanoferrate(Ⅲ) or ferricyanide] is ruby-red. Addition of $Fe[(CN)_6]^{3-}$ to aqueous Fe^{2+} gives the deep blue complex Turnbull's blue and this reaction is used as a qualitative test for Fe^{2+}.

$$3Fe^{2+} + 2Fe[(CN)_6]^{3-} (red) = Fe_3[Fe(CN)_6]_2 (deep\ blue)$$

$K_4Fe[(CN)_6]$ is yellow. Conversely, if $Fe[(CN)_6]^{4-}$ is added to aqueous Fe^{3+}, the deep blue complex Prussian blue is produced. Both Prussian blue and Turnbull's blue are hydrated salts of formula $Fe_4[FeCN)_6]_3 \cdot xH_2O$ ($x=14$).

Complexes of Fe(Ⅲ) with SCN^- are intense red. This serves as a sensitive qualitative and quantitative test for ferric ion.

$$Fe^{3+} + xSCN^- = Fe[(NCS)_x]^{3-} (intense\ red)$$

Ion(Ⅲ) ion also has a strong affinity for F^-

$$Fe^{3+} + 6F^- = FeF_6^- (colorless)$$

Cobalt

Cobalt belongs to the group Ⅷ. It is a hard, lustrous, silver-grey metal. Cobalt in small amounts is essential to many living organisms, including humans. It is a central component of the vitamin cobalamin, or vitamin B-12.

The +2 and +3 oxidation states are most prevalent, and most of the simple compounds of cobalt are cobalt(Ⅱ) compounds.

Blue $CoCl_2$ turns to pink on exposure to moisture and readily form hydrates. Because of this dramatic color change and the ease of the hydration/dehydration reaction, cobalt chloride is used as an indicator for water.

$$CoCl_2 \rightleftharpoons CoCl_2 \cdot H_2O \rightleftharpoons CoCl_2 \cdot 2H_2O \rightleftharpoons CoCl_2 \cdot 6H_2O$$
blue   blue-purple   magenta   pink

All compounds containing cobalt in the +3 oxidation state are relatively unstable, while they are stabilized by complex ion formation.

In the presence of ammonia or amines, cobalt(Ⅱ) is readily oxidised by atmospheric oxygen to give a variety of cobalt(Ⅲ) complexes. For example,

$$4[Co(NH_3)_6]^{2+} (brown\ yellow) + O_2 + 2H_2O = 4[Co(NH_3)_6]^{3+} (deep\ brown) + 4OH^-$$

Complexion $[Co(CN)_6]^{4-}$ can even be oxidized by water to form cobalt(Ⅲ) complex.

$$2[Co(CN)_6]^{4-} + 2H_2O = 2[Co(CN)_6]^{3-} + H_2 \uparrow + 2OH^-$$

Cobalt(Ⅱ) complexes are usually either octahedral or tetrahedral. For example, $[Co(NCS)_6]^{4-}$ and $[CoCl_4]^{2-}$. $[Co(NCS)_6]^{4-}$ ion is blue, which is stable in organic solvent, such as acetone, propane, etc.

Nickel

Nickel belongs to the group Ⅷ. It is a silvery-white and hard metal. This metal is relatively resistant to corrosion. Nickel plays numerous roles in the biology of microorganisms and plants, In fact urease (an enzyme which assists in the hydrolysis of urea) contains nickel.

The most common oxidation state of nickel is $+2$. Nickel(Ⅲ) compounds are relatively strong oxidizing agents.

An aqueous solution of nickel (Ⅱ) salts has a characteristic emerald green color.

Green precipitate $Ni(OH)_2$ forms, when strong base is added to Ni^{2+} salt solution.

$$Ni^{2+} + 2OH^- = Ni(OH)_2 \downarrow$$

Aqueous ammonia also gives the same precipitate, but with excess ammonia this precipitate dissolves to give a deep blue solution.

$$Ni^{2+} + 2NH_3 \cdot H_2O \longrightarrow Ni(OH)_2 \downarrow (green) + 2NH_4^+$$
$$Ni(OH)_2 + 6NH_3 \cdot H_2O(excess) = [Ni(NH_3)_6]^{2+} (deep\ blue) + 2OH^- + 6H_2O$$

There are many other stable complexes of nickel, just like $[Ni(CN)_4]^{2-}$ and $[NiCl_4]^{2-}$.

Copper and Zinc

Copper is the elements of group ⅠB, whose compounds have similar chemical properties.

Copper exhibits two common oxidation states $+1$ (cuprous) and $+2$ (cupric). In aqueous solution, the $+2$ oxidation state is more common.

Zinc is a white, lustrous metal, which behaves as an excellent reducing agent. This element is exclusively $+2$ oxidation state.

(1) Oxides and hydroxides

The color of Cu oxides is as follows. CuO (black), Cu_2O (red), ZnO (white).

$Cu(OH)_2$ and $Zn(OH)_2$ are both amphoteric hydroxides, particularly $Zn(OH)_2$. They can dissolve in excess concentrated KOH and NaOH forming coordination compounds.

$$Cu(OH)_2 (blue) + 2OH^- = [Cu(OH)_4]^{2-} (dark\ blue)$$
$$Zn(OH)_2 (white) + 2OH^- = [Zn(OH)_4]^{2-} (colorless)$$

Addition of concentrated aqueous ammonia to a solution of an Cu^{2+} salt first gives a blue precipitate of $Cu(OH)_2$, then with excess concentrated aqueous ammonia, this precipitate dissolves to give a deep blue solution containing the complex ion $[Cu(NH_3)_4]^{2+}$.

$$Cu(OH)_2 + 4NH_3 = [Cu(NH_3)_4]^{2+} (dark\ blue) + 2OH^-$$

However, precipitate $Zn(OH)_2$ can not dissolve in excess concentrated ammonia solution.

(2) Salts

The potential diagram of copper is as follows.

$$Cu^{2+} \xrightarrow{0.157V} Cu^+ \xrightarrow{0.52V} Cu$$

Cu^+ readily disproportionates in solution

$$2Cu^+ \rightleftharpoons Cu^{2+} + Cu$$

Because copper(Ⅰ) is readily oxidized to copper(Ⅱ) under aqueous solution, copper(Ⅰ) is by far the more common.

Copper(Ⅰ) fluoride is not known. CuCl, CuBr and CuI are white solids, and they are made by reduction of Cu(Ⅱ) in presence of Cl^-, Br^-, I^-.

CuI precipitate forms when any Cu(Ⅱ) salt is added to KI solution.

$$2Cu^{2+} + 4I^- = 2CuI \downarrow + I_2$$

Copper or other reagents react with Cu^{2+} ion to produce the insoluble copper(Ⅰ) salts. For example

$$CuCl_2 + Cu = 2CuCl \downarrow$$

Most zinc salts are highly soluble.

(3) Coordination Compounds

Cu^{2+} can react with chloride ions or ammonia, CN^- or SCN^- etc. forming different complex ions with various colors, such as, $[CuCl_4]^{2-}$ (little yellow), $[Cu(NH_3)_4]^{2+}$ (dark blue).

There are some common complexes of Cu^+, such as, $[CuX_2]^-$ (colorless), $[Cu(CN)_2]^-$ (colorless). They are stable only when excessive ligand exists in the solution. In dilute solution they decompose and CuX is precipitated.

$$2[CuCl_2]^- \xrightarrow{Dilution} 2CuCl \downarrow + 2Cl^-$$

Silver and Mercury

Silver is an important element of group IB in the second transition series, and mercury is the element of group IIB in the third series. But their compounds have similar chemical properties.

(1) Oxides and hydroxides

Ag_2O, HgO and Hg_2O, which are the oxides of Ag and Hg are insoluble. Ag_2O is brown, while HgO is red or yellow.

AgOH, $Hg(OH)_2$ and $Hg_2(OH)_2$, which are hydroxides of Ag and Hg are amphoteric, while alkalinity is relatively strong. And they are so unstable that they will decompose to the oxides as soon as they precipitate from the solutions.

$$2Ag^+ + 2OH^- = 2AgOH \downarrow \qquad 2AgOH = Ag_2O \downarrow + H_2O$$
$$Hg^{2+} + 2OH^- = Hg(OH)_2 \downarrow \qquad Hg(OH)_2 = HgO \downarrow + H_2O$$
$$Hg_2^{2+} + 2OH^- = Hg_2(OH)_2 \downarrow \qquad Hg_2(OH)_2 = HgO \downarrow + Hg + H_2O$$

(2) Conversion of Hg^{2+} and Hg_2^{2+}

The chemistry of Hg(I) is that of the Hg_2^{2+}, which contains an Hg-Hg single bond. Hg(I) compounds can be made by the action of Hg metal on Hg(II) compounds.

$$Hg^{2+} + Hg \rightleftharpoons Hg_2^{2+}$$

Reagents that can form insoluble Hg(II) salts or stable Hg(II) complexes will upset equilibrium and decompose Hg(I) salts. For example, ammonia reacts with Hg_2Cl_2 forming a white precipitate Hg_2NH_2Cl, and Hg_2NH_2Cl further disproportionates to white $HgNH_2Cl$ and black Hg. These reactions can be use to identify Hg_2^{2+} ions.

$$Hg_2Cl_2 + 2NH_3 = Hg_2NH_2Cl \downarrow (white) + NH_4Cl$$
$$Hg_2NH_2Cl = HgNH_2Cl \downarrow (white) + Hg \downarrow (black)$$

Addition of excess OH^-, S^{2-}, I^- and CN^- to Hg(I) salts can also result in formation of Hg and HgO, HgS, $[HgI_4]^{2-}$ and $[Hg(CN)_4]^{2-}$.

$$Hg_2^{2+} + 4CN^- = [Hg(CN)_4]^{2-} + Hg$$

(3) Coordination compounds

The common coordination compounds of Ag conclude $[Ag(NH_3)_2]^+$, $[Ag(CN)_2]^-$, $[Ag(SCN)_2]^-$, $[Ag(S_2O_3)_2]^{3-}$ and $[Ag(Cl)_2]^-$.

Ag^+ ion can react with Cl^-, Br^- and I^- forming precipitate AgCl, AgBr and AgI. It also can react with NH_3, $S_2O_3^{2-}$ or CN^- forming complex ion $[Ag(NH_3)_2]^+$, $[Ag(S_2O_3)_2]^{3-}$ and $[Ag(CN)_2]^-$. According to the K_{sp} of insoluble substance and the K_s of complex ion, there is a competition between the precipitants and the ligands. If NaCl solution, stronger ammonia water, KBr solution, $Na_2S_2O_3$ solution, KI solution and KCN solution are added to AgNO₃ solution one by one, white precipitate will first be formed. Then it is dissolved, light yellow precipitate will be obtained, and it will dissolved, yellow precipitate will be seen, at last it disappears.

$$Ag^+ + Cl^- = AgCl \downarrow \text{ (white)}$$
$$AgCl + 2NH_3 = [Ag(NH_3)_2]^+ + Cl^-$$
$$[Ag(NH_3)_2]^+ + Br^- = AgBr \downarrow \text{ (light yellow)} + 2NH_3$$
$$AgBr + 2S_2O_3^{2-} = [Ag(S_2O_3)_2]^{3-} + Br^-$$
$$[Ag(S_2O_3)_2]^{3-} + I^- = AgI \downarrow \text{ (yellow)} + 2S_2O_3^{2-}$$
$$AgI + 2CN^- = [Ag(CN)_2]^- + I^-$$

(4) Silver mirror reaction

Ag^+ can be reduced by glucose in an aqueous ammonia solution. This reaction can be used to identify Ag^+.

Key Words

lithium ['liθiəm] 锂
sodium ['səudjəm, -diəm] 钠
potassium [pə'tæsjəm] 钾
rubidium [ru:'bidiəm] 铷
cesium ['si:ziəm] 铯
francium ['frænsiəm] 钫
alkali metal ['ælkəlai 'metl] 碱金属
diagonal relationship [dai'ægənl ri'leiʃənʃip]
 对角线规则
ambient temperature ['æmbiənt 'tempritʃə(r)]
 环境温度，室温
simple oxide ['simpl 'ɔksaid] 简单氧化物
peroxide [pə'rɔksaid] 过氧化物
decomposition [ˌdi:kɔmpə'ziʃən] 分解
electrolysis [ilek'trɔlisis] 电解
electropositive [iˌlektrəu 'pɔzətiv] 电正性
molten ['məultən] 熔融的
ignite spontaneously [ig'nait spɔn'teinjəsli]
 自燃

magnesium [mæg'ni:zjəm] 镁
calcium ['kælsiəm] 钙
barium ['bɛəriəm] 钡
alkaline earths metals 碱土金属
amphoteric [ˌæmfə'terik] 两性的
alloy ['ælɔi] 合金
radiocontrast agent 放射性对比剂
aluminum [ə'lju:minjəm] 铝
polymorphic [ˌpɔli'mɔ:fik] 多晶型的
light-emitting diode (LED) 发光二极管
chromium ['krəumjəm] 铬
manganese ['mæŋgəni:z] 锰
iron ['aiən] 铁
cobalt [kə'bɔ:lt] 钴
nickel ['nikl] 镍
copper ['kɔpə] 铜
zinc [ziŋk] 锌
silver ['silvə] 银
mercury ['mə:kjuri] 水银，汞

Exercises

1. All of the group ⅠA and ⅡA metals are produced by electrolysis of molten salts. Why?

2. List three properties that can illustrate the diagonal relationship (the similarity) between lithium and magnesium.

3. Calcium ion makes water "hard", but Na^+ ion does not. Suggest an explanation for this observation.

4. Magnesium metal burns in air to give a white ash. When this material is dissolved in water, the order of ammonia can be detected. Write out balanced chemical equations for this observation.

5. Write complete and balanced equation for reaction between the following substances.

(1) $Na_2O_2 + H_2O \longrightarrow$
(2) $Na + H_2O \longrightarrow$
(3) $KO_2 + CO_2 \longrightarrow$
(4) $K_2CO_3 + CO_2 + H_2O \longrightarrow$
(5) $BeO + NaOH \longrightarrow$
(6) $MgH_2 + H_2O \longrightarrow$
(7) $CaCO_3 + CO_2 + H_2O \longrightarrow$
(8) $Al + H_2O + NaOH \longrightarrow$

6. Explain why Al(Ⅲ) is the only stable oxidation state of aluminum in its compounds?

7. Which properties are considered characteristic of the transition metals?

8. What is meant by the term lanthanide contraction? What properties of the transition elements are affected by the lanthanide contraction?

9. What accounts for the fact that chromium exhibits several oxidation states in its compounds, whereas aluminum exhibits only the +3 oxidation state?

10. Finish the following reaction equations.
(1) $Cr_2O_7^{2-} + H_2S + H^+ \longrightarrow$
(2) $Pb^{2+} + Cr_2O_7^{2-} \longrightarrow$
(3) $NaBiO_3 + Mn^{2+} + H^+ \longrightarrow$
(4) $Fe^{3+} + F^- \longrightarrow$
(5) $Ni(OH)_2 + NH_3 \cdot H_2O(excess) \longrightarrow$
(6) $Cu^{2+} + I^- \longrightarrow$
(7) $CuCl_2 + Cu \longrightarrow$
(8) $Cu^{2+} + NH_3 (excess) \longrightarrow$
(9) $AgBr + S_2O_3^{2-} \longrightarrow$
(10) $HgI_2 + I^- \longrightarrow$

11. Explain the following phenomena.
(1) MnO_4^- acts with SO_3^{2-} solution in acidic, neutral and basic conditions.
(2) $CoCl_2$ is exposed to moisture.
(3) Ammonia reacts with Hg_2Cl_2.

12. Identify the following ions.
    Mn^{2+}, $Cr_2O_7^{2-}$, Fe^{3+}, Cu^{2+}, Zn^{2+}.

13. In the following pair substances, separate one from the other.
(1) Ag^+ and Hg_2^{2+}
(2) Mg^{2+} and Cu^{2+}
(3) CuS and HgS
(4) AgCl and AgI

II

Inorganic Chemical Experiments

Chapter 1 Basic Techniques of Experimental Chemistry

1.1 Safety

The chemistry laboratory is really not a dangerous place, but it demands a reasonable prudence on the part of an experimenter to keep it safe. In the following paragraphs, the more important precautions are discussed.

Eye Protection

The eyes are particularly susceptible to permanent damage by corrosive chemicals as well as by flying fragments.

During doing hazardous experiments, follow all directions carefully, and take care not to endanger your neighbor in particular. For example, when heating a test tube, do not point its mouth toward anyone. Report any accident immediately to your instructor. In case of injury to the eye, flood the eye with lots of water immediately, and continue to rinse for at least 10 mins. If an eyewash fountain is not immediately available, use a rubber tube connected to a faucet. All injuries involving the eyes should be referred to a physician at once.

Poisonous Chemicals

Figure 1-1 Proper way to test the odor of a substance

Most of the chemicals you will work with are poisonous to some degree. It is obvious that you should never taste a chemical unless specifically directed to do so. However, there are more subtle ways of being poisoned. One of these is breathing toxic vapors. Be careful to work in ventilating hood whenever instructed to do so. Even such common substances as carbon tetrachloride, benzene, and mercury are poisonous and potentially dangerous. Avoid prolonged exposure to these liquids or the accompanying vapors. Since heating favors the vapor state, these and other poisonous liquids should be heated only in a hood.

Occasionally, you will be directed to test the odor of a substance. The proper way to do this is to waft a bit of the vapor toward your nose as shown in Figure 1-1. Do not stick your nose in and inhale vapor directly from the test tube.

A possible poisoning hazard, frequently overlooked, is contamination through the hands. Some poisons, e.g., benzene are rapidly absorbed through the skin. All poisons can stick to the hands and eventually end up in the mouth. Immediately scrub your hands thoroughly after exposure to hazardous chemicals, and get into the habit of always washing your hands before leaving the laboratory.

Essential Precautions

Follow all directions with utmost care, especially those having to do with hazardous conditions. Do not perform any unauthorized experiment. If you want to change or supplement the assigned material, first consult your instructor and get his permission. Irresponsible behavior will result in immediate expulsion from the laboratory.

In using chemical reagents, double-check the label to make sure you are not using the wrong chemical. Serious explosions have frequently resulted from such errors.

Smoking is not permitted in the laboratory.

1.2 Recording Results

The chemist considers his notebook as one of his most valuable possessions. It summarizes the work he has done and the results he has obtained. Loss of a notebook can be catastrophic. Furthermore, the notebook record should be kept so that it will be meaningful at a later date both to its author and to other chemists.

In this course, you are asked to keep your record in the laboratory manual. Data should be recorded as soon as they are obtained. Writing should be concise and legible. Qualitative data, which are just as important as quantitative data, should not be omitted. A few key words, perhaps with a sketch, will often suffice. In recording quantitative data, make sure the numbers are labeled with both units and property being measured. Pay strict attention to significant figures so that the numbers give maximum information.

In calculating results from experimental data, it is good practice to show specifically the method of calculation. This can be done either by giving a sample calculation using numbers labeled with units or by writing out a general expression indicating how the experimental data are mathematically combined. For example, in an experiment to determine density, we can write either

$$\text{Density of unknown} = \frac{27.1 \text{ g}}{22.6 \text{ mL}} = 1.20 \text{ g/mL}$$

Or

$$\text{Density of unknown} = \frac{\text{mass of unknown}}{\text{volume of unknown}}$$

It is useful to show the detailed arithmetic, which is better done on scratch paper or in some inconspicuous place in the record book.

1.3 Weighing

One of the most common operations in experimental chemistry is the determination of mass or weight. The mass of an unknown is usually determined by comparing it with a known mass. There are three kinds of balances in common use. Here we will only discuss platform balance.

One model of platform balance is shown in Figure 1-2. It is the most common type of platform balance used in chemical laboratory. When this type of platform balance is used, place the sample on the left pan, while weights are placed on the right pan. The unknown on the left pan is balanced by placing large weights on the right pan. Then sliding the weights attached to the crossbeams until the pointer makes equidistant swings to the right, and left of the center point on the scale.

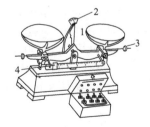

Figure 1-2　Platform balance
1—scale pan; 2—dial;
3—balance screw; 4—vernier

Platform balances, when used properly, are capable of measuring mass to a precision of 0.1 g. To prevent corrosion of the pans, chemical should never be placed directly on the metal surface. Always use a container or, for solids, a piece of weighing paper creased upward along the two diagonals in order to prevent spillage.

1.4 Concerning Liquids

Cleaning Glassware

In spite of the fact that there are elaborate recipes for making cleaning solutions, the most effective way to clean glassware is still to use soap, water, and a brush. Get into the habit of washing your equipment after each use to have ready for the next experiment. Do not forget to rinse off the soap with lots of water. Invert the glassware so that it will drain dry.

Chemists use a variety of glassware to measure the volume of reagents. The specific type of glassware used in any situation depends on how accurately or precisely the volume needs to be known. In general, glassware is divided into two broad categories, glassware for approximate measurements and glassware for accurate and precise measurements.

Glassware for Approximate Measurements

Five common types of glassware are used for the approximate measurement of volume, reagent bottles, beakers, Erlenmeyer flasks, graduated cylinders, and disposable pipettes. A reagent bottle (see Figure 1-3) is the least accurate as they seldom have any marks to indicate approximate volume. They are used to prepare solutions with an approximate concentration. Adding 0.1 moles of a reagent to a 1 L bottle and adding water to the top of the bottles rounded shoulder will produce a solution that is approximately 0.1 mol/L.

Figure 1-3   Reagent bottle

Beakers and Erlenmeyer flasks (see Figure 1-4) usually have several graduation marks on their sides. These marks provide an approximate measurement of volume, They are usually accurate to within $\pm 10\%$ of the flask's maximum volume. For example, adding water to the 100 mL mark on a 250 mL beaker will yield a net volume between 75 mL and 125 mL.

A graduated cylinder provides a more accurate measure of volume than a beaker or Erlenmeyer flask. Typically, a graduated cylinder is accurate to within $\pm 5\%$ of the cylinder's maximum volume. When delivering 5 mL using a 10 mL graduated cylinder, the actual volume is probably between 4.5 mL and 5.5 mL.

A disposable pipette (i.e., disposable dropper) is a useful way to add a reagent whose volume is given in drops. A commonly used estimate is that 20 drops is roughly equivalent to 1 mL.

In general, the precision for all these types of glassware is better than their respective accuracies, although their precision is seldom an issue.

To measure the amount of a liquid sample, it is usually more convenient to determine its volume rather than its mass. One of simplest devices for doing this is the graduated cylinder

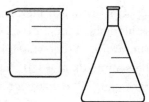

Figure 1-4   Beaker and Erlenmeyer flask

(see Figure 1-5). The correct way to read the volume of liquid is to hold the cylinder vertical at eye level and look at the top surface (meniscus) of the liquid (see Figure 1-6). The meniscus is curved, with a rather flat part in the center. By noting the position of this flat part relative to the calibration marks on the cylinder, you can determine the volume of a liquid to approximately 0.1 mL with the cylinder shown. It is important to keep graduated cylinders clean for at least two reasons, dirt may chemically contaminate an experiment, and its presence in the cylinder may throw off accurate determination of volume. (Glassware which is dirty does not drain properly so that the delivered volume is not equal to that indicated by the calibration marks.)

Figure 1-5 Graduate cylinder

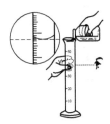

Figure 1-6 Correct ways to read the volume of liquid in graduate cylinder

Glassware for Accurate and Precise Measurements

Sometimes we need to know a reagent's exact volume. This is the case we worry both about the volume's accuracy (How close is it to 10 mL?) and its precision (How much variation might we expect in the delivered volume from one trial to the next?). Three types of glassware are commonly used when we need accurate and precise measurements of volume, volumetric flasks (i.e., measuring flask), volumetric pipettes, and burettes. In general, the precision of these types of glassware is better than their respective accuracies.

1. Volumetric Flask

When filled to its calibration mark, a volumetric flask (see Figure 1-7) is designed to contain a specified volume of solution, usually to within $\pm(0.03 \sim 0.2)\%$ of the stated value, depending on the size of the volumetric flask (although accuracy can be improved through calibration, typically by determining the mass of water contained within the flask and converting to volume using water's known temperature dependent density). A volumetric flask with a capacity of less than 100 mL, generally measures volume to the hundredths of a milliliter, whereas volumetric flasks of 100 mL or greater capacity measure volumes to the tenth of a milliliter. For example, a 10 mL volumetric flask contains 10.00 mL, but a 250 mL volumetric flask holds 250.0 mL. This is an important issue to consider when keeping track of significant figures.

Note the use of the word contain in describing the properties of a volumetric flask. This description is important. Although a 100 mL volumetric flask will contain exactly 100.0 mL (± 0.1 mL), it cannot be used to deliver 100.0 mL to another container. This apparent contradiction results from the fact that you can never completely transfer a liquid from one container to another, some liquid, even if it is only a drop, always remains behind.

Because a volumetric flask contains a solution of known volume, it is very useful when preparing a solution whose exact con-

Figure 1-7 Volumetric flask

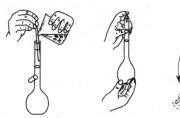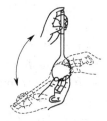

Figure 1-8 Operation of volumetric flask

centration must be known. Shown as Figure 1-8, dissolve completely a known amount of reagent in some solvent, and then transfer it to a clean volumetric flask with a glass rod. After that, use additional solvent in several portions to transfer the remaining reagent in the beaker to the flask. Note that the whole solvent volume added should be less than the volume of the volumetric flask. The final adjustment of volume to the flask's calibration mark is made dropwise using a disposable pipette or a solvent dispensing bottle. To complete the mixing process, the volumetric flask is inverted and shaken at least 10 times.

2. Burette

Another common device for measuring liquid volume is the burettes (Figure 1-9). A burette is also used to deliver a variable volume of solution. In general, burettes are constructed so that it is possible to measure volume more precisely than with a graduated cylinder. The most commonly used burette in the laboratory is a 50 mL burette, which has major markings for every 1 mL increment and minor markings for every 0.1 mL increment. The accuracy of measuring volume is approximately 0.1%. Thus, for a 50 mL burette, volumes are accurate to approximately ±0.05 mL.

Figure 1-9 Acidic burette (a) and basic burette (b)

An acidic burette [Figure 1-9(a)] is a long, narrow tube with graduated markings and a stopcock on the bottom end. The topmost mark is labeled with a volume of zero, with additional markings increasing in volume as they go down the burette. The basic burette illustrated [Figure 1-9(b)] can be made by attaching a rubber connector containing a glass bead to the constricted tip of a gas-measuring tube. A nozzle made by drawing down a piece of glass tubing is inserted at the lower end. Pouring liquid through a funnel into the top fills the burette. When the side of the rubber at the glass bead is pinched, some liquid can be drained out to free the tip of all air. It is not necessary for the meniscus to be at the zero mark, since the volume delivered from a burette can be obtained simply by taking the difference between the initial and final readings. In reading a burette, it is imperative that the eye must be at the level of the meniscus (Figure 1-10). For colorless liquids like water, the bottom of the meniscus is read. For dark liquids, it is generally more convenient to read the top (i.e., the place where the liquid contacts the glass). Keep your burette cleaned, otherwise, it will not drain properly, as you can easily tell by noting whether drops of liquid are left hanging from the walls. Burettes equipped with stopcocks should be rinsed with water and filled with distilled water for storage. Rinsing is especially important if strong base has been used. For accurate use of burette, it is necessary that the volume of liquid held below the calibration marks be the same for the initial and final readings. To achieve this, make sure that there is no air trapped in the connector or tip. A good test is to twist the tip upward so that a trapped air bubble may rise and then be squeezed out (see Figure 1-11). Furthermore, make sure that no drop is left hanging from the burette tip either before or after delivering liquid. After delivering a volume, always touch the tip to the wall of the receiver.

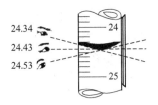

Figure 1-10 Reading a burette (24.43 mL is right)

Figure 1-11 Drive away air bubbles in the connector or tip

3. Pipette

A volumetric pipette is used to deliver a specified volume of solution. Several different styles of volumetric pipettes are available, but the most common (and the most accurate) is the transfer pipette (also called pipette). Pipettes consist of a long tube with a bulge in the middle and a single calibration mark [Figure 1-12(a)]. The accuracy of a transfer pipette is similar to that of a volumetric flask of equal volume. Thus, for example, a 100 mL transfer pipette will deliver 100.0 mL of solution (± 0.1 mL). As with a volumetric flask, accuracy can be improved by calibrating with water.

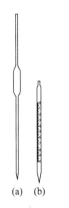 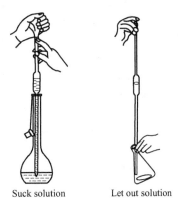

(a) (b)

Suck solution   Let out solution

Figure 1-12 Two styles of pipette

Figure 1-13 Operation of pipette

Sometimes it is also necessary to obtain samples of a decimal volume, for example, 4.50 mL. This also can be done by another type of pipette. As shown in Figure 1-12(b), this type of pipette consists of glass pipe having a delivery tip at the lower end and an opening for suction filling at the upper end. Successive marks engraved on the stem, which are similar to the marks of burette, indicate the height to which the liquid level must be draun to get exactly the volume of liquid.

As shown in Figure 1-13, the proper way to use a pipette is to draw the liquid up past the ring mark using a rubber suction bulb, detach the bulb, and quickly place a fingertip over the suction end to hold the suction. Then, by slightly rolling the fingertip aside, allow excess liquid to drain out of the delivery tip until the meniscus descends to the ring mark. The tip of the pipette is then transferred to a receiving vessel, and the liquid is allowed to drain out. Usually 10 sec more are counted off to the receiver wall to get off the last drop. Pipettes are available in various sizes, 1, 5, 10, 25, 50 mL and are generally calibrated to deliver that volume at a standard temperature of 25℃。

Handling Reagent Bottles

When adding liquids directly from bottles, you should observe certain elementary pre-

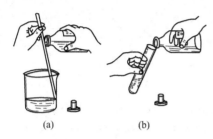

Figure 1-14 Proper way to pour from a reagent bottle

cautions. First, double-check the label to make sure you have the right reagent. Second, never set the stopper down so as to contaminate the inner surface. The proper way to pour from a reagent bottle to a beaker is shown in Figure 1-14(a). Note that a stirring rod is held against the mouth of the bottle so as to permit liquid to flow down along it without splashing into the receiver. As shown in Figure 1-14(b), liquid can be poured directly to a test tube without a stirring rod. Finally, when done, replace the stopper, and set the bottle back on the reagent shelf. Under no circumstances should you pour chemicals back into a reagent bottle or, for that matter, place anything - e. g., a medicine dropper - in it.

Decantation and Filtration

A common problem in the laboratory is to separate a liquid from a solid. This should be done in two stages. First, decant, i. e., let the solid settle, and carefully pour off most of the liquid down a stirring rod (Figure 1-15). Second, proceed to filter. When only a few milliliters of liquid remain in the beaker, swirl it gently and pour the mixture into the funnel. Rinse down the sides of the beaker with a stream of distilled water from your wash bottle, swirl, and pour into the funnel. Repeat until beaker and stirring rod are clean. Allow filter to drain.

Figure 1-15 Operation of decantation

The filter paper is prepared for insertion into the funnel by folding it in half along one diameter of the filter paper and then in quarters, as shown in Figure 1-16. For rapid filtration, the filter paper should fit snugly in the funnel so that air cannot be drawn down between the paper and the funnel. Tear off a corner of the folded paper as shown in the fourth sketch of Figure 1-16 to get a tighter seal. Insert the paper into the funnel, and pour some solvent (usually water) through it. It is then possible to press the wet paper gently against the funnel wall. When the funnel is properly sealed, liquid will be retained in the funnel stem, and its weight will help pull liquid through the paper. If liquid is not retained in the stem, either, the paper is not sealed or the stem is dirty, in which case it should be cleaned (use a pipe cleaner or a twisted strip of cheesecloth).

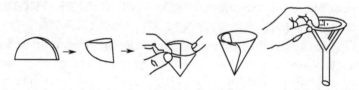

Figure 1-16 Preparation of filter paper

When pouring a solid-liquid mixture into a filter, it is a good idea to pour down a stirring rod, as shown in Figure 1-17. However, be careful not to punch a hole in the filter paper with the end of the glass rod. To prevent splashing and have the bottom tip of the funnel touch the receiver wall. Do not let the level of the liquid in the funnel go above the top of the paper.

Suction Filtration

Filtration is frequently a slow operation. It can be speeded up by use of suction filtration. Suction can be applied by means of a water aspirator (Figure 1-18), which is a gadget that fits on a water faucet and uses a fast stream of water to suck in air through a side arm. Because the suction is quite strong, it is necessary to use a special funnel, called a Büchener funnel, to support the filter paper. Otherwise, the paper might break through. Put the paper in the funnel, and wet it first with solvent. Then decant most of the liquid into the center of the filter paper. Pour it through slowly so the precipitate does not get a chance to seep under the filter paper.

Figure 1-17　Atmospheric filtration

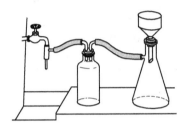

Figure 1-18　Suction filtration

Figure 1-19　Electric centrifuge

Centrifugation

Rapid separation of solids and liquids can also be done in centrifuge, an electric example of which is shown in Figure 1-19. The electric centrifuge consists of two parts. a fixed base which houses an electric motor and a freely rotating head which accommodates text tubes in an inclined position. When the head is rotated at high speed, the denser phase in the test tubes (usually the solid phase) separates to the part of the test tube farthest from the axis of rotation. The liquid phase, which corresponds to the filtrate, can be poured off, or it can be sucked up into a collector, e. g. , medicine dropper or pipette. Washing of precipitates can be done by adding wash solution to the solid, stirring the mixture with a glass rod, and repeating the centrifugation.

When operating a centrifuge, the following rules should be observed.

(1) The weight load in the rotating head must be symmetrically distributed about the axis of rotation. If you put your mixture in one slot, you must place an identical test tube with an approximately equal volume of water in the slot diametrically opposite.

(2) Allow the centrifuge to rotate about one minute for most efficient separation. The optimum time is shorter for centrifuges rotating faster than 1700 r/min. Keep your hands away from the top of the centrifuge while it is rotating.

(3) Allow the rotating head to come to a full stop before trying to remove your sample. You can help show the rotation by gently pressing your hands to the sloping sides of the cone. However, do not brake too hard, otherwise, you will stir up the sediment again.

Toward Understanding pH

We use the words acidic and basic as extremes that describe solutions as hot and cold are extremes that describe temperature. Just as mixing hot and cold water evens out the temperature, mixing acids and bases can cancel their extreme effects and are considered neutral.

The pH scale can tell if a liquid is more acid or more basic, just as the Fahrenheit or Celsius scale is used to measure temperature. The range of the pH scale is from 0 to 14 and from very acidic to very basic. A pH of 7 is neutral. A pH less than 7 is acidic and greater than 7 is basic. The lower the number, the more acidic is the substance.

Each whole pH value below 7 is ten times more acidic than the next higher value. For example, a pH of 4 is ten times more acidic than a pH of 5, and a hundred times (10×10) more acidic than a pH of 6. This holds true for pH values above 7, each of which is ten times more basic (also called alkaline) than the next lower whole value. An example would be, a pH of 10 is ten times more alkaline than a pH of 9.

pH Indicator

A pH indicator is a weak acid or a weak base. Indicators have a very useful property, they change color depending on the pH of the solution they are in. You may have noticed that red cabbage changes color when you slice it up and cook it. When you buy the red cabbage it is actually blue. It will be quite blue when you slice it up and add some tap water. That is because your tap water is very slightly alkaline. If you add some vinegar it will turn red. This is because there is a pigment (colored chemical) in the red cabbage that acts as a pH indicator.

This color change is not at a fixed pH, but rather, it occurs gradually over a range of pH values. This range is termed the color change interval. pH indicator is defined by a useful pH range. For example phenolphthalein changes from colorless at 8.0 to pink at 10.0. And bromthymol blue has a useful range from 6.0 (yellow) to 7.6 (blue).

pH indicators are very special chemicals, they will change color if you change the pH (by adding acid or alkali). Perhaps we should call them pH indicators because there are other kinds of indicators.

Liquid pH indicators are used to test other solutions. A few drops of the right indicator added to an unknown solution can tell you its pH value. Chemists use pH indicators in a common laboratory procedure called titration. Here, an unknown substance is measured by carefully adding a solution of known concentration until a neutral point is reached. The neutral point is indicated by the color change of a pH indicator mixed in with the unknown solution.

Indicators change color at different pH values. Litmus changes at pH 7, while phenolphthalein changes from pink to colorless at pH 9 (See Table 1-1).

Table 1-1 Indicator color at different pH values

Indicator	pH	Color in Acid	Color in Alkali
Litmus	7.0	Red	Blue
Phenolphthalein	9.7	Colorless	Red/Pink
Methyl Orange	3.7	Red	Yellow
Bromophenol Blue	4.0	Yellow	Blue

pH Test Paper

How are pH values determined? Simple. One can use a special pH paper (called Hydrion^{TM} pH Paper or pH test paper), which is impregnated with pH indicators. Immerse a dry glass rod in the solution with unknown pH. Then attach it quickly to a piece of pH test paper. It turns a specific color depending upon the pH value of the substance. The color of the test strip is matched to a color chart, which gives the pH value.

pH Meter

In recent years, the demand for measuring pH has grown dramatically. This includes environmental, agricultural, wastewater, pharmaceutical, and educational applications, to name just a few.

A pH meter can also be used to measure the pH of soil, water, or other substances. It is an instrument that has a probe, which is inserted into a soil or liquid sample. It gives a "readout" concerning the pH of the substance tested. Knowing the pH of the soil can help a farmer know what soil type is best in which to grow particular plants, vegetables, or flowers.

Electronic, benchtop meters are available that read pH to resolutions of 0.001. Portable, "pocket size" meters suitable for a wide array of testing needs are now with accuracies of 0.1 or 0.01 pH. These easy-to-use meters feature built-in memorized buffer values for quick, "automatic" calibration and automatic temperature compensation that eliminates errors in pH measurement caused by solution temperature variations.

A pH meter measures the pH of a solution utilizing a glass electrode. The electrode is an ion sensitive electrode that will ideally respond to one specific ion, in this case H^+. It is made of very thin glass that allows H^+ ions to pass through it. A typical modern pH meter has a glass and reference electrode in one tube. The meter measures electrical potential and converts this data into a pH reading for a sample.

Before taking a pH measurement, the meter requires periodic calibration using standard value calibration solutions (sometimes called buffer solutions). Typical standards are pH 4.01, pH 7.01 and pH 10.01. The probe is immersed in the standard value calibration solution. The knobs on the box are used to adjust the displayed pH value to the known pH of the solution, thus calibrating the meter. Now the new-style pH meter is more convenient and accurate when calibrating the instrument.

To use the pH meter, the water sample is placed in the cup, and the glass probe at the end of the retractable arm is placed in the water. Inside the thin glass bulb at the end of the probe there are two electrodes that measure voltage. One electrode is contained in a liquid that has a fixed acidity, or pH. The other electrode responds to the acidity of the water sample. A voltmeter in the probe measures the difference between the voltages of the two electrodes. The meter then translates the voltage difference into pH and displays it on the little screen on the main box.

Chapter 2　Typical Chemical Laboratory Apparatus

alcohol burner ['ælkəhəl 'bə:nə] 酒精灯		dropping board ['drɔpiŋ bɔ:d] 点滴板	
asbestos gauze [æz'bestəs gɔ:z] 石棉网		Erlenmeyer flask ['ə:lənmaiə flɑ:sk] 锥形烧瓶	
beaker ['bi:kə] 烧杯	1000mL	evaporating dish [i'væpəreitiŋ diʃ] 蒸发皿	
Buchner funnel ['buknə 'fʌnəl] 布氏漏斗		filter paper ['filtə 'peipə] 滤纸	
burette [bjʊ'ret] 滴定管（酸式，碱式）	酸式　碱式	funnel ['fʌnəl] 漏斗	
		graduated cylinder ['grædjueitid 'silində] 量筒	
centrifugal tube [sen'trifjugəl 'tju:b] 离心试管		hair brush [hɛə brʌʃ] 毛刷	
crucible ['kru:sibl] 坩埚		iron clamp ['aiən klæmp] 铁夹	
crucible tong ['kru:sibl tɔŋ] 坩埚钳		iron clamp ring ['aiən klæmp riŋ]铁环 ring stand [riŋ stænd]铁架	
drop bottle [drɔp 'bɔtl] 滴瓶		long neck funnel [lɔŋ nek 'fʌnəl] 长颈漏斗	

Chapter 2　Typical Chemical Laboratory Apparatus

续表

名称	图	名称	图
measuring flask ['meʒəriŋ flɑ:sk] 容量瓶		test-tube [test 'tju:b] 试管	
medicine dropper ['medsin 'drɔpə] 滴管		test-tube clamp [test 'tju:b klæmp] 试管夹	
mortar ['mɔ:tə] 研钵 pestle ['pestl] 研棒		test-tube stand [test 'tju:b stænd] 试管架	
narrow-mouth(ed) bottle ['nærəu-mauθ 'bɔtl] 细口瓶		triangle tie ['traiæŋgl tai] 三脚架	
pipette [pi'pet] 移液管，吸量管	移液管　吸量管	washing bottle ['wɔʃiŋ 'bɔtl] 洗瓶	
		watch glass [wɔtʃ glɑ:s] 表面皿	
round bottom flask [raund 'bɔtəm flɑ:sk] 圆底烧瓶		water extractor ['wɔ:tə iks'træktə] 干燥器	
rubber suction bulb ['rʌbə 'sʌkʃən bʌlb] 吸耳球		water-bath ['wɔ:tə bɑ:θ] 水浴锅	
separating funnel ['sepəreitiŋ 'fʌnəl] 分液漏斗		weighing bottle ['weiiŋ 'bɔtl] 称量瓶	
spatula ['spætjulə] 药勺		wild-mouth bottle [waild-mauθ 'bɔtl] 广口瓶	
suction bottle ['sʌkʃən 'bɔtl] 吸滤瓶		wire triangle ['waiə 'traiæŋgl] 泥三角	

Chapter 3　Experiments

3.1　Some Elementary Operation

Objectives

(1) To master the correct operation of platform balance, glassware cleaning, heating solution in test tube and graduated cylinder.

(2) To master the way of transferring reagents, filtration and rinsing of precipitate and preparation of lotion.

(3) To master the operation of alcohol burner.

Apparatus, Reagents and Materials

1. Apparatus

Glassware, platform balance, filter paper, alcohol burner.

2. Reagents and Materials

Solids　$K_2Cr_2O_7$

Acids　H_2SO_4 1 mol/L, H_2SO_4 concentrated

Salts　$PbAc_2$ 0.1 mol/L

Procedure

1. Glassware Cleaning

Clean all the glassware equipments.

2. Heat up Solution in Test Tube

Place distilled water into a test tube. Hold the tube with a test tube clamp and heat up the water with alcohol burner to boiling (Figure 3-1). Do not point the mouth of the test tube toward anyone.

3. Operation of Graduated Cylinder

Take 2 mL of water with a graduated cylinder into a test tube (Figure 3-2), noting the liquid height. So the volume of liquid reagents can be estimated without graduated cylinder. Then take 3 mL and 5 mL of water in the same way. Note the correct way to determine the solution volume in graduated cylinder (Figure 3-3).

Figure 3-1　Heat up solution
in test tube

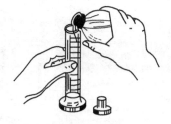

Figure 3-2　Take solutions with
a graduated cylinder

4. Preparation of Lotion

Weight out 5 g of solid $K_2Cr_2O_7$ using a dry watch glass, transfer the solid into a 250 mL beaker and then add 100 mL of concentrated sulfuric acid. Mix them up to form a chromic acid lotion. After the liquid cools down, transfer it to a reagent bottle.

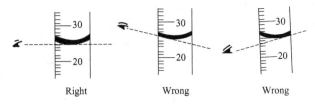

Figure 3-3 Determine the solution volume in the graduated cylinder

5. Filtration and Rinse of Precipitate

Take 20 mL of 0.1 mol/L Pb(Ac)$_2$ solution and pour it into a small beaker. Add 2 mL of 1 mol/L H$_2$SO$_4$ drop by drop, at the same time mix them up with glass rod, observing the white precipitate. After the solution is layered, filtrate the solution with decantation method and rinse the precipitate twice with 10 mL of distilled water per time. Transfer the upper clear layer to a funnel for filtration, then transfer the precipitate (PbSO$_4$) to the filter paper. Force out some water in washing bottle to rinse the precipitate (Figure 3-4) and observe the filtrate's clarity.

Figure 3-4 Rinse the precipitate in the funnel

6. Operation of Alcohol Burner

(1) Light up the alcohol burner only with match or burning batten. To avoid alcohol pouring out and burning, never light up an alcohol burner with another burning one (Figure 3-5).

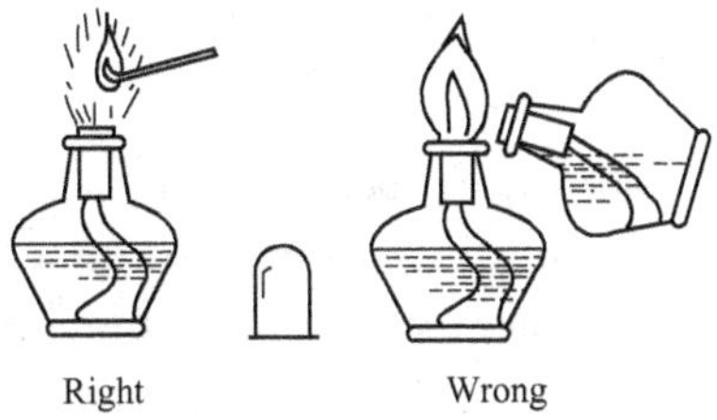

(2) Extinguish the flame with the crown top of burner. Blowing it off is forbidden.

(3) To save alcohol extinguish the flame instantly when the burner is not being used and cover the crown top. Because the lamp wick will absorb moisture without the cap and it will be hard to light it up again.

Figure 3-5 Light up an alcohol burner

(4) The burner top will be quite hot after a long time burning or burning under the asbestos gauze. Move the crown top of burner away when the fire is extinct. Cover the crown top again after the burner top cools down, or alcohol vapor will condense when meeting with cold alcohol burner cap, which leads to the burst of burner top.

(5) If alcohol escape and burn, put out the fire with water because alcohol is water-soluble.

Key Words

agitator [ˈædʒiteitə] 搅拌器
alcohol burner [ˈælkəhɔl ˈbəːnə] 酒精灯
asbestos gauze [æzˈbestɔs gɔːz] 石棉网
beaker [ˈbiːkə] 烧杯
concentrated [ˈkɔnsentreitid] 浓的
crown top of burner [kraun tɔp ɔv ˈbəːnə] 灯帽
decantation [ˌdiːkænˈteiʃən] 倾析
erlenmeyer flask [ˈəːlənmaiə flɑːsk] 锥形烧瓶

filter paper [ˈfiltə ˈpeipə] 滤纸
filtration [filˈtreiʃən] 过滤
funnel [ˈfʌnəl] 漏斗
graduated cylinder [ˈgrædjueitid ˈsilində] 量筒
lamp wick [læmp wik] 灯芯
lead acetate [liːd ˈæsiˌteit] 醋酸铅 PbAc$_2$
lead sulfate [liːd ˈsʌlfeit] 硫酸铅 PbSO$_4$
liquid reagent [ˈlikwid ri(ː)ˈeidʒənt] 液体试剂

lotion [ˈləuʃən] 洗液
pipette [piˈpet] 移液管
platform balance [ˈplætfɔːm ˈbæləns] 台秤
potassium dichromate [pəˈtæsjəm daiˈkrəumeit] 重铬酸钾 $K_2Cr_2O_7$
precipitate [priˈsipiteit] n. 沉淀物 v. 使沉淀
reagent bottle [ri(ː)ˈeidʒənt ˈbɔtl] 试剂瓶
rinse [rins] 清洗

rubber suction bulb [ˈrʌbə ˈsʌkʃən bʌlb] 洗耳球
sucker [ˈsʌkə] 吸管
sulphuric acid [sʌlˈfjuərik ˈæsid] 硫酸
test tube [test ˈtjuːb] 试管
volumetric flask [vɔljuˈmetrik flɑːsk] 容量瓶
watch glass [wɔtʃ glɑːs] 表面皿
washing bottle [ˈwɔʃiŋ ˈbɔtl] 洗瓶

3.2 Acid-Base Titration

Objectives

(1) To master the titration operations.
(2) To measure the concentrations of NaOH and HCl solutions.

Principles

The reaction of an acid and a base to form a salt and water is known as neutralization. In this experiment $H_2C_2O_4$ and HCl are used as the acids. The following are reactions of $H_2C_2O_4$ and HCl with NaOH respectively.

$$H_2C_2O_4 + 2NaOH = Na_2C_2O_4 + 2H_2O$$
$$HCl + NaOH = NaCl + H_2O$$

Titration is the process of measuring the volume of one reagent to react with a measured volume or weight of another reagent. Figure 3-6 shows the operation using basic pipette [Figure 3-6(a)] and acidic pipette [Figure 3-6(b)]. In this experiment we will determine the concentration of a NaOH solution by titrating it with a standard acid ($H_2C_2O_4$) solution of known concentration. Then an HCl solution of unknown concentration will be titrated with the NaOH solution.

In the titration, the point of neutralization, which is called the end-point, is observed when an indicator, which is placed in the solution being titrated, changes color. The indicator selected is one that changes color when the stoichiometric quantity of base (according to the chemical equation) has been added to the acid. A solution of phenolphthalein, an organic compound, is used as the indicator in this experiment. Phenolphthalein is colorless in acid solution but changes to pink when the solution becomes slightly alkaline. When the number of moles of sodium hydroxide added is equal to the number of moles of HCl originally present, the reaction is complete. The next drop of sodium hydroxide added changes the indicator from colorless to pink.

Use the following relationships in your calculations, according to the above reaction equation

(1) Calculations of the concentration of NaOH

$$2c(H_2C_2O_4)V(H_2C_2O_4) = c(NaOH)V(NaOH)$$
$$c(NaOH) = 2c(H_2C_2O_4)V(H_2C_2O_4)/V(NaOH)$$

(2) Calculations of the concentration of HCl

$$c(HCl)V(HCl) = c(NaOH)V(NaOH)$$

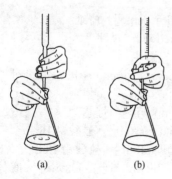

Figure 3-6 Operation of titration

$$c(\text{HCl}) = c(\text{NaOH})V(\text{NaOH})/V(\text{HCl})$$

Apparatus, Reagents and Materials

1. Apparatus

Basic burette, pipette.

2. Reagents and Materials

Acids HCl solution, $H_2C_2O_4$ standard solution

Bases NaOH solution

Indicators Phenolphthalein

Procedure

1. Measuring the concentration of NaOH

Rinse the basic burette thrice with 5 to 10 mL portions of sodium hydroxide solution by holding the burette in a horizontal position, rolling the solution around to wet the whole inner surface and letting out the rinsing through the burette tip. Discard the rinsing. Fill the burette with the base, making sure that the tip is completely filled and contains no air bubbles. You can drive away air bubbles as shown in Figure 1-11. Adjust the level of the liquid in the burette so that the bottom of the meniscus is at exactly 0.00 mL. Record the initial burette reading (0.00 mL) in the space provided on the report sheet. Be careful not to waste the solutions. Do not expose the base unnecessarily to air because the base will absorb CO_2.

Rinse the pipette thrice with 2 to 5 mL portions of the standard oxalic acid solution, running the rinsing through the pipette tip. Discard the rinsing. Draw 20.00 mL of the standard oxalic acid solution into the pipette with a rubber suction bulb, and then move the pipette to the Erlenmeyer flask that is to receive the acid and allow the liquid to drain while holding the pipette in a vertical position. And then add to the flask 2~3 drops of phenolphthalein indicator.

To let out the solution, move the glass bead in the burette with the forefinger and the thumb of your left hand (if you are right handed) to allow the solution to enter the Erlenmeyer flask. As shown in Figure 3-6(a), this procedure leaves your right hand free to swirl the solution and create a whirlpool of solution in the flask during the titration, so that the solutions are mixed completely. At first, put the titrating solution rapidly, and the indicator color change is sharp and conspicuous. But as the quantity of solution gets closer to the endpoint amount (the indicator color changes slowly), continuously reduce the rate of addition, so that the last additions are made drop-by-drop. When the first permanent detectable pale pink color is obtained ("Permanent" means a persistence of the pink color for at least 30 sec of swirling). The endpoint of the titration is reached. Record the final reading of the burette in Data Table 3-1 after allowing 1 min for the walls of the burette to drain.

Run two more similar titrations using phenolphthalein. The difference among the three base volumes added in the titrations should be less than 0.10 mL.

2. Measuring the concentration of HCl

Transfer 20.00 mL of the hydrochloric acid solution (whose exact concentration is unknown) into an Erlenmeyer flask with the pipette. And add 2~3 drops of phenolphthalein indicator into the flask. Titrate it with the NaOH solution according to the above-mentioned operations. Record the volume of the NaOH solution used in Data Table 3-2.

Run two more similar titrations using phenolphthalein. The difference among the three base volumes obtained in the titrations should be less than 0.10 mL.

Data Table 3-1 Measure the concentration of NaOH

Titrations order	1	2	3
Volume of NaOH used/mL			
Volume of $H_2C_2O_4$ used/mL			
Concentration of $H_2C_2O_4$/(mol/L)			
Concentration of NaOH/(mol/L)			
Average concentration of NaOH/(mol/L)			

Data Table 3-2 Measure the concentration of HCl

Titrations order	1	2	3
Volume of NaOH used/mL			
Volume of HCl used/mL			
Concentration of HCl/(mol/L)			
Average concentration of HCl/(mol/L)			

Notes

(1) When the liquid in the pipette is let out, it's better to rest the tip of pipette to the inner wall of the Erlenmeyer flask for 30s, and then take it away.

(2) You can control the dropping velocity a little quicker at the beginning of the titration, at this moment the pink of the indicator will disappear immediately. When near the endpoint, pink disappears slowly, and you must titrate a drop at a time until pink doesn't disappear in half minute, which means the endpoint is located.

(3) After each titration there should be no drop outside the burette tip and the burette tip contains no air bubbles.

(4) The mixed solutions will fade after a long time because CO_2 in the air dissolves in the solution. But it doesn't mean that the neutralization is uncompleted.

(5) During the titration, there is part of the flask wall above the solution, which may retain drops of unreacted titrating solution. To avoid experimental errors the solution on the flask walls should be washed down with distilled water several times before it approaches the endpoint.

Key Words

burette [bjuə'ret] 滴定管
chemical equation ['kemikəl i'kweiʃən] 化学方程式
concentration [ˌkɔnsen'treiʃən] 浓度
distilled water [dis'tild 'wɔːtə] 蒸馏水
end-point 滴定终点
erlenmeyer flask ['əːlənmaiə flɑːsk] 锥形瓶
glass bead [glɑːs biːd] 玻璃珠
hydrochloric acid [ˌhaidrəu'klɔːrik 'æsid] 盐酸
meniscus [mi'niskəs] 新月
neutralization [ˌnjuːtrəlai'zeiʃən] 中和
organic compound [ɔː'gænik 'kɔmpaund] 有机化合物
oxalic acid [ɔk'sælik 'æsid] 草酸
phenolphthalein [ˌfinɔl'fθæliːn] 酚酞
pink [piŋk] 粉红色的
pipette [pi'pet] 移液管，吸量管
rinse [rins] 润洗，刷洗
sodium hydroxide ['səudjəm hai'drɔksaid] 氢氧化钠
stoichiometric quantity [stɔikiə'metrik 'kwɔntiti] 化学计量数
swirl [swəːl] n. 漩涡，涡状形；vt. 使成漩涡

tip [tip] 尖嘴，尖端
titrate ['taitreit] 用滴定法测量

titration [tai'treiʃən] 滴定

3.3 Electrolyte Solutions

Objectives

(1) To understand the dissociation equilibria of weak acids and weak bases, and the principles of their equilibrium shifts.

(2) To prepare buffer solutions and test their properties.

(3) To understand the behavior of salts hydrolysis and the principle of the salts hydrolysis equilibria.

(4) To understand the multi-phase dissociation equilibria of sparingly soluble salts and the principles of their dissociation equilibrium shifts.

Principles

1. Dissociation equilibria of weak electrolytes and their equilibrium shifts

The extent of the ionization of a weak acid or weak base can be reduced by adding a strong electrolyte that provides an ion common to the dissociation equilibrium. This phenomenon is called the common-ion effect. For example, when a strong electrolyte (For example, NaAc, fully dissociated) is added to a weakly dissociated electrolyte (For example, HAc), sodium acetate provides a high concentration of acetate ions (Ac^-) that suppresses the dissociation of HAc and reduces the amount of H^+ (aq), thus making the solution less acidic. The above-mentioned equilibrium will shift toward the formation of HAc.

2. Buffer solution

As it's known, a buffer solution is made of week acid/base and its salts. By choosing the appropriate components, a solution can be buffered at virtually any pH. For example, when sodium acetate is added to acetic acid solution, a buffer solution forms. The addition of ammonium chloride to ammonia solution is a further example.

3. Salt hydrolysis reactions

Some salts, when added to water, react with the water to yield solutions, which are either alkaline or acidic. These salts are said to be hydrolyzed. The addition of hydrolysis products to the obtained solution, restrains the salt hydrolysis reaction. For a given hydrolysis reaction, it will be accelerated as the solution becomes more dilute or at a higher temperature.

4. Dissociation equilibria of sparingly soluble electrolyte compounds and their shifts

For a saturated solution of sparingly soluble salt, there is an equilibrium that exists between the solid compound and the ions in the solution. In a given saturated solution of sparingly soluble salt, the product of the active concentration of the ions raised to the power as indicated by the chemical formula is a constant, which is called the solubility product constant. It should be noted that the equilibrium might be shifted to the resulting in dissolve. On addition of appropriate components, the precipitate may dissolve owing to the reaction of ions particularly with respect to precipitation. In this case, the product of the active concentration of the ions raised to the power is less than the solubility product. For example, the solubility of the salt calcium carbonate $CaCO_3$ is much greater in the presence of an acid than in pure water, owing to the formation of carbonic acid (H_2CO_3) which is formed by the reaction between H^+ and CO_3^{2-} ions. Carbonic acid (H_2CO_3) further decomposes to give the water and carbon dioxide.

5. How to take solutions with droppers (Figure 3-7)

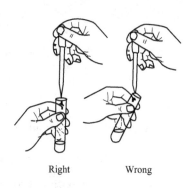

Figure 3-7 Take solutions with droppers

Apparatus, Reagents and Materials

1. Apparatus

Pipette, burette.

2. Reagents and Materials

Solids NH_4Cl

Acids HAc (1 mol/L, 0.1 mol/L), HCl (6 mol/L, 0.1 mol/L), citric acid 0.1 mol/L, $H_2C_2O_4$ 0.1 mol/L

Bases $NH_3 \cdot H_2O$ (2 mol/L, 1 mol/L, 0.1 mol/L, 6 mol/L), NaOH 0.1 mol/L

Salts $MgCl_2$ 0.1 mol/L, Na_3PO_4 0.01 mol/L, NH_4Cl (0.1mol/L, saturated), $CaCl_2$ 0.1 mol/L, NaAc 0.5 mol/L, $Cu(NO_3)_2$, 0.1 mol/L, Na_2HPO_4 0.2 mol/L, $(NH_4)_2C_2O_4$ 0.1 mol/L, $SbCl_3$, 0.1 mol/L, Na_2CO_3 1 mol/L, NH_4Ac 1 mol/L, $FeCl_3$ 1 mol/L, $Bi(NO_3)_3$ 0.1 mol/L

Indicators phenolphthalein indicator, methyl orange indicator, thymol blue, thymol-sulfonphthalein

Procedure

1. Ionization equilibria of week acids/bases

(1) Place 1 mL of distilled water, 2 drops of 2 mol/L $NH_3 \cdot H_2O$ and 1 drop of phenolphthalein indicator into 2 test tubes respectively. Shake the tubes to mix the solution well and observe the solution color. Add a little solid NH_4Cl into one tube and swirl it. Compare the color change in two test tubes. Explain it.

(2) Place 2 mL of 1 mol/L HAc solution and 1drop of methyl orange indicator into a test tube. Observe the solution color. Add a little more solid NaAc, noting the solution color change. Explain it.

(3) Place 5 drops of 0.1 mol/L $MgCl_2$ solution into two test tubes respectively, then put 5 drops of saturated NH_4Cl solution into one of them. Add 5 drops of 2 mol/L $NH_3 \cdot H_2O$ into each tube, noting the solution change in two tubes. Explain the difference.

2. Buffer solution

(1) Place 3 mL of 0.1 mol/L HAc solution and 3 mL of 0.1 mol/L NaAc solution into a test tube to form a buffer solution. Add 5 drops of thymol blue indicator, noting the solution color. Divide the solution into 3 test tubes (tube a, b and c). Add 5 drops of 0.1 mol/L HCl solution into tube a, 5 drops of 0.1 mol/L NaOH solution into tube b and 5 drops of H_2O into tube c. Observe the color change. Add excessive 0.1 mol/L HCl solution into tube a, and excessive 0.1 mol/L NaOH solution into tube b, noting the solution color change. Make a conclusion on it.

Table 3-1 lists the color-changing range of thymol blue indicator.

Table 3-1 The color-changing range of thymol blue indicator

pH value	Less than 2.8	2.8~9.6	More than 9.6
Color	Red	Yellow	Blue

(2) Prepare the buffer solution and observe the color-change range of indicator

Prepare a buffer solution with pH ranging from 2.2 to 5.0 according to the solution volume in the following table.

Place 4 mL of as-prepared buffer solution into 5 Nessler tubes respectively. Line up the tubes with pH value from low to high, then add 1 drop of methyl orange indicator into each tube and mix them well. Observe the solution color and define the color-changing range of methyl orange indicator.

pH value	0.1 mol/L citric acid solution/mL	0.2 mol/L Na_2HPO_4 solution/mL
2.2	19.6	0.40
3.0	15.89	4.11
4.0	12.29	7.71
4.4	11.29	8.71
5.0	9.70	10.30

(3) Buffer capacity of buffer solution

Take 25 mL of 0.1 mol/L $NH_3 \cdot H_2O$ and 25 mL of 0.1 mol/L NH_4Cl solution into a beaker with pipette to make a buffer solution. Measure its pH value with precise pH paper and compare it with the calculated pH value.

Add 0.5 mL of 0.1 mol/L HCl solution into as-prepared buffer solution and measure its pH value with precise pH paper. Add 1 mL of 0.1 mol/L NaOH solution and measure its pH value again. Compare the measured value with the calculated one. Write down the datum in the following table.

Buffer solution	Calculated pH value	Measured pH value
25 mL of 0.1 mol/L $NH_3 \cdot H_2O$		
25 mL of 0.1 mol/L NH_4Cl		
Add 0.5 mL of 0.1 mol/L HCl solution into as-prepared solution		
Add 1 mL of 0.1 mol/L NaOH solution into above solution		

3. Hydrolysis of salts

(1) Obtain 4 test tubes. Add 5 drops of 1 mol/L Na_2CO_3 into the first tube, 5 drops of $FeCl_3$ into the second, 5 drops of NaCl into the third and 5 drops of NH_4Ac solution into the forth. Check their acidity/basicity with pH test paper. Which salts have hydrolyzed? Write down the ionic equation.

(2) Place a little solid $SbCl_3$ [or 2 drops of 0.1 mol/L $Bi(NO_3)_3$ solution] into a test tube containing about 1 mL of water. What will happen? Check its acidity/basicity with pH test paper. Add 6 mol/L HCl solution until solution just becomes clear. Add more water and note the phenomenon. Explain it according to the principle about equilibrium shift.

4. Formation and dissolution of precipitate

Place 1 mL of 0.1 mol/L $CaCl_2$ solution into two test tubes respectively. Add 5 drops of 0.1 mol/L $(NH_4)_2C_2O_4$ solution into one tube (No. 1) and 5 drops of 0.1 mol/L $H_2C_2O_4$ solution into the other (No. 2). Compare the change. Add 5 drops of 6 mol/L HCl solution into No. 1 tube. What happens? Then add a little excessive 6 mol/L $NH_3 \cdot H_2O$ solution into No. 1 tube and note the change. In order to judge the precipitate component [$Ca(OH)_2$ or CaC_2O_4], take another test tube, add equal amount of $CaCl_2$ solution and $NH_3 \cdot H_2O$. Observe the phenomenon and give the answer.

Notes

(1) The dropping board should be cleaned before use.

(2) In Procedure 2 (3) the pH value of the buffer solution should be measured quickly, because ammonia is favor to volatilize.

(3) In Procedure 3 (2) solid or concentrated Bi(NO₃)₃ should be used for obvious phenomenon.

Key Words

acetic acid [ə'si:tik 'æsid]　醋酸
ammonia water [ə'məunjə 'wɔ:tə]　氨水 $NH_3 \cdot H_2O$
ammonium acetate [ə'məunjəm 'æsi,teit]　醋酸铵 NH_4Ac
ammonium chloride [ə'məunjəm 'klɔ:raid]　氯化铵 NH_4Cl
antimony chloride ['æntiməni 'klɔ:raid]　氯化锑 $SbCl_3$
bismuth nitrate ['bizməθ 'naitreit]　硝酸铋 $Bi(NO_3)_3$
calcium chloride ['kælsiəm 'klɔ:raid]　氯化钙 $CaCl_2$
citric acid ['sitrik 'æsid]　柠檬酸
cupric nitrate ['kju:prik 'naitreit]　硝酸铜 $Cu(NO_3)_2$
cupric oxalate ['kju:prik 'ɔksəleit]　草酸铜 CuC_2O_4

disodium hydrogen phosphate [dai'səudiəm 'haidrəudʒən 'fɔsfeit]　磷酸氢二钠 Na_2HPO_4
dropping board ['drɔpiŋ bɔ:d]　点滴板
ferric chloride ['ferik 'klɔ:raid]　氯化铁 $FeCl_3$
magnesium chloride [mæg'ni:zjəm 'klɔ:raid]　氯化镁 $MgCl_2$
methyl orange ['meθil 'ɔrindʒ]　甲基橙
Nessler tube ['neslə tju:b]　奈氏比色管
oxalic acid [ɔk'sælik 'æsid]　草酸 $H_2C_2O_4$
phenolphthalein [finɔl'fθæli:n]　酚酞
phenolphthalein indicator [,finɔl'fθæli:n 'indikeitə]　酚酞指示剂
sodium carbonate ['səudjəm 'kɑ:bəneit]　碳酸钠 Na_2CO_3
sodium phosphate ['səudjəm 'fɔsfeit]　磷酸钠 Na_3PO_4
thymol blue ['θaiməul blu:]　百里酚蓝

3.4　Preparation of Officinal Sodium Chloride and Its Purity Examination

Objectives

(1) To master the principle and ways of impurities removal.
(2) To practice the methods of purity examination.

Principles

1. Purification

In this experiment, officinal NaCl with high purity and high quality will be refined from raw salt by physical and chemical treatments. The major impurities in the raw salt are some insoluble substance, such as mud, sand etc. and some soluble ions such as Br^-, SO_4^{2-}, I^-, K^+, Ca^{2+}, Mg^{2+}, Fe^{2+}, etc.

As sodium chloride is a soluble crystalline solid, there are some general methods to remove the impurities.

(1) Insoluble substance such as mud and sand etc. can be removed by filtration.

(2) Soluble impurities, such as SO_4^{2-}, Ca^{2+}, Mg^{2+}, Fe^{2+} and few heavy metals ions, can be removed by precipitation and filtration. For example, SO_4^{2-} ions can be precipitated by superfluous Ba^{2+} and $BaSO_4$ can be removed by suction filtration. The carbonates and hydroxides of Ca^{2+}, Mg^{2+}, Fe^{2+}, Ba^{2+} and some heavy metals ions play important roles in the removal of these cations.

(3) Soluble impurities, such as Br^-, I^-, K^+, can be removed by recrystallization.

As NaCl crystallizes from the solution, these ions are excluded from growing crystal lattice, giving a pure solid. When separating the solid from the solution by suction filtration, these soluble ions staying in the solution can be removed.

2. Examination

In this experiment, you should finally obtain officinal NaCl with high purity and high quality. Therefore the lower the concentrations of impurities are, the better the product should be. You can determine the concentrations of impurities qualitatively by comparing your samples with the standard samples prepared by the lab instructor. Although the concentrations of heavy metals ions may be very small, it is essential to examine their concentrations, which affect the quality of the product strongly.

(1) The examinations of Ca^{2+}, Mg^{2+}, SO_4^{2-}, Ba^{2+}, K^+ will be achieved by turbidimetry. According to the turbidimetry method, the intensity of turbidity in a liquid is proportional to the concentration of precipitate (or fine particles suspended in solution), only if the thickness (or path length) of the solution is constant. Therefore you will prepare the sample solutions through precipitating these ions with specific precipitants, and match these sample solutions against the standard solutions containing known concentrations of these ions. Then you can determine the concentrations of these ions qualitatively. The less intensity of turbidity of sample solution than that of the standard solution means the higher quality of sample.

(2) The examinations of the ions of heavy metals such as Pb, Bi, Cu, Hg, Sb, Sn, Co, Zn etc. will be achieved by colorimetric method. These heavy metals can be colored by sulfide ion under the specified test conditions. Colorimetric analysis will determine small concentrations of substances where the substance is colored, or reacts with a suitable reagent to produce a colored substance by comparing with standard sample. When matching visually the color of a solution containing the constituent to be determined against the color of standard solutions containing known concentrations of that constituent, you can determine the concentrations of these heavy metals qualitatively. The less intensity of color of sample solution than that of the standard solution means the less concentration of these metals.

Apparatus, Reagents and Materials

1. Apparatus

Electric cooker, evaporating dish, Buchner funnel, beakers (250 mL), platform balance, colorimetric cylinders.

2. Reagents and Materials

Solids   raw sodium chloride

Acids H_2SO_4 6 mol/L, HCl (concentrated, 1/50 mol/L)

Bases   NaOH (2 mol/L, 1/50 mol/L), Na_2CO_3 saturated

Salts $BaCl_2$ 25%, KI 10%, KBr 10%, Ca^{2+} test solution (TS), Mg^{2+} TS, $(NH_4)_2C_2O_4$ TS, NH_4Cl TS

Indicators   bromthymol blue

Others   starch-KI test paper

Procedure

1. Purification of salt

(1) Obtain 50.00 g of raw salts using a platform balance, and put the salt into a beaker. Add enough water into the beaker. Place a piece of asbestos gauze on an electric cooker

and put the beaker on the asbestos gauze. Then heat the mixture to a gentle boiling. Stir with a glass-stirring rod. Continue to heat the solution until all of the salts are dissolved.

(2) Add 25% $BaCl_2$ drop by drop to the beaker on stirring until the precipitate is completed. Put the solution without disturbance for a moment while the solution is still hot.

(3) Set up a suction filtration with a clean Buchner funnel and a clean suction bottle (Your laboratory instructor will provide you with instructions for operating suction filtration). Decant and filter the hot solution by suction filtration.

(4) Wash the beaker and transfer the filtrate to the beaker. Discard the precipitate on the filter paper. Continue heating the beaker to a gentle boiling. Add saturated solution of Na_2CO_3 drop by drop to the beaker and then adjust the pH between 10 and 11 with NaOH solution until the precipitate is completed. Repeat step 3.

(5) Discard the precipitate on the filter paper and transfer the filtrate to a clean evaporating dish. Adjust the pH of the filtrate between 3 and 4 with HCl solution. Heat and evaporate the solution until dense mash occurs and a large amount of NaCl crystals precipitate.

(6) Filtrate the hot mash by suction filtration. The suction should be continued for minutes to partially dry the crystals. Discard the filtrate in the suction bottle after measure its exact volume. Wash the evaporating dish and then use a glass rod to carefully remove the crystals (the product) from the filter paper and to transfer the crystals to the evaporating dish.

(7) Dry the product by heating. Then cool down the product. Measure the mass of the product with a platform balance. Calculate the productivity of NaCl. The purity of the product needs to be examined afterwards.

2. Examination

(1) Clarity of the product dissolved in water

Dissolve 5 g of products in 25 mL of distilled water. Observe the solution carefully. The solution should be colorless and transparent.

(2) Acidity or basicity

NaCl, a salt produced by strong acid and base, should be neutral in water. But it may contain a little acid or base during the process of purification. So the acidity or basicity of qualified product should vary in a narrow scope. In this experiment, bromothymol blue will be used as indicator because the color change scope of this indicator is pH6.6 ~ pH7.6 with the color from yellow to blue.

Obtain 5 g of product dissolved in 50 mL of fresh distilled water and add two drops of bromothymol blue into the solution. If the solution presents yellow, add 1/50 mol/L NaOH solution until the color of the solution turns just to faint blue. If the amount of NaOH solution used does not exceed 0.1 mL, the product is qualified. If the solution presents blue, add 1/50 mol/L HCl solution until the color of the solution turns just to faint yellow. If the amount of HCl solution does not exceed 0.2 mL, the product is qualified.

(3) Bromide and Iodide

Dissolve 1g of product in a test tube with 3 mL of distilled water. Add 1 mL of chloroform into the test tube. Then add dilute chlorine (Cl_2) TS dropwise with constant shaking. The layer of chloroform should not present amaranth, yellow or orange color.

Control test Add 1 mL of 10% iodide TS and bromide TS into two tubes respectively, the following step is similar to that of the test solution. The color of chloroform in the tube containing the iodide TS is amaranth while the other tube is yellow or orange.

$$2Br^- + Cl_2 \longrightarrow Br^2 + 2Cl^- \qquad 2I^- + Cl_2 \longrightarrow I^2 + 2Cl^-$$

(4) Barium ion

Dissolve 4g of products in 20 mL of distilled water. Filter if necessary. Divide the solution into two equal portions. Add 2 mL of diluted H_2SO_4 into one portion and 2 mL of distilled water into the other portion. Put aside these two solutions for two hours. Then observe them. The two solutions should be equally clear and transparent.

(5) Calcium and Magnesium ions

Dissolve 4 g of products in 20 mL of distilled water. Add 2 mL of ammonia TS into the solution and divide the solution into two equal portions. Add 1 mL of ammonium oxalate $[(NH_4)_2C_2O_4]$ TS into one portion, 1 mL of disodium hydrogen phosphate (Na_2HPO_4) TS and a few drops of ammonium chloride TS into the other portion. Put aside these two solutions for five minutes. The two solutions should not be turbid within 5 minutes.

Control test Calcium Take 1 mL of calcium TS into a test tube, alkalized slightly with ammonia TS.

Add 1 mL of ammonium oxalate TS, white crystal is precipitated.

$$Ca^{2+} + C_2O_4^{2-} \longrightarrow CaC_2O_4 \downarrow \text{ (white)}$$

Magnesium Take 1 mL of magnesium TS into a test tube, add a few drops of ammonia TS and ammonium chloride. Then add, drop by drop, disodium hydrogen phosphate TS, white precipitate is separated from the solution.

$$Mg^{2+} + HPO_4^{2-} + NH_4^+ + OH^- \longrightarrow MgNH_4PO_4 \downarrow \text{ (white)} + H_2O$$

(6) Sulfate

Put 1 mL of standard potassium sulfate solution into a 50 mL colorimetric cylinder. Dilute the solution with water to about 35 mL. Add 5 mL of 1 mol/L hydrochloric acid and 5 mL of barium chloride TS. Dilute the solution with water to 50 mL and mix well.

Place 5 g of products into another 50 mL colorimetric cylinder. Add about 35 mL of distilled water and then add 5 mL of 1 mol/L HCl into the cylinder. Filter if necessary. Add 5 mL of $BaCl_2$, dilute the solution into the graduate line of 50 mL. Shake and make the solution uniform.

Put aside the two cylinders for ten minutes. View downward and compare the sample solution with the standard solution prepared by the lab instructor. The intensity of turbidity of your sample solution should be less than that of the standard solution.

Standard potassium sulfate Dissolve 0.01813 g of potassium chloride with distilled water in a 1000 mL volumetric flask, dilute with water to volume and mix well. This solution contains the equivalent of 0.1 mg of sulfate per mL.

(7) Iron ion

Dissolve 5 g of products with 35 mL of distilled water in a 50 mL colorimetric cylinder. Add 5 mL of 1 mol/L HCl, 30 mg ammonium persulfate and 5 mL of ammonium thiocyanate (NH_4SCN) TS into the solution. Dilute the solution to the graduate line of 50 mL. Shake and make the solution uniform. Compare the intensity of color of sample solution with that of the standard solution prepared by the lab instructor. The intensity of color of your sample solution should be less than that of the standard solution.

The equation of reaction between iron ions and thiocyanate ions is as follows.

$$Fe^{3+} + 6SCN^- \longrightarrow [Fe(SCN)_6]^{3-} \text{ (sanguine color)}$$

Standard ion solution Dissolve 863.0 mg of ferric ammonium sulfate in water, add 2 mL of dilute hydrochloric acid, and dilute with water to 1000 mL. Pipette 10 mL of this solution into a 100 mL volumetric flask, add 0.5 mL of dilute hydrochloric acid, dilute with water to volume, and mix well. This solution contains the equivalent of 0.01 mg of iron per mL.

(8) Potassium ion

Place 5 g of products into a 50 mL colorimetric cylinder. Add about 20 mL of distilled water to dissolve the product. Add 2 drops of 2 mol/L acetic acid (CH_3COOH) to adjust the solution pH between 5 and 6. Then add 2 mL of 0.1 mol/L sodium tetraphenylboron $[NaB(C_6H_5)_4]$ and dilute the solution to 50 mL. Shake and make the solution uniform. Compare the sample solution with the standard solution prepared by the lab instructor. The intensity of turbidity of your sample solution should be less than that of the standard solution.

The equation of reaction between potassium ions and $B(C_6H_5)_4^-$ ions is as follows.

$$K^+ + B(C_6H_5)_4^- \longrightarrow KB(C_6H_5)_4 \downarrow (\text{white})$$

Standard potassium sulfate solution Weigh accurately 2.2280 g of potassium sulfate previously dried at 105℃. Dissolve it with water in 1000 mL volumetric flask, dilute to volume, and mix. The obtained solution contains the equivalent of 1 mg of potassium per mL.

Sodium tetraphenylboron solution Triturate 1.5 g of sodium tetraphenylboron with 10 mL of water, then add 40 mL of water, triturate again and filter.

(9) Heavy Metals ions

Obtain two 50 mL colorimetric cylinders.

Into a colorimetric cylinder add 1 mL of standard lead solution (0.01 mg Pb/mL), add 2 mL of dilute acetic acid (CH_3COOH) and dilute with water to 25 mL. Place 5 g of products into the other colorimetric cylinder. Add 23 mL of distilled water to dissolve the product. Add 2 mL of dilute acetic acid. Then add 10 mL of H_2S solution into two cylinders respectively. Shake them and make the solution uniform. The cylinders should stand for 10 minutes in dark place. Compare the color of two solutions. The color of your sample solution should not be darker than that of the standard solution. That means the content of heavy metal doesn't exceed the prescriptive limit.

One of the relative equations of reactions is as follows.

$$Pb^{2+} + S^{2-} \longrightarrow PbS \downarrow$$

Notes

(1) Don't add too much water, otherwise you will spend too much time in evaporating the solution.

(2) When remove the impurities by adding precipitate agents, don't let solution boil for a long time. Add a little water if NaCl crystallizes on the solution surface.

(3) During evaporation, the crystal membrane of NaCl on the surface of the condensed solution should be ruptured with a glass rod, otherwise the crystal will splatter everywhere.

(4) Some soluble impurities, such as Br^-, I^- and K^+, will be removed away together with the mother liquid. So the solution should not be evaporated to dryness. Reserve about 30 mL of liquid before finishing the evaporation.

(5) Do Procedure 2 (4) first, for the solutions should be left alone for two hours.

(6) The solution prepared in Procedure 2 (1) will be further used in Procedure 2 (2).

(7) In Procedure 2 (2), add NaOH if the solution presents yellow, or add HCl if the solution presents blue.

(8) Use the colorimetric tube correctly.

Key Words

amaranth [ˈæmərænθ] 紫红色
ammonium persulfate [əˈməunjəm pəˈsʌlfeit] 过硫酸铵
ammonium oxalate [əˈməunjəm ˈɔksəleit] 草酸铵 $(NH_4)_2C_2O_4$
ammonium thiocyanate [əˈməunjəm ˌθaiəuˈsaiəneit] 硫氰酸铵 NH_4SCN
bromothymol blue [ˌbrəuməˈθaiməl bluː] 溴麝香草酚蓝
colorimetric method [ˌkʌləriˈmetrik ˈmeθəd] 比色法
colorimetric cylinder [ˌkʌləriˈmetrik ˈsilində] 比色管
dense mash [dens mæʃ] 稠厚糊状
disodium hydrogen phosphate [daiˈsəudiəm ˈhaidrəuʤən ˈfɔsfeit] 磷酸氢二钠 Na_2HPO_4
electric cooker [iˈlektrik ˈkukə] 电炉

evaporate [iˈvæpəreit] （使）蒸发
filtrate [ˈfiltreit] v. 过滤,筛选 n. 滤出液
impurity [imˈpjuəriti] 杂质
officinal [ɔfiˈsainl] n. 成药; adj. 药用的
platform balance [ˈplætfɔːm ˈbæləns] 托盘天平
precipitant [priˈsipitənt] 沉淀剂
recrystallization [riˈkristəlaizeiʃən] 重结晶
sanguine [ˈsæŋgwin] 血红色,血红色的
saturated [ˈsætʃəreitid] 饱和的
sodium tetraphenylboron [ˈsəudjəm ˌtetrəˌfiːnilˈbɔrɔn] 四苯硼钠
starch-KI test paper 淀粉碘化钾试纸
suction bottle [ˈsʌkʃən ˈbɔtl] 抽滤瓶
suction filtration [ˈsʌkʃən filˈtreiʃən] 抽滤
turbidimetry [təːbidiˈmetri] 比浊法
turbidity [təːˈbiditi] 浑浊

3.5 Oxidation-Reduction Reactions

Objectives

(1) To study the reactions of some oxidizing agents and reducing agents, and to gain practice in writing and balancing redox reactions.

(2) To determine the order of some metals in the table of electrode potential.

(3) To study the influence of temperature, pressure and ion concentrations on redox reactions.

(4) To study the principle of the electrochemical cell and the electrolytic cell.

Principles

Usually, a redox reaction is written as two half-reactions, showing the movement of electrons from one chemical species to another. It is possible to build an apparatus where the two half-reactions take place at separate sites, with electrons being transferred indirectly. Such an apparatus is known as an electrochemical or voltaic cell.

Each half-reaction has an electrode potential. The potential difference between an anode and a cathode can be measured by a voltage measuring device. But the absolute potential of the anode and cathode cannot be measured directly. Defining a standard electrode, all other potential measurements can be made against this standard electrode. Tabulating all available potentials, then we could predict comparative strengths of oxidizing and reducing agents. Generally, the most negative values indicate the strongest reducing agent, while the largest positive number for standard electrode potentially indicates the strongest oxidizing agent.

So each half-reaction has a standard electrode potential ($E_{\text{cathode}}^{\ominus}$ and $E_{\text{anode}}^{\ominus}$). If the overall reaction potential ($E_{\text{cell}}^{\ominus}$) is positive, the redox reaction proceeds to the right. If the overall reaction potential is negative, the redox reaction proceeds to the left.

$$E_{\text{cell}}^{\ominus} = E_{\text{cathode}}^{\ominus} - E_{\text{anode}}^{\ominus} \tag{3.1}$$

A device that generates electrical energy from chemical energy is called an electrochemical cell. An electron flow is produced by spontaneous oxidation-reduction reactions taking place at the electrodes.

A device converts electrical energy to chemical energy, which is called electrolytic cell, and the process of conversion is called electrolysis. It consists of a vessel with two electrodes dipped in an electrolyte. The electrode attached to the negative terminal of an external source of energy (Battery) is called the cathode and the one, which connects to the positive terminal, is called the anode.

Apparatus, Reagents and Materials

1. Apparatus

A piece of copper, lead wire, a piece of copper, a piece of zinc, lead wire, a piece of zinc, salt bridge

2. Reagents and Materials

Solids   Pb, Zn, CH_3CSNH_2, $(NH_4)_2S_2O_8$

Acids HNO_3 (2 mol/L, 1∶10), HCl concentrated, H_2SO_4 (3 mol/L, 1 mol/L), HAc 6 mol/L, H_2S solution

Bases   NaOH(40%, 6 mol/L)

Salts $KMnO_4$ 0.01 mol/L, $FeSO_4$ 0.5 mol/L, Na_2SO_3 0.1 mol/L, $Na_2C_2O_4$ 0.1 mol/L, NaBr 1 mol/L, $K_2Cr_2O_7$ 0.1 mol/L, NaI 1 mol/L, $Fe_2(SO_4)_3$ 0.025 mol/L, $Pb(NO_3)_2$ 1 mol/L, $CuSO_4$ 1 mol/L, Na_2SO_4 0.5 mol/L, $ZnSO_4$ 1 mol/L, $MnSO_4$ 0.2 mol/L, $AgNO_3$ 0.1 mol/L

Others red litmus-paper, H_2O_2 3%, $CHCl_3$

Procedure

1. Oxidizing Agents and Reducing Agents

(1) Add 5 drops of 0.01 mol/L $KMnO_4$ and 3 drops of 3 mol/L H_2SO_4 into each of two test tubes. Then add 1~2 drops of 3% H_2O_2 into one test tube. Add 2~3 drops of 0.5 mol/L $FeSO_4$ into the other test tube. Write down the equations and point out the oxidizing and reducing agents in the reactions.

(2) Add 3 drops of 0.1 mol/L $K_2Cr_2O_7$ and 5 drops of 3 mol/L H_2SO_4 into a test tube. Shake and add a little solid CH_3CSNH_2. Write down the equations and point out the oxidizing and reducing agents in the reactions.

2. Order of Zinc, Lead, Copper in the Table of Standard Electrode Potentials

Put some zinc granules with clean surface into one test tube containing 3 mL of 0.5 mol/L $Pb(NO_3)_2$ solution and into the other test tube containing 3 mL of 1 mol/L $CuSO_4$ solution respectively. Record the observations and write down the equations.

Place some Pb granules instead of Zn into 1 mol/L $ZnSO_4$ and $CuSO_4$ solution respectively. Record the observations and write down the equations.

Determine the order of zinc, lead, copper in the table of Standard Electrode Potentials and explain. So you can rank the reducing strengths of the metals.

3. Influence of Concentration, Acidity and Temperature on the Redox Reactions

(1) Influence of concentration

Place some Zn granules into each of two test tubes, Add 2 mL of concentrated nitric acid into one test tube and 2 mL of 2 mol/L nitric acid into the other.

Note the changes in two test tubes. Observe the difference in reaction velocities and products of these two test tubes. The main product of concentrated nitric acid being reduced can be determined by its color while the product of dilute nitric acid being reduced can be determined by the NH_4^+ in solution.

Identification of NH_4^+ in solutions with a gas chamber. Place 5 drops of solution identified in a big watch glass, and then add 3 drops of 6 mol/L NaOH solution. Obtain a small watch glass, adhere a wet red litmus paper to the watch glass. Cover the big watch glass with the small watch glass to form a sealed-room and observe the color change of the litmus-paper after ten minutes.

(2) Influence of acidity

Add 0.5 mL of 1 mol/L NaBr solution into each of two test tubes. Put 10 drops of 3 mol/L H_2SO_4 into one test tube and add 1 drop of 0.01 mol/L $KMnO_4$. Place 10 drops of 6 mol/L HAc into the other test tube and add 1 drop of 0.01 mol/L $KMnO_4$. Compare the velocity of the color disappearance in these two test tubes. Write down the equations and explain.

(3) Influence of temperature

Add 2 mL of 0.1 mol/L $Na_2C_2O_4$ solution, 0.5 mL of 3 mol/L H_2SO_4 and 1 drops of 0.01 mol/L $KMnO_4$ into each of two test tubes. Heat one test tube in a beaker containing 80℃ warm water and remain the other test tube unheated. Observe the speed of the color fading away. Write down the equation and explain.

(4) Influence of acidity and basicity on the products of redox reaction

Add 10 drops of 0.1 mol/L Na_2SO_3 solution into each of three test tubes. Add 10 drops of 1 mol/L H_2SO_4 into one test tube, 10 drops of distilled water into the second and 10 drops of 40% NaOH into the third. Shake and then add 3 drops of 0.01 mol/L $KMnO_4$ into each test tube. Observe the reactions and write down the equations taking into account that the reducing products of $KMnO_4$ are Mn^{2+} in acid medium, MnO_2 in neutral medium and MnO_4^{2-} in alkaline medium.

4. Influence of Catalysts on the Redox Reaction

Place 2 drops of 0.1 mol/L $MnSO_4$ solution and 1 mL of 3 mol/L H_2SO_4 into a test tube. Shake and mix them well. Add a spoon of solid $(NH_4)_2S_2O_8$, shake sufficiently until the solid dissolves completely. Divide the solution into two portions. Add 1~2 drops of 0.1 mol/L $AgNO_3$ into one portion, and heat it up in a water bath. Put aside the test tube for a little while, observing the color change. Write down the reaction equation and compare the phenomenon with the other portion. What's the difference?

5. Selection of Oxidizing Agent

Which oxidizing agent can satisfy the requiring that oxidizes I^- into I_2, while does not oxidize Br^- to Br_2 in the mixture of NaBr and NaI solution?

(1) Obtain two test tubes. Add 10 drops of 1 mol/L NaBr into one test tube and 1 mol/L NaI into the other. Then add 10 drops of 3 mol/L H_2SO_4, 1 mL of $CHCl_3$ and 2~3 drops of 0.01 mol/L $KMnO_4$ into each test tube. Shake and observe the color change in the

chloroform layers.

(2) Replace $KMnO_4$ with $Fe_2(SO_4)_3$ solution and repeat Procedure 5 (1). Observe the change in the tubes and write down the equations.

Select the suitable oxidizing agent according to your results and explain it.

6. Electrochemical Cell (or Voltaic Cell) and Electrolytic Cell

Obtain three 50 mL beakers. Add 25 mL of 1 mol/L $ZnSO_4$ solution into one beaker, $CuSO_4$ into the second and 50 mL of 0.1 mol/L Na_2SO_4 into the third. Assemble the apparatus as shown in Figure 3-8.

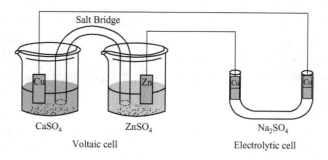

Figure 3-8 The apparatus

Add 3 drops of phenolphthalein indicator into the Na_2SO_4 solution around the cathode. Put the apparatus for a little while and observe the color change of solution around the cathode. Explain the phenomenon and write down the half-reaction.

Notes

(1) Sulfuric acid, sodium hydroxide and hydrogen peroxide are corrosive. Avoid spills and wash with water immediately in case of spills.

(2) Sodium sulfite solution is strongly basic and partially decomposes to give poisonous sulfur dioxide gas.

(3) In Procedure 2, owing to the low reaction rates, the tubes should be placed in the test tube rack for a while without any disturbance, then observe the phenomenon. The remainders should not be poured into the basin.

(4) In Procedure 3 (1), the reaction between concentrated HNO_3 and Zn will produce NO_2 that is toxic and volatile, so work in the ventilating hood as directed.

Key Words

ammonium persulfate [ə'məunjəm pə'sʌlfeit] 过（二）硫酸铵 $(NH_4)_2S_2O_8$
anode ['ænəud] 阴极，负极
cathode ['kæθəud] 阳极，正极
electrochemical cell [i,lektrəu'kemikəl sel] 原电池
electrode potential [i'lektrəud] 电极电位
electrolytic cell [i,lektrəu'litik sel] 电解池
hydrosulphuric acid [,haidrəusʌl'fjuərik 'æsid] 氢硫酸

potential difference [pə'tenʃ(ə)l 'difərəns] 电位差
red litmus paper [red 'litməs 'peipə] 红色石蕊试纸
shake [ʃeik] 振荡
Standard Hydrogen Electrode ['stændəd 'haidrəudʒən i'lektrəud] 标准氢电极
thioacetamide [,θaiəuæsi'tæmid] 硫代乙酰胺 CH_3CSNH_2
watch glass [wɔtʃ glɑːs] 表面皿

3.6 Coordination Compounds

Objectives

(1) To understand the formation and properties of coordination compounds.

(2) To master the applications of coordination compounds.

Principles

The formation of coordination compound is reversible. For example, when Cu^{2+} reacts with NH_3, tetraamminecopper (II) ion, $[Cu(NH_3)_4]^{2+}$, will be formed. The chemical equation for this reaction is as following.

$$Cu^{2+} + 4NH_3 \longrightarrow [Cu(NH_3)_4]^{2+}$$

(1) If the concentration of ligand (such as NH_3) is increased, the above-mentioned equilibrium will shift toward the formation of complex ion. Conversely, decreasing the concentration of ligand will cause the shift of the equilibrium toward the dissociation of complex ion.

(2) If the central ion can react with a precipitant forming insoluble substance, the complex ion will be destroyed. Contrarily, the reaction of a ligand with metal ion forming soluble complex ion will dissolve the slightly soluble salt of the metal ion. If a ligand is added prior to the precipitant, the deposition will be prevented. The competition between the metal ion and the precipitant/ligand depends on the relative magnitude of K_{sp} for slightly soluble substance and K_s for complex ion.

(3) If the ligand is a week base (such as NH_3), or a week acid group [such as CN^-, $S_2O_3^{2-}$, ethylenediaminotetraacetate (EDTA)], the addition of strong acid will cause the dissociation of complex ion.

(4) When two different ligands exist in the solution of a metal ion, a more stable complex ion will be formed. For example, when Fe^{3+} is added to the solution containing F^- and SCN^-, $[FeF_6]^{3-}$ will form, while $[Fe(NCS)_6]^{3-}$ will not.

(5) Addition of ligand to form complex ion will change the electronic structure of central metal ion, which can further change the redox property of the metal. For example, the reducibility of $[Co(CN)_6]^{4-}$ is much stronger than that of Co^{2+}.

(6) $[Cr(OH)Y]^{2-}$ will be formed when excessive $NH_3 \cdot H_2O$ is added to the solution of the complex ion prepared by the reaction between Cr^{3+} and EDTA.

(7) Some constants

① The solubility products of slightly soluble silver salts

Slightly soluble silver salts	AgCl	AgBr	AgI
K_{sp}	1.8×10^{-10}	5.3×10^{-13}	8.5×10^{-17}

② The stability constants of complex ions containing silver

Complex ions containing silver	$[Ag(NH_3)_2]^+$	$[Ag(S_2O_3)_2]^{3-}$	$[Ag(CN)_2]^-$
K_s	1.5×10^7	2.4×10^{13}	1.3×10^{21}

③ Standard electrode potential

$$E^{\ominus}_{Co^{3+}/Co^{2+}} = +1.842 \text{V} \qquad E^{\ominus}_{I_2/I^-} = +0.5345 \text{V}$$

Apparatus, Reagents and Materials

1. Apparatus

Beakers (50 mL), glass rod, test-tubes (10×100), test-tubes (10×500)

2. Reagents and Materials

Solids $CuSO_4 \cdot 5H_2O$, NaF

Acids   HCl 1 mol/L

Bases $NH_3 \cdot H_2O$ (2 mol/L, concentrated), NaOH 1 mol/L, Na_2CO_3 0.1 mol/L

Salts Na_2S 0.1 mol/L, $BaCl_2$, 0.1 mol/L, KSCN 1 mol/L, NaCl 0.1 mol/L, KBr 0.1 mol/L, EDTA-2Na 0.1 mol/L, $K_4P_2O_7$ solution, $Na_2S_2O_3$ 0.5 mol/L, KCN 0.5 mol/L, $AgNO_3$ 0.1 mol/L, KI 0.1 mol/L, sodium citrate 1 mol/L, $FeCl_3$ 0.1 mol/L, $CuSO_4$ 0.1 mol/L, $CaCl_2$ 0.1 mol/L, $CoCl_2$ 0.5 mol/L, $CrCl_3$ 0.1 mol/L

Others C_2H_5OH 95%, H_2O_2 30%, starch solution, acetone, solution of iodine, phenolphthalein indicator, pH test paper or red litmus test paper

Procedure

1. Formation and Properties of Tetraamminecopper (II) Sulfate $\{[Cu(NH_3)_4]SO_4\}$

(1) Formation

Weigh out 2.5g of $CuSO_4 \cdot 5H_2O$ in a small beaker and add 10 mL of water. Stir until the solid has completely dissolved, then add 5 mL of stronger ammonia water and mix the reagents. Add isovolumic ethyl alcohol, stirring to mix them well. Leave the beaker for 2~3 mins, then filter the solution using suction. Wash the solid $\{[Cu(NH_3)_4]SO_4 \cdot H_2O\}$ twice in the funnel by a small amount of ethanol. Record the characters of the product. The relative equation is as follows.

$$CuSO_4 + 4NH_3 \rightleftharpoons [Cu(NH_3)_4]SO_4$$

(2) Properties

① Dissolve a small amount of product in several drops of water, note the solution color. Then add more water, observe the change.

② Dissolve a small amount of product in several drops of water, add excessive 1 mol/L HCl solution drop by drop. Observe the change. Then add excessive stronger ammonia water, note the color change. Explain it.

Discuss the formation and dissociation of coordination compound according to these phenomena.

③ Dissolve a small amount of product in several drops of water, divide the solution into 3 test tubes.

(a) Add 0.1 mol/L sodium carbonate solution into No. 1 tube, observing whether cupric subcarbonate deposits or not.

(b) Add 0.1 mol/L sodium sulfide solution into No. 2 tube, observing whether copper sulfide occurs or not.

Discuss the concentration change of Cu^{2+} in the solution according to these phenomena.

(c) Add 0.1 mol/L barium chloride solution into No. 3 tube, observing whether barium sulfate precipitates or not.

Discuss the site of Cu^{2+} and SO_4^{2-} in the coordination compound.

④ Place some product into a dry test-tube, smell and check whether it has ammonia taste. Put a wet pH test paper or red litmus test paper near the top of test tube and heat the tube slightly. Observe and write down the following. (a) the color change of test paper. (b) the color of remaining solid. (c) the odor. (d) the equation.

Is NH_3 a constituent of the coordination? Is that bonding stable? Explain your answer.

2. Stability of Coordination Compound

Estimate the phenomena in the following steps according to the K_{sp} of AgCl, AgBr, AgI and the K_s of $[Ag(NH_3)_2]^+$, $[Ag(S_2O_3)_2]^{3-}$, $[Ag(CN)_2]^-$. Verify your estimation by tests.

(1) Add 3~4 drops of 0.1 mol/L $AgNO_3$ solution into a test-tube, then add isovolumic 0.1 mol/L NaCl solution.

(2) Add 2 drops of stronger ammonia water into solution (1).

(3) Add 2 drops of 0.1 mol/L KBr solution into solution (2).

(4) Add 2 drops of 0.5 mol/L $Na_2S_2O_3$ solution into solution (3).

(5) Add 2 drops of 0.1 mol/L KI solution into solution (4).

(6) Add 2 drops of 0.5 mol/L KCN solution into solution (5).

Record the phenomena of each step and write down the equations.

3. Influence of Solution pH on the Coordination Equilibrium

(1) pH change during the formation of coordination compound

Take two test tubes. Add 2 mL of 0.1 mol/L $CaCl_2$ solution into one test-tube and 2 mL of 0.1 mol/L EDTA-2Na (Na_2H_2Y) solution into the other. Add 1 drop of phenolphthalein indicator into each tube and 2 mol/L ammonia water to adjust the solution color until the color just becomes red. Mix two solutions. What color change will happen to the solution? Write down the equation. In what case, the solution pH will decrease during the formation of coordination compound?

(2) Effect of solution pH on coordination equilibrium

① Coordination of citric acid to Fe^{3+}

Place 1 mL of 0.1 mol/L $FeCl_3$ solution into a test-tube, then add 1 mL of 1 mol/L sodium citrate solution, observing the color change of solution. Divide the solution into 3 portions, one for control. Add 1 mol/L NaOH solution and 1 mol/L HCl solution into the other two tubes respectively. Compare the color in 3 tubes [Fe^{3+} is tan, citrate complex of Fe^{3+} is bright yellow or yellowish green, $Fe(OH)_3$ is red-brown]. Discuss the stability of ferric citrate in acidic and basic solutions.

② Effect of solution pH on the formation of $[Fe(NCS)_6]^{3-}$

Place 4~5 drops of 0.1 mol/L $FeCl_3$ solution into a test-tube, add 1 mL of 0.01 mol/L KSCN solution. Divide the solution into 2 tubes, add several drops of 1 mol/L HCl solution into one tube and 1 mol/L NaOH solution into the other. Observe the color change. Discuss the stability of $[Fe(NCS)_6]^{3-}$ in acidic and basic solutions.

4. Activity of Coordination Compound

Add 2 mL of 0.1 mol/L EDTA-2Na solution into ten drops of 0.1 mol/L $CrCl_3$ solution and mix them well. Observe the color change in the tube. Does any coordination compound form in the tube (Cr^{3+}-EDTA, modena)? Heat the solution to the boiling point and keep for several minutes. Does any coordination compound form in the tube? Add 1~2 drops of 2 mol/L ammonia water. Observe whether a green deposit of $Cr(OH)_3$ forms in the tube or not. Discuss why it is not easy for Cr^{3+} to react with EDTA to form coordination compound. Is it because of the instability of the coordination compound?

5. Change of Oxidation-Reduction Ability by Coordination

(1) Add 30% H_2O_2 solution into 4~5 drops of $CoCl_2$ solution. Observe and write down the phenomenon. Is H_2O_2 able to oxidize Co^{2+} to Co^{3+} (brown color)?

(2) Add excessive stronger ammonia water into 4~5 drops of $CoCl_2$ solution until the

precipitate has dissolved. Observe the color change. Does the color change when H_2O_2 is added? Acidify the solution with 6 mol/L HCl and observe the color change. What charge does Co have now? What effect does the formation of ammonia complex have on the reducibility of Co^{2+}?

(3) Put 1 mL of $CoCl_2$ solution into a test tube, add 5 drops of the starch solution. Add the solution of iodine drop by drop, observe and record the color change. Place 2 drops of $CoCl_2$ solution into another tube, add excessive KCN solution until the precipitate has completely dissolved. Then add 5 drops of the starch solution, subsequently add the iodine solution drop by drop. Observe the color change and write down the equations. Discuss the effect of KCN on the reducibility of Co^{2+}.

6. Coordination Masking: Mask of Fe^{3+} by F^-

Place several drops of 0.1 mol/L $FeCl_3$ solution into a test tube, add several drops of 0.1 mol/L KSCN solution and some solid NaF. Mix them well and observe the color. Write down the equations.

Place several drops of 0.5 mol/L $CoCl_2$ solution into another tube, add several drops of 1 mol/L KSCN solution. Then add isovolumic acetone. The existence of Co^{2+} can be verified if blue color occurs in the tube. Does the blue color fade if solid NaF is added?

Design a method for checking Co^{2+} when Fe^{3+} also exits.

7. Formation of Chelate: Preparation of Tetrapyrophosphatecopper(II) Ion

Place 2 drops of 0.1 mol/L $CuSO_4$ solution into a test tube and add $K_4P_2O_7$ solution drop by drop until the precipitate has completely dissolved. A dark blue transparent solution will form in the test tube.

$$2Cu^{2+} + P_2O_7^{4-} = CuP_2O_7 \downarrow$$
$$CuP_2O_7 + 3P_2O_7^{4-} = 2[Cu(P_2O_7)_2]^{6-}$$

The structure of tetrapyrophosphatecopper (II) ion is as following.

$$\begin{bmatrix} \begin{array}{c} O & & O \\ | & & | \\ O-P=O & & O-P=O \\ | & & | \\ O & Cu & O \\ | & & | \\ O=P-O & & O-P-O \\ | & & | \\ O & & O \end{array} \end{bmatrix}^{6-}$$

Notes

(1) Be careful when using KCN because it is extreme toxic substance. The solution should be recollected after the experiment.

(2) When synthesizing $[Cu(NH_3)_4]SO_4$, ammonia water should be added after all $CuSO_4 \cdot 5H_2O$ has dissolved completely. In this step, stronger ammonia water should be used.

Key Words

copper sulfate pentahydrate ['kɔpə 'sʌlfeit ˌpentə'haidreit] 水合硫酸铜 $CuSO_4 \cdot 5H_2O$

cupric subcarbonate ['kju:prik 'sʌbˌka:bəneit] 碱式碳酸铜

ethylenediaminotetraacetate ['eθili:n ˌdai ˌæminəuˌte'træsiteit] 乙二胺四乙酸 EDTA

potassium pyrophosphate [pə'tæsjəm ˌpairəu'fɔsfeit] 焦磷酸钾 $K_4P_2O_7$

potassium rhodanate [pə'tæsjəm 'rəudəˌneit] 硫氰化钾 KSCN

sodium citrate ['səudjəm 'sitrit] 枸橼酸钠，柠檬酸钠

sodium thiosulfate ['səudjəm ,θaiəu'sʌlfeit]
硫代硫酸钠 $Na_2S_2O_3$

solution of iodine [sə'lju:ʃən ɔv 'aiədi:n]
碘液

3.7 The Halogens

Objectives

(1) To test the physical and chemical properties of the halogens, hydrogen halides, oxyacids and oxysalts of the halogens.

(2) To master the common principles and methods of preparing the halogens in the laboratory, and study the common methods for identifying halogen ions.

Principles

The elements of group ⅦA, the halogens, include fluorine, chlorine, bromine, and iodine. Halogens have the electron configuration ns^2np^5, where n can be 2, 3, 4, 5, or 6. They are one electron short of a noble-gas configuration, and they can accept readily one electron. So all the halogens are very active nonmetals and are most often found in the -1 oxidation number. Fluorine is the most electronegative element and is found only in 0 and -1 oxidation states. Other halogens, when combined with something more electronegative than themselves, exhibit positive oxidation states up to $+7$.

Chlorine, bromine, and iodine can be prepared by oxidizing their halides.

Because of their high electronegativities, the halogens tend to gain electrons from other substances and thereby serve as oxidizing agents. The oxidizing ability of the halogens, which is indicated by their standard reduction potentials, decreases going down the group ($F_2 > Cl_2 > Br_2 > I_2$). As a result, a given halogen is able to oxidize the anions of the halogens below it in the family. For example, Cl_2 will oxidize Br^- and I^-, but not F^-.

But halogen anions are all reducing agents. The reactivity of halide anions changes by the converse trend: $I^- > Br^- > Cl^- > F^-$.

Since all the halogen molecules are nonpolar molecules, so they are very soluble in nonpolar solvents (organic solvents). Besides, iodine dissolves readily in potassium iodide solution, and forms KI_3.

Silver halides dissolve neither in water nor in dilute nitric acid, but the silver salts of CO_3^{2-}, PO_4^{3-}, CrO_4^{2-} can dissolve in nitric acid solution. In order to avoid the disturbance from other anions, the halide anions should be precipitated to silver halides in nitric acid solution.

Since the solubility of silver halides in ammonia solution is different, so changing the concentration of ammonia solution can separate the mixture of halide anions. For example, $(NH_4)_2CO_3$ is usually used to dissolve AgCl and then separate it from AgBr and AgI. The reaction equations are shown below.

$$(NH_4)_2CO_3 + H_2O \longrightarrow NH_4HCO_3 + NH_3 \cdot H_2O$$
$$AgCl + 2NH_3 \cdot H_2O \longrightarrow [Ag(NH_3)_2]^+ + Cl^- + 2H_2O$$

The acid strengths of the oxyacids increase with increasing oxidation state of the central halogen atom. The oxyacids of the halogens are rather unstable. They generally decompose (sometimes explosively) on attempts to isolate them. All of the oxyacids are strong oxidizing agents. The oxyanions, formed on removal of H^+ from the oxyacids, are generally more stable than the oxyacids. Hypochlorite salts are used as bleaches and disinfectants because of the powerful oxidizing capabilities of the ClO^- ion. Sodium chlorite, is used as a bleaching agent. Chlorate slats are similarly very reactive. For example, potassium chlorate is used in making matches and fireworks.

Apparatus, Reagents and Materials

1. Apparatus

Centrifugal tubes, centrifugal machine

2. Reagents and Materials

Solids I_2, red phosphorus, Zn, KCl, KBr, KI, MnO_2, $KClO_3$

Acids H_2SO_4 (3 mol/L, 18 mol/L), HCl 12 mol/L, HNO_3 6 mol/L

Salts   KBr 0.1 mol/L, KCl 0.1 mol/L, KI, 0.1 mol/L, $AgNO_3$ 0.1 mol/L, NaClO solution, I_2 solution

Others   blue litmus test paper, $Pb(Ac)_2$ test paper, starch-KI test paper, filter paper, chloroform

Procedures

1. Reaction of Iodine with Metals and Nonmetals

(1) Reaction of iodine solution with zinc powder

Add a spoon of zinc powder into a test tube containing 1 mL of iodine solution and shake it sufficiently. Take another test tube added with 1 mL of iodine solution as control. Observe the color change of the iodine solution (If the phenomenon is not obvious, heat slightly). Write down the reaction equation, and explain it.

(2) Reaction of iodine with red phosphorus

Mix a little solid iodine and red phosphorus in a test tube and add 1~2 drops of water (If the red phosphorus is moist, don't add water). Heat it in a water bath for a little while, then the reaction occurs intensively. Put a piece of wet blue litmus test paper near the top of the test tube to detect the formation of HI. Note the phenomenon, and write down the reaction equations. First, write down the reaction equation of forming PI_3, and then the hydrated reaction equation of PI_3.

2. Comparison of the Oxidizabilities of Chlorine, Bromine, Iodine

(1) Comparing the oxidizability of chlorine and iodine

Add 1 drop of 0.1 mol/L KI solution into a test rube, and then dilute it with distilled water to about 1 mL. Add chlorine water drop by drop, observe the color change. Then add 1 mL of chloroform, shake and observe the color change in the chloroform layer. After that, add excessive chlorine water (or vent gaseous chlorine) into the solution until the color of the chloroform layer disappears. Explain the phenomenon, and write down the reaction equations.

(2) Comparing the oxidizability of bromine and iodine

Add several drops of bromine water into about 1 mL of 0.1 mol/L KI solution, then add several drops of starch solution. Record the phenomenon and write down the reaction equations.

(3) Comparing the oxidizability of chlorine and bromine

Add several drops of chlorine water into about 1 mL of 0.1 mol/L KBr solution, observe the color change, and then add 1 mL of chloroform. Shake and observe the color change in the chloroform layer. Explain the phenomenon, and write down the reaction equation.

Summarize above experimental results and give the sequence of the oxidizability strength of chlorine, bromine and iodine. Explain it with standard electrode potentials.

3. Preparation of Halogens

Take 3 dry test tubes. Put some KCl crystals into the first, some KBr crystals into the second and some KI crystals into the third. Then add 2 mL of 3 mol/L H_2SO_4 and a little solid MnO_2 into each tube. Put a piece of starch-KI test paper near the mouth of the test tube containing KCl to identify whether the emitted gas is chlorine. Add 1 mL of chloroform into the other two test tubes respectively, observe the color of the chloroform layer, and write down the reaction equations.

4. Comparing the reducing ability of hydrogen halides

(1) To a dry test tube, add several KCl granules, and then add 2~3 drops of concentrated H_2SO_4. Observe the change, and put a piece of blue litmus test paper near the mouth of the test tube to detect the emitted gas (HCl).

(2) To a dry test tube, add several KBr granules, and then add 2~3 drops of concentrated H_2SO_4. Observe the change, and put a piece of paper stained with I_2 solution near the top of the test tube to identify the escaped gas (SO_2).

(3) To a dry test tube, add several KI granules, and then add 2~3 drops of concentrated H_2SO_4. Observe the change, and put a piece of lead acetate test paper near the top of the test tube to check the emitted gas (H_2S).

Summarize the reaction products in the above three test tubes, and then range the sequence of the reducibility strength of chloride, bromide and iodide. Write down the corresponding reaction equations, and explain the results with standard electrode potentials.

5. Oxidizability of Hypochlorites and Chlorates

(1) Add 1 mL of 0.1 mol/L KI solution and 1 mL of chloroform into a test tube, and then add 1~2 drops of sodium hypochlorite (NaClO). Shake sufficiently and observe the color in the chloroform layer. And then add excessive NaClO solution drop by drop, keep shaking until the color of chloroform layer disappears. Record the phenomena, and write down the corresponding reaction equations.

(2) Dissolve a little $KClO_3$ crystals in 1~2 mL of water, and then add 10 drops of 0.1 mol/L KI solution. Divide the obtained solution into two portions. Acidify one portion with 1 mol/L H_2SO_4 and take the other as control. After a while, observe the change. Compare the oxidizability of chlorates in neutral and acidic solution.

6. Separation and Identification of the Solution Containing the Mixture of Cl^-, Br^- and I^-

(1) Formation of AgX precipitate

Add 3 drops of 0.1 mol/L NaCl, 3 drops of 0.1 mol/L KBr and 3 drops of 0.1 mol/L KI into a centrifugal tube and acidify the mixed solution with 2 drops of 6 mol/L HNO_3. And then add 0.1 mol/L $AgNO_3$ solution drop by drop until the deposition finishes. Separate the precipitate from the solution with centrifugation, discard the solution. Wash the precipitate twice using 4~5 drops of distilled water per time. After stirring, centrifuge and discard the solution (taken out with a capillary syringe).

(2) Dissolution of AgCl and identification of Cl^-

Add 2 mL of 12% $(NH_4)_2CO_3$ solution into the above-obtained silver halide precipitate, stirring sufficiently. Then centrifuge and transfer the clear solution {$[Ag(NH_3)_2]Cl$} into a test tube (Figure 3-9). Acidify it with 6 mol/L HNO_3. There exists Cl^- in the solution if white AgCl is formed. The precipitate should be left to identify Br^- and I^- in next step.

(3) Identification of Br^- and I^-

Figure 3-9 Suck the upper clear solution

Add 5 drops of water and a little zinc powder into the centrifugal tube containing the precipitate obtained in Procedure 6 (2) and stir sufficiently. After silver halides are completely reduced, i. e. the precipitations looks black, centrifuge. Suck the clear liquid into another centrifugal tube (or a test tube), and then add 10 drops of chloroform. After that add the chlorine water. Shaking sufficiently after every drop is added. Observe the color change in the chloroform layer. There exists I^- if the chloroform layer is amaranth because of the formation of I_2. Then add chlorine water until amaranth fades (I^- is oxidized to colorless $NaIO_3$). There exists Br^- if the chloroform layer is orange or gold yellow.

The relative reaction equations are as following.

$$2AgBr + Zn \rightleftharpoons Zn^{2+} + 2Br^- + 2Ag \downarrow$$
$$2AgI + Zn \rightleftharpoons Zn^{2+} + 2I^- + 2Ag \downarrow$$
$$2I^- + Cl_2 \rightleftharpoons I_2 + 2Cl^-$$
$$I_2 + 5Cl_2 + 6H_2O \rightleftharpoons 2HIO_3 + 10HCl$$
$$2Br^- + Cl_2 \rightleftharpoons Br_2 + 2Cl^-$$

Notes

(1) In this experiment, both PI_3 and Cl_2 gases are toxic and irritating. Inbreathing them will quickly irritate throat and cause cough and breathlessness. So the experiments of preparing PI_3 and Cl_2 gases must be done in the stink cupboard, and the laboratory should be ventilated.

(2) When you identify the gases escaping in the reactions, you must wet the test papers before put it near the mouth of the test tubes, and should not throw them into the test tubes.

(3) The method of centrifuge and the way of separating liquid and precipitate can be referenced in the experimental technique.

Key Words

blue litmus test paper [bluː'litməs test 'peipə] 蓝色石蕊试纸
centrifugal tube [sen'trifjugəl 'tjuːb] 离心试管
centrifugal machine [sen'trifjugəl mə'ʃiːn] 离心机
centrifuge ['sentrifjuːdʒ]
Chlorate ['klɔːrit] 氯酸盐

stink cupboard [stiŋk 'kʌbəd] 通风橱
hypochlorite [ˌhaipəu'klɔːrait] 次氯酸盐
hypochlorous acid [ˌhaipəu'klɔːrəs 'æsid] 次氯酸
sodium hypochlorite [ˌsəudjəm ˌhaipəu'klɔːrait] 次氯酸钠 NaClO
red phosphorus [red 'fɔsfərəs] 红磷

3.8 Chromium, Manganese and Iron

Objectives

(1) To learn the formation and the bulk properties of the compounds of Cr, Mn and Fe with important valence states.

(2) To master the oxidizability and reducibility of the compounds containing chromium and manganese, and the effect of medium on redox properties and products.

(3) To understand the hydroxide of Fe (II) and Fe (III), the redox property of ferric salts and the formation of coordination compounds.

(4) To be familiar with the identification of chromium, manganese, iron.

Principles

Chromium, manganese and iron belong to group VIB, VIIB and VIII respectively. They all have variable oxidation states. Cr (III) and Cr (VI) are stable. Mn (II), Mn (IV), Mn (VI) and Mn (VII) are important, while Fe (II) and Fe (III) are common.

1. Chromium

Chromium is a shiny metal and has very little tendency to corrode. Solutions of chromium compounds or complex ions are richly colored. Chromium dissolves slowly in dilute hydrochloric or sulfuric acid, liberating hydrogen and forming a blue solution of the Cr (II) or chromous ion. Normally, however, you do not observe this blue color, because the Cr (II) ion is rapidly oxidized in air to the violet-colored chromium (III) or chromic ion. When hydrochloric acid solution is used, the solution appears green as a result of the formation of complex ions containing chloride coordinated to chromium, for example, $[Cr(H_2O)_4Cl_2]^+$.

(1) Cr(II) is instable and shows relatively strong reducibility. It can be obtained by reducing Cr (III) and Cr (VI) with some reducers (for example, Zn).

(2) Cr (II) is a kind of amphoteric substance. In basic solution, Cr (III) salt can be oxidized by some strong oxidants, for example, Na_2O_2 and H_2O_2, to bright yellow chromate.

(3) Chromium is frequently encountered in aqueous solution in the +6 oxidation state. In basic solution the yellow chromate ion, CrO_4^{2-}, is the most stable. In acidic solution the dichromate ion, $Cr_2O_7^{2-}$ is formed.

(4) Chromate and dichromate are both strong oxidants and can be reduced to Cr (III) with green or blue color. In acidic solution, $Cr_2O_7^{2-}$ can react with H_2O_2 and form chromium peroxide, $CrO(O_2)_2$, which looks blue in ethylene. This reaction can identify chromium.

2. Manganese

Manganese can exist in all oxidation states from +2 to +7, although +2 and +7 are the most common. Mn (II) is stable in acidic solution. It forms an extensive series of salts with all of the common anions. Mn^{2+} can react with $NaBiO_3$ and be oxidized to amaranthine $KMnO_4$ in HNO_3 solution. This reaction can be used to identify Mn^{2+}. Mn (IV), such as MnO_2, is an intermediate oxidation state, it behaves as both an oxidizing agent and a reducing agent. Manganate (VI) can be prepared by the reaction of MnO_2 and oxidizer, such as $KClO_3$, in strong basic case with heating. The green MnO_4^{2-} is stable only in basic solution, and it will disproportionate under neutral or weak basic condition. Manganese (VII) is found in the intensely purple MnO_4^- ion, which behaves as a strong oxidizing agent and its reduction product changes with the medium.

3. Iron

Iron is a kind of moderately active metal. It exists in aqueous solution in either the +2 (ferrous) or +3 (ferric) oxidation state. Iron can react with non oxidizing acids such as dilute sulfuric acid or acetic acid to form Fe^{2+}. However, in the presence of air, Fe^{2+} tends to be oxidized to Fe^{3+}.

(1) Hydroxides

Iron (II) hydroxide is white and slightly soluble, which can be oxidized by air. Iron (III) hydroxide [Fe(OH)$_3$], which is red brown-colored and slightly soluble, can be precipitated from a solution of Fe^{3+} ion by addition of base. It is amphoteric, while alkalinity is relatively strong. So it dissolves in strongly acidic solution but not in basic solution.

(2) Oxidizability and reducibility of ferric salts

The Fe (II) salts shows quite strong reducibility. It can be oxidized to Fe (III) by $KMnO_4$ in acidic case. The ferric (III) salts have faintish oxidizability and can be reduced to Fe (II) only by stronger reducer, for example, KI.

(3) The relatively strong coordination ability of iron

Most iron complex has a coordination number of 6, such as $Fe[(CN)_6]^{4-}$, $Fe[(NCS)_6]^{3-}$, etc. The formation of iron complex can also be used to identify Fe (II) and Fe (III).

Apparatus, Reagents and Materials

1. Apparatus

Centrifugal machine

2. Reagents and Materials

Solids   Zinc powder, MnO_2, $NaBiO_3$, $FeSO_4 \cdot 7H_2O$

Acids   HCl (2 mol/L, 6 mol/L, concentrated), HNO_3 (6 mol/L, concentrated), H_2SO_4 (2 mol/L, 1 mol/L, 6 mol/L concentrated)

Bases   NaOH (2 mol/L, 6 mol/L, 40%), $NH_3 \cdot H_2O$ (2 mol/L, 6 mol/L)

Salts $CrCl_3$ 0.1 mol/L, $K_2Cr_2O_7$ 0.2 mol/L, Na_2SO_3 0.1 mol/L, $FeCl_3$ (0.1 mol/L), $MnSO_4$ (0.1 mol/L, 0.5 mol/L), Na_2CO_3 0.1 mol/L, KI 0.1 mol/L, $K_4[Fe(CN)_6]$ 0.1 mol/L, $K_3[Fe(CN)_6]$ 0.1 mol/L

Others H_2O_2 3%, ethylene, solution of starch, starch-KI paper

Procedure

1. Chromium

(1) Formation of Cr (II)

Put 10 drops of 0.1 mol/L $CrCl_3$ solution into a test tube, add 10 drops of 6 mol/L HCl solution and some zinc granules. Warm slightly the tube until plentiful gas escape from the tube. Note that the solution color changes from green (Cr^{3+}) to blue (Cr^{2+}). Take out the upper clear layer and add several drops of concentrated HNO_3. Observe the change in the solution.

(2) Synthesis and properties of chromic hydroxide [Cr(OH)$_3$]

Add, drop by drop, 2 mol/L NaOH solution into 10 drops of $CrCl_3$ solution until Cr(OH)$_3$ precipitates. Observe the precipitate color. Test the amphotericity of Cr(OH)$_3$ and write down the equation.

(3) Oxidation of Cr (III) and identification of Cr^{3+}

Add excessive 6 mol/L NaOH solution into 1~2 drops of 0.1 mol/L $CrCl_3$ solution until the precipitate has completely dissolved. Then add 3 drops of H_2O_2 solution and heat the test tube. Observe the color change, explain the phenomenon and write down the equation. Add 10 drops of ethylene after the test tube cools to room temperature. Then acidify the solution with 6 mol/L HNO_3 solution. Swirl the tube. There exits Cr^{3+} in the solution if the ethylene layer looks blue.

(4) Transition between the chromate and the dichromate

Add a little 2 mol/L NaOH solution into 5 drops of $K_2Cr_2O_7$ solution and observe the color change. Acidify the solution with 1 mol/L H_2SO_4 solution, noting the color change. Explain it and write down the equation.

(5) Oxidizability of Cr (VI)

① Add 5 drops of 1 mol/L H_2SO_4 solution into 5 drops of 0.1 mol/L $K_2Cr_2O_7$ solution, then add 0.1 mol/L Na_2SO_3 solution drop by drop. Observe the change and write down the equation.

② Add 15 drops of concentrated HCl solution into 5 drops of 0.5 mol/L $K_2Cr_2O_7$ solution, then heat the test tube. Check the gas with wet starch-KI paper. Observe the color change of the paper and the solution. Explain the phenomenon and write down the equation.

2. Manganese

(1) Synthesis and properties of the hydroxide of Mn (II)

Add 2 mol/L NaOH solution into 5 drops of $MnSO_4$ solution slowly and with great care until the deposition has finished. Divide the solution (including the precipitate) into two portions, one for testing amphotericity of manganese hydroxide [$Mn(OH)_2$] quickly. Swirl the other test tube in air, note the color change of the precipitate and explain it.

(2) Synthesis and properties of Mn (III) compounds

Add 1 mL of concentrated sulfuric acid into 10 drops of 0.5 mol/L $MnSO_4$ solution, cooling down by cold water. Then add 2~3 drops of 0.01 mol/L $KMnO_4$ solution. Observe the formation of deep red-colored Mn^{3+} (If the acidity is not enough MnO_2 will form).

Neutralize the deep red solution with 0.1 mol/L Na_2CO_3 solution, and Mn^{3+} will be converted to Mn^{2+} and MnO_2 obtaining a rusty precipitate.

(3) Formation of Mn (IV) compounds

Add 0.01 mol/L $MnSO_4$ solution into 10 drops of 0.01 mol/L $KMnO_4$ solution drop by drop. Observe the formation of MnO_2.

(4) Formation of Mn (V) compounds

Add a drop of 0.01 mol/L $KMnO_4$ solution into 1 mL of 40% NaOH solution, swirl the test tube and observe the character light blue of Mn (V).

(5) Formation of Mn (VI) compounds

Add 20 drops of 40% NaOH solution into 10 drops of 0.01 mol/L $KMnO_4$ solution, then add some solid MnO_2. Heat and shake the solution and leave it for a while. After centrifugation and sedimentation, the character color of Mn (VI) will form in the upper clear layer.

Take out the upper layer, acidify it with 6 mol/L H_2SO_4 solution. Observe the color change and the formation of precipitate. According to these experiments discuss the stability of Mn with different valence charges.

(6) Oxidation of MnO_2

Add 10 drops of concentrated HCl solution into a test tube containing a small amount of MnO_2 powder. Warm it slightly and test whether there is chlorine escaping with starch-KI paper.

(7) Relationship between the reduction products of potassium hypermanganate and the medium

Obtain 3 test tubes. Place 5 drops of 0.01 mol/L $KMnO_4$ solution into each tube and then add 2 drops of 2 mol/L H_2SO_4 solution into the first, 2 drops of water into the second and 2 drops of 6 mol/L NaOH solution into the third. After that, add several drops of 0.1 mol/L Na_2SO_3 solution into each tube and observe the change. Write down the equations and explain the relationship between the reduction products of potassium hypermanganate and the medium.

(8) Identification of Mn^{2+}

Put 5 drops of 0.1 mol/L $MnSO_4$ solution into a test tube, then add 5 drops of 6 mol/L HNO_3 solution and a little solid $NaBiO_3$. The exiting of Mn^{2+} will be verified if the upper layer looks purple after centrifugation and sedimentation.

3. Synthesis and Properties of the Hydroxides of Iron

(1) Synthesis and properties of Fe(II) hydroxide

Add 1~2 drops of 2 mol/L H_2SO_4 solution into 2 mL of distilled water, then heat the solution to boiling and keep for a while (why?). Then add a small amount of solid $FeSO_4 \cdot 7H_2O$. At the same time, heat 1 mL of 2 mol/L NaOH solution to the boiling point in another test tube (why?). Add the NaOH solution into the $FeSO_4$ solution rapidly and do not swirl the tube. Observe the change. Leave the tube for a while and then observe the color change of the precipitate.

(2) Synthesis and properties of Fe(III) hydroxide

Add 2 mol/L NaOH solution into 5 drops of 0.1 mol/L $FeCl_3$ solution, observe the color and appearance of the precipitate. Add concentrated HCl solution, then observe the change.

4. Reducibility of the Ferric Salts

(1) Reducibility of the ferrous salts

Add 1~2 drops of 2 mol/L H_2SO_4 solution into 10 drops of fresh $FeSO_4$ solution, then add a small amount of 0.01 mol/L $KMnO_4$ solution. Observe the color fading.

(2) Oxidizability of ferric salts

Add 0.1 mol/L KI solution into 10 drops of 0.1 mol/L $FeCl_3$ solution drop by drop and observe the color change. Then add 1 drop of starch solution, observe the change.

5. Coordination Compounds of Iron

(1) Coordination compounds of Fe(II)

Add several drops of 2 mol/L NaOH solution into 10 drops of 0.1 mol/L $K_4[Fe(CN)_6]$ solution. Does any precipitate of $Fe(OH)_2$ occur in the tube? Explain this phenomenon. Place 5 drops of 0.1 mol/L $FeCl_3$ solution into another test tube, add 2 drops of 0.1 mol/L $K_4[Fe(CN)_6]$ solution and observe the change in the tube.

(2) Coordination compounds of Fe(III)

Add several drops of 2 mol/L NaOH solution into a test tube containing 10 drops of 0.1 mol/L $K_3[Fe(CN)_6]$ solution. Does any precipitate of $Fe(OH)_3$ occur in the tube? Explain this phenomenon.

Place a little solid $FeSO_4 \cdot 7H_2O$ into a test tube and dissolve it in 10 drops of water. Add 2 drops of $K_3[Fe(CN)_6]$ solution and observe the change in the tube.

Notes

(1) The color of the solution containing Cr^{3+} changes with the concentration of Cr^{3+} or the corresponding anions.

(2) In Procedure 1(2), NaOH should be added slowly for excessive NaOH will dissolve the precipitate.

(3) In Procedure 1(3), add more ethylene for obvious phenomenon.

(4) When preparing the hydroxide of Mn^{2+}, heat the two solutions to boiling point before they are mixed.

(5) Be careful when operating the centrifugal machine.

Key Words

acidify [əˈsidifai] 酸化
amaranthine [ˌæməˈrænθain] 紫红色的
chromic hydroxide [ˈkrəumik haiˈdrɔksaid] 氢氧化铬
chromium chloride [ˈkrəumjəm ˈklɔ:raid] 氯化铬 $CrCl_3$
chromium peroxide [ˈkrəumjəm pəˈrɔksaid] 过氧化铬
disproportionate [ˌdisprəˈpɔ:ʃənit] 歧化
Ethylether [ˈeθili:θə] 乙醚
ferric salts [ˈferik ˈsɔ:lts] 铁盐
hydrated ferrous sulfate [ˈhaidreitid ˈferəs ˈsʌlfeit] 水合硫酸亚铁 $FeSO_4 \cdot 7H_2O$
identification [aiˌdentifiˈkeiʃən] 鉴定
manganese sulfate [ˈmæŋgəni:z ˈsʌlfeit] 硫酸锰 $MnSO_4$
manganic hydroxide [mænˈgænik haiˈdrɔksaid] 氢氧化锰

potassium dichromate [pəˈtæsjəm daiˈkrəumeit] 重铬酸钾 $K_2Cr_2O_7$
potassium ferricyanide [pəˈtæsjəm ˌferiˈsaiənaid] 六氰合铁（Ⅲ）酸钾
red potassium prussiate [red pəˈtæsjəm ˈprʌʃiit] 赤血盐 $K_3[Fe(CN)_6]$
potassium ferrocyanide [pəˈtæsjəmferəuˈsaiənaid] 六氰合铁（Ⅱ）酸钾
yellow prussiate of potash [ˈjeləu ˈprʌʃiit ɔv ˈpɔtæʃ] 黄血盐 $K_4[Fe(CN)_6]$
potassium hypermanganate [pəˈtæsjəmˌhaipəˈmæŋgəneit] 高锰酸钾 $KMnO_4$
sedimentation [ˌsedimenˈteiʃən] 沉降
sodium bismuthate [ˈsəudjəm ˈbizməθeit] 硝酸铋 $NaBiO_3$
sodium sulfite [ˈsəudjəm ˈsʌlfait] 亚硫酸钠 Na_2SO_3

3.9 Copper, Silver, Zinc and Mercury

Objectives

(1) To understand the relative activity of copper, silver, zinc and mercury, and the acidity/basicity of their hydroxides.

(2) To understand the change of valence state of copper, silver, zinc and mercury, and the preparation and properties of their coordination compounds.

(3) To be familiar with the identifying of the ions of copper, silver, zinc and mercury, and the silver mirror reaction.

Principles

Copper, silver, zinc and mercury are important elements of group ⅠB and group ⅡB in the period table. We will discuss the four elements' properties together because of the familiar properties of their compounds.

1. Copper and zinc

Copper exhibits two common oxidation states in aqueous solution, +1 (cuprous) and +2 (cupric). The +2 oxidation state is more common. Salts of Cu (Ⅱ) are water soluble with a few exceptions, and many form blue solutions because of hydration.

Zinc is a white, lustrous metal. Zn is exclusively +2 oxidation state and forms colorless salts containing Zn^{2+} ions.

(1) Oxides and hydroxides

The color of Cu oxides is as follows, CuO (black), Cu_2O (red), ZnO (white).

$Cu(OH)_2$ and $Zn(OH)_2$ are both amphoteric hydroxides, particularly $Zn(OH)_2$. They can dissolve in excess hydroxide because coordination compounds are formed. $Cu(OH)_2$ and $Zn(OH)_2$ readily lose water by heating to form CuO and ZnO.

(2) Inversion of Cu^{2+} and Cu^+

Salts of Cu (I) are often water insoluble and are mostly white in color. Cu^+ readily disproportionates in solution, $2Cu^+ \rightleftharpoons Cu^{2+} + Cu$. Furthermore Cu (I) is readily oxidized to Cu (II) under most solution conditions, the +2 oxidation state is by far the more common.

Cu^{2+} is a relative weak oxidizing reagent, which will be reduced by copper or other reagent and form the insoluble copper (I) salt. For example, $CuCl_2 + Cu \rightleftharpoons 2CuCl\downarrow$ When KI is added to a Cu^{2+} solution, Cu^{2+} will be reduced by I^- and then cuprous iodide, a white precipitate, is separated out. The reaction equation is as follows, $2Cu^{2+} + 4I^- \rightleftharpoons 2CuI\downarrow + I_2$

(3) Coordination compounds

Cu^{2+} can react with chloride ions or ammonia and form different complex ions with various stabilities. Cu^{2+} can also react with halide ions (except F^-), CN^- or SCN^- etc., and form the copper (I) complexes, $[CuX_2]^-$. They are stable only when excessive ligand exists in the solution. On dilution they decompose and CuX is precipitated.

(4) Identifying Cu^{2+} and Zn^{2+}

(a) Identifying Cu^{2+}

There are two methods for identifying Cu^{2+}. One applies the reaction of copper (II) with potassium ferrocyanide {$K_4[Fe(CN)_6]$, yellow prussiate of potash} forming a reddish brown precipitate, $Cu_2[Fe(CN)_6]$. The other is based on the reaction of Cu^{2+} with aqueous ammonia forming a dark blue copper (II) complexes, $[Cu(NH_3)_4]^{2+}$.

(b) Identifying Zn^{2+}

Zn^{2+} can react with colorless solution of dithizone to give a pink chelate.

2. Silver and Mercury

Ag^+ and Hg^{2+} are both colorless. Usually, their compounds are also colorless. However, they have 18 electronic shields, long ionic radii, strong polarizing power. As a result they are apt to deform. They will polarize strongly with other anions that are easy to deform. Therefore, many compounds they form are dark-colored and sparingly soluble in water.

AgS (black)   AgI (yellow) Ag_2O (brown)
HgS (black) HgI_2 (red)   HgO (red or yellow)

(1) Oxides and hydroxides

The hydroxides of Ag and Hg, AgOH, $Hg(OH)_2$ and $Hg_2(OH)_2$, are amphoteric aptly to basic. And they are so unstable that they will decompose to the oxides as soon as they precipitate from the solutions.

(2) Conversion of Hg^{2+} and Hg_2^{2+}

Hg_2^{2+} salt will disproportionate in some conditions. For example, ammonia reacts with Hg_2Cl_2 forming Hg_2NH_2Cl, a white precipitate. Hg_2NH_2Cl further disproportionates to white $HgNH_2Cl$ and black Hg. This reaction shows black color for separation of Hg, which can be use to identify Hg_2^{2+} ions. Hg_2I_2 (green yellow) do almost the same in excess KI solution, forming $[HgI_4]^{2-}$ and Hg.

(3) Coordination compounds

Ag^+ complexes are linear and sp orbital hybridized with coordination number two. Hg^{2+} form tetrahedral complexes by sp^3 hybrid with coordination number four. Ag^+ reacts with excess ammonia to form $[Ag(NH_3)_2]^+$, however, Hg^{2+} ions form $HgNH_2Cl$, a white precipitate.

Many of the complexes of Ag (Ⅰ) and Hg (Ⅱ) with a certain ligand usually contain a mixture of the various complex ions. Only when the concentration of ligand is very low or relatively high is a single species present.

(4) Silver mirror reaction

Ag^+ will be reduced by glucose in an aqueous ammonia solution. This reaction can be used to identify Ag^+.

Apparatus, Reagents and Materials

1. Apparatus

Centrifugal machine, dropping board, alcohol burner

2. Reagents and Materials

Solids   Cu

Acids   HCl (2 mol/L, concentrated)

Bases   NaOH (1 mol/L, 2 mol/L, 6 mol/L), $NH_3 \cdot H_2O$ (2 mol/L, 6 mol/L)

Salts $CuSO_4$ 0.1 mol/L, $ZnSO_4$ 0.1 mol/L, KI (0.1 mol/L, saturated), KSCN saturated, $K_4[Fe(CN)_6]$ 10%, $CuCl_2$ 1 mol/L, $AgNO_3$ 0.1 mol/L, $Hg(NO_3)_2$ 0.1 mol/L, $Hg_2(NO_3)_2$ 0.1 mol/L, $HgCl_2$ 0.1 mol/L, KBr 0.1 mol/L, $Na_2S_2O_3$ 0.1 mol/L

Others   starch solution, dithizone solution (solvent: CCl_4), glucose solution 10%

Procedure

1. Copper and Zinc

(1) Hydroxides of copper and zinc

Add 6 drops of 0.1 mol/L $CuSO_4$ solution into 3 test tubes respectively, add 2 mol/L NaOH solution into each one and observe the formation of precipitate. Then add 2 mol/L HCl solution into the 1st test tube and excessive 6 mol/L NaOH solution into the 2nd test tubes, and heat the 3rd test tube. Observe the phenomenon in each test tube.

Take 2 test tubes and place 5 drops of 0.1 mol/L $ZnSO_4$ solution into each tube. Then add 2 mol/L NaOH solution respectively and note the formation of precipitate. Add 2 mol/L HCl solution into one tube and excessive 6 mol/L NaOH solution into the other, comparing the phenomenon.

(2) Coordination compounds of copper and zinc

Take 2 test tubes. Place 10 drops of 0.1 mol/L $CuSO_4$ solution into one tube and 10 drops of 0.1 mol/L $ZnSO_4$ solution into the other. And then add 6 mol/L $NH_3 \cdot H_2O$ respectively to prepare complex ions. Add 1 mol/L NaOH solution into 2 tubes and note whether precipitate forms again.

(3) Oxidizability of Cu (Ⅱ) and coordination compounds of Cu (Ⅰ)

(a) Take 2 centrifugal tubes, and add 5 drops of 0.1 mol/L $CuSO_4$ solution and 20 drops of 0.1 mol/L KI solution into each tube. Isolate the precipitate with centrifugation. Test whether there is I_2 in the clear upper layer. Rinse the precipitate twice with distilled water and observe its color.

Add saturated KI solution into one tube until the precipitate just dissolves. Dilute the resulting solution with distilled water, observe the formation of precipitate again.

Add saturated KNCS solution into the other tube until the precipitate just dissolves. Precipitate forms again when the solution is diluted. Explain it.

(b) Add 10 drops of 1 mol/L $CuCl_2$ solution, 3~4 drops of concentrated HCl and a little copper powder into a test tube. Heat the solution to boiling until it looks dirt yellow. Stop heating, take a little solution and dilute it. See if there is white precipitate. Explain it and write down the equation.

(4) Identification of Cu^{2+} and Zn^{2+}

(a) Identifying Cu^{2+} Add 2 drops of 0.1 mol/L $CuSO_4$ solution into a dropping board, then add 2 drops of 20% $K_4[Fe(CN)_6]$ solution. The formation of reddish brown precipitate $\{Cu_2[Fe(CN)_6]\}$ indicates the existence of Cu^{2+}.

(b) Identifying Zn^{2+} Place 2 drops of 0.1 mol/L $ZnSO_4$ solution into a test tube, then add 6 mol/L NaOH solution until the precipitate dissolves. Add 10 drops of dithizone solution. The pink color of aqueous solution shows the existence of Zn^{2+}.

2. Silver and Mercury

(1) Synthesis and properties of the oxides and hydroxides of Ag and Hg

Obtain 3 test tubes. Place 5 drops of 0.1 mol/L $AgNO_3$ solution into the first, 5 drops of $Hg(NO_3)_2$ solution into the second and 5 drops of $Hg_2(NO_3)_2$ solution into the third. Then add several drops of 2 mol/L NaOH solution into each tube. Observe the precipitate appearance. Explain the components of the solids in each tube.

(2) Reaction with ammonia water

(a) Put 5 drops of 0.1 mol/L $AgNO_3$ solution into a test tube, then add 2 mol/L ammonia water drop by drop. Observe the precipitate. Subsequently, add 2mol/L ammonia water drop by drop and observe the dissolution of solid. Treat the obtained solution with 2 drops of 2 mol/L NaOH and observe if any precipitate forms again. Explain it and write down the equation.

(b) Place 5 drops of 0.1 mol/L $HgCl_2$ solution into a test tube, then add 2 mol/L ammonia water drop by drop. Observe the precipitate. Subsequently, add excessive 2 mol/L ammonia water and observe if the precipitate dissolves. Explain it and write down the equation.

(c) Place 5 drops of 0.1 mol/L $Hg_2(NO_3)_2$ solution into a test tube, then add several drops of 6 mol/L ammonia water. Observe the precipitate. Subsequently, add excessive 6 mol/L ammonia water and observe if the precipitate dissolves. Explain it and write down the equation.

Conclude the difference between the reactions of silver and mercury salts with ammonia water.

3. Other Coordination Compounds of Ag and Hg

(1) Place 5 drops of 0.1 mol/L $AgNO_3$ solution into a test tube, then add 10 drops of 0.1 mol/L KBr solution. Observe the precipitate color and isolate it by centrifugation. Discard the solution. Add 0.1 mol/L $Na_2S_2O_3$ solution into the tube drop by drop. Stir it and see if the solid dissolves. Explain it and write down the equations.

(2) Put 5 drops of 0.1 mol/L $Hg(NO_3)_2$ solution into a test tube, then add 10 drops of 0.1 mol/L KI solution. Observe the precipitate color and then add excessive KI solution to see if the solid dissolves. Explain it and write down the equations.

(3) Place 5 drops of 0.1 mol/L $Hg_2(NO_3)_2$ solution into a test tube, then add 10 drops of 0.1 mol/L KI solution. Observe the precipitate color and then add excessive KI solution to note the change. Explain it and write down the equations.

4. Formation of Silver Mirror

Place 2 mL of 0.1 mol/L $AgNO_3$ solution into a clean test tube, and then add 2 mol/L ammonia water drop by drop until the precipitate just dissolves. Add 3 drops of 10% glucose solution into the tube. Heat the test tube with a water bath. In a while, a shining mirror of silver is formed on the inner surface of the tube. How to clean the silver mirror?

Notes

(1) Be careful with using $HgCl_2$, which is toxic.

(2) A little longer time should be spent on boiling the mixture of $CuCl_2$ solution with copper powder. Otherwise, the phenomenon is not obvious sometimes.

Key Words

chelate ['ki:leit] 螯合物
dehydrate [di:'haidreit] 脱水
disproportionate [,disprə'pɔ:ʃənit] 歧化
dithizone ['daiθaizəun] 二苯硫腙，双硫腙
glucose ['glu:kəus] 葡萄糖
mercuric chloride [mə:'kjuərik 'klɔ:raid] 氯化汞 $HgCl_2$

mercuric nitrate [mə:'kjuərik 'naitreit] 硝酸汞 $Hg(NO_3)_2$
mercurous chloride ['mə:kjurəs 'klɔ:raid] 氯化亚汞 Hg_2Cl_2
mercurous nitrate ['mə:kjurəs 'naitreit] 硝酸亚汞 $Hg_2(NO_3)_2$
reddish brown ['rediʃ braun] 红褐色

3.10 Determination of the Ionization Constant of HAc by pH Meter

Objectives

(1) To determine the ionization constant of HAc and further master its principles.

(2) To use the pH meter correctly.

Principles

In the aqueous solution of acetic acid (HAc), a weak acid, there exists ionization equilibrium: $HAc \rightleftharpoons H^+ + Ac^-$

The expression of its ionization constant is as following.

$$K_{HAc} = [H^+][Ac^-]/[HAc] \quad (3.2)$$

Where K_{HAc} is the ionization constant of HAc, $[H^+]$, $[Ac^-]$ and $[HAc]$ are the equilibrium concentration of H^+, Ac^- and HAc respectively. Other relative equations are as follows.

$$pH = -\lg[H^+] \quad (3.3)$$

$$[H^+] = [Ac^-] \quad (3.4)$$

$$[HAc] = c_{HAc} - [H^+] \quad (3.5)$$

In this experiment, the total concentration of HAc (c_{HAc}) can be obtained by titrating HAc solution with standard NaOH solution. The pH value of the HAc solution can be determined by pH meter, which can be used to calculate $[H^+]$ [equation (3.3)]. According to the equation (3.2) to (3.5), series of K_{HAc} value can be calculated. Therefore the ionization constant of HAc is equal to their average.

Apparatus, Reagents and Materials

1. Apparatus

pH meter (PHSJ-4A), acidic burette (50 mL), basic burette (50 mL), beakers (100 mL), glass rod, Erlenmeyer flasks (100 mL)

2. Reagents and Materials

Acids　　HAc about 0.1 mol/L

Bases　　NaOH (0.1000 mol/L, calibrated)

Others　　standard buffer solution (pH=4.00), phenolphthalein indicator

Operation of pH meter

1. Compound Electrode of the pH Meter

The pH meter has two electrodes in one tube, glass and reference electrode, which is called compound electrode. A saturated KCl and HCl solution is contained in a tube, which is inside of an outer tube that will have contact with the solution to be measured. This outer tube has a double glass bulb with Na^+, which makes an ion specific electrode. When measuring the pH of a solution, a salt bridge forms. The Na^+ ion, not H^+, crosses the glass membranes of the pH electrode and allows for a change in free energy, which is measured by the pH meter as the concentration of H^+.

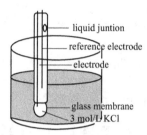

Figure 3-10　Compound electrode

As shown in Figure 3-10, the glass bulb at the bottom of the electrode is a thin glass membrane which releases ions into the solution inside the bulb when protons dock to the outside of the glass membrane. This membrane should be completely submerged in your solution and not be in contact with the wall of your solution-container. It is also sensitive to mechanical disturbances, and it should therefore not be bounced off the bottom of your beaker or be hit by an out-of-control magnetic stir bar. The small fiber or ceramic "patch" above the glass bulb is a liquid junction, which allows contact between your solution and the reference electrode. This "patch" must therefore also be submerged in your solution during measurements.

2. Buffer Solution

A buffer solution is used as the standardization solution, because its pH is known and it will maintain its pH in case of contamination as long as it is not excessive. A buffer solution of pH = 7 is commonly used although the instrument should be standardized in the pH region of the unknown solutions.

3. Calibrating the pH Meter

(1) Turns the meter on

(2) Connect the electrode to the Twist-Lock input located at the top of the meter. Line up the arrow, line on the electrode's Twist-Lock connector and push until it locks in place. To disconnect, twist the connector ring in the arrow direction and pull apart.

(3) Remove the protective end cover from the electrode. Before first using your pH electrode or whenever the electrode is dry, soak it several hours in an electrode filling or storage solution (3 mol/L KCl solution) or in a buffer for pH electrodes.

(4) As shown in Figure 3-11, rinse the bulb with distilled water from a wash bottle into an empty beaker (a). Blot the electrode dry with filter paper (b). Do not rub! This procedure should be carried out whenever the electrodes are transferred from one solution to another to minimize the chance of contamination.

(5) Begin to calibrate the pH meter by setting it in a buffer solution of pH = 7.00 (or

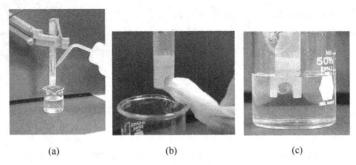

(a)　　　　　　　　(b)　　　　　　　　(c)

Figure 3-11　The proper way to measure the pH of the solution
(a) Rinse the electrode;
(b) Blot the electrodes dry with filter paper;
(c) Immerse the end of the probe completely in the solution

6.86).

(6) Lower the electrodes into the standardization solution carefully so they do not strike the bottom of the beaker and break. Immerse the end of the probe completely in the buffer solution [Figure 3-11(c)]. If an experiment requires the use of a magnetic stirring bar in the solution whose pH is being measured, be careful that the bar does not hit the electrodes.

(7) Gently stir the solution in a circular motion to create a homogeneous sample. Press the "cal" key. When the reading stabilizes (i. e. an "eye" occurs on the liquid crystal screen), press the "enter" key.

(8) Do a two-point calibration using the instruction manual for the specific pH meter. Rinse the electrode between immersions in buffers. Some studies may allow for a one-point calibration with pH 7 buffer. Follow instruction manual for the specific pH meter for a one-point calibration. Lower the electrode into the 2nd buffer solution, repeat step 7.

(9) When you do two-point calibration, after calibration with pH 7 buffer is finished, the meter should be calibrated with a basic solution (usually, a pH=10 buffer solution) if the sample to be tested is basic or with an acidic solution (usually, a pH=4 buffer solution) and if the sample is acidic. Continue to calibrate the meter according to the pH of the sample to be measured.

(10) Finally, rinse the electrode with pH = 7 solution and measure the pH of this solution. If the meter does not read "7.00", re-calibrate the instrument.

4. Measuring the Solution pH

(1) After calibrating the instrument, rinse the bulb with distilled water from a wash bottle into an empty beaker. Blot the electrodes dry with filter paper as above-mentioned.

(2) Place the electrode in unknown solution. Gently stir the solution in a circular motion to create a homogeneous sample. Press the "pH" key. When a stable pH reading is obtained on the liquid crystal screen, record the reading.

(3) If the pH of more than one solution is to be measured, repeat Step 1 for each solution. Note that the solution with highest pH value should be measured first, subsequently measure other solutions with decreasing pH value.

(4) When the use of the pH meter is completed, proceed to Step 5.

(5) Remove electrodes from last solution. Rinse electrodes and blot dry. Cap the end of

the electrode with its plastic cup containing 3 mol/L KCl.

(6) Turn the meter off and/or disconnect line cord and clean work space.

Be cautious when interpreting the results of an experiment, several factors may affect the function of the pH meter.

Procedure

1. Calibration of the Concentration of HAc (c_{HAc})

Transfer 25.00 mL of HAc solution with unknown concentration from an acidic burette into a 100 mL Erlenmeyer flask, and then add 2 drops of phenolphthalein. Titrate this HAc solution with 0.1 mol/L NaOH standard solution until endpoint (as shown in Figure 3-12), which means that the first appearance of pink color lasts for at least 30 sec with swirling. Repeat the above-mentioned titration until the difference of the base volumes added in the titration is not above 0.05 mL. Record the base volume added, and calculate the concentration of the HAc solution. Write down the datum in the Data Table 1.

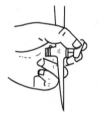

Rotate stopcock with left hand   Titrate solution with acidic burette

Figure 3-12 Operation of acidic burette

2. Preparation of HAc Solutions with Different Concentration

Take two burettes and five dry 100 mL beakers numbered from 1 to 5.

Transfer exactly 48.00 mL of calibrated HAc solution from one burette (burette a) into No. 1 beaker, and swirl the solution.

Transfer exactly 24.00 mL of calibrated HAc solution from burette a into No. 2 beaker and 24.00 mL of distilled water from the other burette (burette b). Mix the solution well.

Prepare the other three HAc solutions with difference concentrations according to the following table using the same procedure. Calculate the concentration of each HAc solutions, and write down the datum in the Data Table 3-3.

No.	Volume of calibrated HAc/mL	Volume of water/mL	c_{HAc}/(mol/L)	pH of HAc
1	48.00	0		
2	24.00	24.00		
3	12.00	36.00		
4	6.00	42.00		
5	3.00	45.00		

3. Measurement of the pH of the HAc Solutions (No. 1~5) with the pH Meter

Measure the pH of the HAc solutions (No. 1~5) with the pH meter. The solution with highest pH value should be measured first, subsequently measure other solutions with decreasing pH value. Record the datum in the Data Table 3-4. Then calculate K_{HAc} value by equation 1~4.

Data Record and Result

Data Table 3-3 The concentration of the HAc solution

Titration No.		1	2	3
Concentration of standard NaOH solution/(mol/L)				
Volume of used HAc solution/mL				
Volume of used NaOH solution/mL				
Concentration of HAc/(mol/L)				
Concentration of HAc solution/(mol/L)	Measured value			
	Average			

Data Table 3-4 The ionization constant of HAc (K_{HAc})   (Tem: ℃)

No.	Volume of HAc /mL	Volume of water /mL	[HAc] /(mol/L)	pH	[H^+] /(mol/L)	K_{HAc}
1						
2						
3						
4						
5						

Notes

(1) The main sensible part of a compound electrode is a glass bulb in its upper part. The bulb is so thin that be sure not to touch it with any hard substance. When the bulb is broken, this compound electrode will be invalid completely.

(2) The apparatus should be standardized by standard buffer solutions.

(3) The solution with highest pH value should be measured first, subsequently measure other solutions with decreasing pH value.

(4) The electrode should be rinsed with distilled water before it is immersed in another solution.

(5) After the experiment is finished, rinse the electrode with distilled water, then coated it with a plastic cup containing 3 mol/L KCl.

Key Words

calibration [ˌkæliˈbreiʃən] 标定

electromagnetic agitator [ilektrəuˈmægnitik ˈædʒiteitə] 电磁搅拌器

ionization constant [ˌaiənaiˈzeiʃən ˈkɔnstənt] 电离常数

3.11 Preparation of Ammonium Iron (Ⅱ) Sulfate

Objectives

(1) To understand the principle of preparation of double salt.

(2) To practice the basic operations skills of heating on water bath, filtrations under constant pressure and vacuum, evaporation, concentrating, crystallization, and drying, et al.

(3) To learn to test the purity in the products with visual colorimetric assay.

Principles

The Mohr's salt, namely ammonium iron (Ⅱ) sulfate [$(NH_4)_2SO_4 \cdot FeSO_4 \cdot 6H_2O$] is a transparent and bluish monoclinic crystal. It is more stable than other ferrous salts and difficult to oxidize in the air. Furthermore, it is soluble in water but insoluble in ethanol. Therefore, it is usually used as a primary standard substance in quantitative analysis.

Similar with most double salts, $(NH_4)_2SO_4 \cdot FeSO_4$ has a lower solubility in water than either of its components [$(NH_4)_2SO_4$ and $FeSO_4$]. Thus the precipitate or crystals of $(NH_4)_2SO_4 \cdot FeSO_4$ can be obtained by mixing the two highly concentrated solutions of $FeSO_4$.

Firstly, the $FeSO_4$ solution is prepared by reacting iron powder with diluted H_2SO_4 solution. In the following step, $(NH_4)_2SO_4$ powder is added to the solution and dissolved completely by heating. After the solution is concentrated and cooled to room temperature, the $(NH_4)_2SO_4 \cdot FeSO_4$ crystals will precipitate. The related reaction equations are as follows:

$$Fe + H_2SO_4 = FeSO_4 + H_2 \uparrow$$
$$FeSO_4 + (NH_4)_2SO_4 + 6H_2O = (NH_4)_2SO_4 \cdot FeSO_4 \cdot 6H_2O$$

Since the Fe^{3+} can react with excess SCN^- to produce a blood red complex $[Fe(SCN)_6]^{3-}$, the content of the Fe^{3+} impurity in the product can be determined by visual colorimetry. The color of the solution will become darker with the increase of the concentration of the $[Fe(SCN)_6]^{3-}$ complex. And the solutions having similar complex concentrations will show similar colors. Consequently, a series of $[Fe(SCN)_6]^{3-}$ standard solutions of concentration gradient and the sample solution should be firstly prepared by adding KSCN. Then compare the colors of the sample solution and the $[Fe(SCN)_6]^{3-}$ standard solutions, and find out the standard solution which exhibits color close to that of the sample solution. The content of Fe^{3+} in the product can be finally figured out based on the Fe^{3+} concentration of the related standard solution. The grade of the product can be evaluated according to the content of the Fe^{3+} impurity (The product is classified into grade Ⅰ, Ⅱ, and Ⅲ according to the upper limits of Fe^{3+} contents of 0.05, 0.10 and 0.20 mg.).

Apparatus, Reagents and Materials

1. Apparatus

Platform balance, water circulating vacuum pump, conical flask (150 mL×3), beakers (or water bath), graduated cylinder (25 mL×3, 10 mL×3), Buchner funnel, filter flask, evaporation dish, visual colorimetric tubes (25 mL)

2. Reagents and Materials

Solids   iron powder, $(NH_4)_2SO_4$

Acids H_2SO_4 3 mol/L, HCl 3 mol/L

Bases Na_2CO_3 solution 10%

Salts   KSCN solution 25%, Fe^{3+} standard solution

Others   ethanol 25%, filter papers

Procedure

1. Degreasing of the Iron Powder

Weigh 4.0 g of iron powder and transfer it to a conical flask, add 20 mL of 10% Na_2CO_3 solution and heat this mixture on a water bath for 10 minutes. Discard the solution by decantation and wash the iron powder with distilled water.

2. Preparation of $FeSO_4$

Add 25 mL of 3 mol/L H_2SO_4 solution to the above conical flask filled with the degreased iron powder. Then heat it in a water bath for 30 minutes (operating in the drought cupboard). Keep shaking the conical flask to accelerate the reaction while heating, and supplement appropriate amount of distilled water if more solution is evaporated. When the solution almost stops bubbling, filter it while the solution is hot. Transfer the filtrated solution to an evaporation dish. And wash the residues in conical flask with 1 mL of 3 mol/L H_2SO_4 and subsequent distilled water, then filter the obtained solution and combine the filtrate to the evaporation dish (Note since the Fe^{2+} is stable in strong acid solution, the additional H_2SO_4 can avoid the transformation from Fe^{2+} to Fe^{3+}). Collect the iron residue and dry it with filter paper. Weigh the residue and calculate the consumption of the iron in the reaction.

3. Preparation of $(NH_4)_2SO_4 \cdot FeSO_4 \cdot 6H_2O$

Weigh 9.5 g of $(NH_4)_2SO_4$ (C.P.) and add it to the $FeSO_4$ solution in the evaporation dish. Heat the evaporation dish on water bath and keep stirring to dissolve the $(NH_4)_2SO_4$ completely. Continue the heating to concentrate the solution, until the crystal films appear on the surface of the solution. Stop heating and spontaneously cool the solution to room temperature, then $(NH_4)_2SO_4 \cdot FeSO_4 \cdot 6H_2O$ will crystallize. Filter the solution with vacuum filtration and wash the crystals twice with small amount of ethanol. Transfer the crystals to a piece of filter paper and press them with another one to absorb the solution, then drying them in the air. Weigh the product and calculate the yield rate.

4. Testing the Product

Weigh 1.0 g of the product and transfer to a 25 mL colorimetric tube. Then dissolve it with 15 mL of distilled water, and add 2 mL of 3 mol/L HCl and 1 mL of 25% KSCN. Subsequently, dilute it to 25 mL with distilled water and shake up and down the tube to get a uniform solution. Finally, determine the grade of the product by comparing it with the standard solutions.

Recording and Treating Data

Date　　　　Temperature　　　　℃　　　Humidity

(1) Preparation of Ammonium Iron (Ⅱ) Sulfate

The weight of the iron powder m_1(Fe)=_____ g;

The weight of the iron residue m_2(Fe)=_____ g;

$m[(NH_4)_2SO_4 \cdot FeSO_4 \cdot 6H_2O]$ =_____ g.

The yield rate of $(NH_4)_2SO_4 \cdot FeSO_4 \cdot 6H_2O$ (%) = $\dfrac{\text{weight of product (g)}}{\text{yield of theoretical (g)}} \times 100\%$ = _____

(2) Testing the purity of the product

Fe^{3+} standard solution	Grade(Ⅰ)	Grade(Ⅱ)	Grade(Ⅲ)
Grade of Ammonium Iron(Ⅱ)Sulfate			

*Preparation of the $[Fe(SCN)_6]^{3-}$ standard solutions

Transfer 5.00 mL of 0.01 mg/mL Fe^{3+} standard solution prepared before experiment to a colorimetric tube with volumetric pipette, add 2 mL of HCl solution and 1 mL of KSCN solution. Thus the grade Ⅰ standard solution (containing 0.05 mg of Fe^{3+}) is obtained after the mixed solution is diluted to 25 mL with distilled water. And the grade Ⅱ (containing 0.10 mg of Fe^{3+}) and grade Ⅲ (containing 0.20 mg of Fe^{3+}) solutions can be prepared respectively in the same way.

Key Words

preparation [,prepəˈreiʃən] 制备
colorimetry [kʌləˈrimitri] 比色法
concentration [,kɔnsenˈtreiʃən] 浓缩
degreasing [diːˈɡriːziŋ] 脱脂（法）

evaporation [i,væpəˈreiʃən] 蒸发
crystal [ˈkristl] 结晶
impurity [imˈpjuəriti] 杂质
filtration [filˈtreiʃən] 过滤

3.12 Preparation and Content Determination of Zinc Gluconate

Objectives

(1) To learn to prepare zinc gluconate.
(2) To learn to determine the content of zinc salts.

Principles

Zinc is one of the essential microelements in human body. And zinc deficiency may lead to various diseases. Zinc-containing drugs such as oral administered zinc sulfate ($ZnSO_4$) used to be employed to therapy zinc deficiency. However, oral $ZnSO_4$ should be taken after meal, and the absorption of which by gastrointestinal tract is very low. Moreover, $ZnSO_4$ can react with gastric juice to generate zinc dichloride ($ZnCl_2$), which causes harm to gastric mucosa. Therefore, zinc gluconate has taken the place of $ZnSO_4$ and become primary Zn supplement drug recently.

Zinc gluconate can be prepared by reacting calcium gluconate with equimolar of $ZnSO_4$. And the resultants are zinc gluconate and calcium sulfate ($CaSO_4$). After filtering the $CaSO_4$ precipitate and concentrating the filtrate, the zinc gluconate is crystallized by adding 95% ethanol to the solution. The related reaction equation is as follow.

$$Ca(C_6H_{11}O_7)_2 + ZnSO_4 = Zn(C_6H_{11}O_7)_2 + CaSO_4 \downarrow$$

Complexometric titration can be used to determine the content of zinc gluconate. The zinc gluconate is titrated with EDTA standard solution under the condition of weak-base NH_3-NH_4Cl. Finally, calculate the content of zinc gluconate according to the titrant consumption amount.

Apparatus, Reagents and Materials

1. Apparatus

Analytical balance, water bath, water circulating vacuum pump, electric heater, acidic burette, volumetric pipet, volumetric flask, evaporation dish, conical flask, beakers (250 mL), graduated cylinder (100 mL), Buchner funnel, filter flask, temperature meter

2. Reagents and Materials

Solids calcium gluconate, $ZnSO_4 \cdot 7H_2O$

Others ethanol 95%, EDTA standard solution 0.05 mol/L, NH_3-NH_4Cl buffer solution (pH=10), eriochrome black T indicator

Procedure

1. Preparation of Zinc Gluconate

Measure 80 mL of distilled water and transfer it into a 250 mL beaker. Then heat the water to 80~90 ℃ and add 13.4 g of $ZnSO_4 \cdot 7H_2O$ to it. Stir the solution to dissolve the

$ZnSO_4 \cdot 7H_2O$ completely. Keep stirring and slowly add 20 g calcium gluconate to the above solution at the constant temperature of 90 ℃. Continue to react for 20 mins at this constant temperature.

Filter the precipitate with double layers of filter paper by vacuum filtration while the solution is hot. Discard the precipitate and transfer the filtrate into an evaporation dish. Heat the dish on boiling water bath and concentrate the solution to about 20 mL (Thick sate solution, suction filtrate again if there is precipitate).

Cool the solution to room temperature and add 20 mL of 95% ethanol to it while stirring. At this moment, a large amount of colloidal zinc gluconate is precipitated. After the stewing, discard the ethanol by decantation. Add 20 mL of 5% ethanol to the colloidal precipitate once, and adequately stir the solution to make the precipitate crystallize. The crude product is obtained after vacuum filtration. Weigh it and calculate its yield rate.

Add 20 mL of distilled water to the crude product and heat it to 90 ℃ to dissolve the precipitate. Then filter the solution while it is hot. After cooling the solution to room temperature, add 20 mL of 95% ethanol and sufficiently stir it to make zinc gluconate crystallize completely. Filter the product by vacuum filtration and dry it at 50 ℃. Finally, weigh the refined product and calculate its yield rate.

2. Determination of the Content

Accurately weigh 1.6 g of zinc gluconate and prepare 100 mL of zinc gluconate with volumetric flask. Then transfer 25.00 mL of the solution into a 250 mL conical flask, followed by adding 10 mL of NH_3-NH_4Cl buffer solution and 4 drops of eriochrome black T indicator. The solution is titrated with 0.05 mol/L EDTA standard solution to the end point that it turns blue. Calculate the content of zinc based on the following equation.

$$Zn\% = \frac{(cV)_{EDTA} \times 65}{\frac{1}{4} \times W_s} \times 100\% \tag{3.6}$$

Appendix

Decantation is a process for the separation of mixtures, carefully pouring a solution from a container in order to leave the precipitate (sediments) in the bottom of the original container. Usually a small amount of solution must be left in the container, and care must be taken to prevent a small amount of precipitate from flowing with the solution out of the container. It's generally used to separate a liquid from an insoluble solid.

Key Words

zinc gluconate ['ziŋk 'gluːkəuˌneit] 葡萄糖锌
precipitate [pri'sipiteit] 沉淀
complexometric titration [ˌkɔmpleks'ɔitrik tai'treiʃən] 配位滴定
filter paper ['filt 'peipə] 滤纸
vacuum filtration ['vækjuəm fil'treiʃən] 减压过滤
standard solution ['stændəd sə'ljuːʃən] 标准溶液
yield rate [jiːld reit] 产率

Appendix

Appendix A The base units of SI system

Physical Quantities	Name of unit	Abbreviation
Mass	kilogram	kg
Length	meter	m
Time	second	s
Electric current	Ampere	A
Temperature	Kelvin	K
Luminous intensity	candela	cd
Amount of substance	mole	mol

Appendix B Relative atomic mass (International atomic weight in 1997)

Atomic Number	Symbol	Name	Atomic Weight	Atomic Number	Symbol	Name	Atomic Weight
1	H	Hydrogen	1.007 94(7)	34	Se	Selenium	78.96(3)
2	He	Helium	4.002 602(2)	35	Br	Bromine	79.904(1)
3	Li	Lithium	6.941(2)	36	Kr	Krypton	83.80(1)
4	Be	Beryllium	9.012 182(3)	37	Rb	Rubidium	85.467 8(3)
5	B	Boron	10.811(7)	38	Sr	Strontium	87.62(1)
6	C	Carbon	12.010 7(8)	39	Y	Yttrium	87.905 85(2)
7	N	Nitrogen	14.006 74(7)	40	Zr	Zirconium	91.224(2)
8	O	Oxygen	15.999 4(3)	41	Nb	Niobium	92.906 38(2)
9	F	Fluorine	18.998 4032(5)	42	Mo	Molybdenum	95.94(1)
10	Ne	Neon	20.179 7(6)	43	Tc	Technetium	(98)
11	Na	Sodium	22.989 770(2)	44	Ru	Ruthenium	101.07(2)
12	Mg	Magnesium	24.305 0(6)	45	Rh	Rhodium	102.905 50(2)
13	Al	Aluminium	26.981 538(2)	46	Pd	Palladium	106.42(1)
14	Si	Silicon	28.085 5(3)	47	Ag	Silver	107.868 2(2)
15	P	Phosphorus	30.973 761(2)	48	Cd	Cadmium	112.411(8)
16	S	Sulfur	32.066(6)	49	In	Indium	114.818(3)
17	Cl	Chlorine	35.452 7(9)	50	Sn	Tin	118.710(7)
18	Ar	Argon	39.948(1)	51	Sb	Antimony	121.760(1)
19	K	Potassium	39.098 3(1)	52	Te	Tellurium	127.60(3)
20	Ca	Calcium	40.078(4)	53	I	Iodine	126.904 47(3)
21	Sc	Scandium	44.955 910(8)	54	Xe	Xenon	131.29(2)
22	Ti	Titanium	47.867(1)	55	Cs	Caesium	132.905 45(2)
23	V	Vanadium	50.941 5(1)	56	Ba	Barium	137.327(7)
24	Cr	Chromium	51.996 1(6)	57	La	Lanthanum	138.905 5(2)
25	Mn	Manganese	54.938 049(9)	58	Ce	Cerium	140.116(1)
26	Fe	Iron	55.845(2)	59	Pr	Praseodymium	140.907 65(2)
27	Co	Cobalt	58.933 200(9)	60	Nd	Neodymium	144.24(3)
28	Ni	Nickel	58.693 4(2)	61	Pm	Promethium	(145)
29	Cu	Copper	63.546(3)	62	Sm	Samarium	150.36(3)
30	Zn	Zinc	65.39(2)	63	Eu	Europium	151.964(1)
31	Ga	Gallium	69.723(1)	64	Gd	Gadolinium	157.25(3)
32	Ge	Germanium	72.61(2)	65	Tb	Terbium	158.925 34(2)
33	As	Arsenic	74.921 60(2)	66	Dy	Dysprosium	162.50(3)

Appendix B (Continued)

Atomic Number	Symbol	Name	Atomic Weight	Atomic Number	Symbol	Name	Atomic Weight
67	Ho	Holmium	164.930 32(2)	90	Th	Thorium	232.038 1(1)
68	Er	Erbium	167.26(3)	91	Pa	Protactinium	231.035 88(2)
69	Tm	Thulium	168.934 21(2)	92	U	Uranium	238.028 9(1)
70	Yb	Ytterbium	173.04(3)	93	Np	Nepturnium	(237)
71	Lu	Lutetium	174.967(1)	94	Pu	Plutonium	(244)
72	Hf	Hafnium	178.49(2)	95	Am	Americium	(243)
73	Ta	Tantalum	180.947 9(1)	96	Cm	Curium	(247)
74	W	Tungsten	183.84(1)	97	Bk	Berkelium	(247)
75	Re	Rhenium	186.207(1)	98	Cf	Californium	(251)
76	Os	Osmium	190.23(3)	99	Es	Einsteinium	(252)
77	Ir	Iridium	192.217(3)	100	Fm	Fermium	(257)
78	Pt	Platinum	195.078(2)	101	Md	Mendelevium	(258)
79	Au	Gold	196.966 55(2)	102	No	Nobelium	(259)
80	Hg	Mercury	200.59(2)	103	Lr	Lawrencium	(260)
81	Tl	Thallium	204.383 3(2)	104	Rf	Rutherfordium	(261)
82	Pb	Lead	207.2(1)	105	Db	Dubnium	(262)
83	Bi	Bismuth	208.980 38(2)	106	Sg	Seaborgium	(263)
84	Po	Polonium	(210)	107	Bh	Bohrium	(264)
85	At	Astatine	(210)	108	Hs	Hassium	(265)
86	Rn	Radon	(222)	109	Mt	Meitnerium	(268)
87	Fr	Francium	(223)	110	Uun		(269)
88	Ra	Radium	(226)	111	Uuu		(272)
89	Ac	Actinium	(227)	112	Uub		(277)

Appendix C Selected Thermodynamic Data (298.15K)

Substances	State	$\Delta_f H_m^\ominus/(kJ/mol)$	$\Delta_f G_m^\ominus/(kJ/mol)$	$S_m^\ominus/(J \cdot K^{-1}/mol)$
Ag	c	0.0	0.0	42.6
AgCl	c	−127.0	−109.8	96.3
AgBr	c	−100.4	−96.9	107.1
AgI	c	−61.8	−66.2	115.5
$AgNO_3$	c	−124.4	−33.4	140.9
Ag_2O	c	−31.1	−11.2	121.3
Al	c	0.0	0.0	28.3
Al_2O_3	c	−1675.7	−1582.3	50.9
$AlCl_3$	c	−704.2	−628.8	109.3
B_2O_3	c	−1273.5	−1194.3	54.0
Ba	c	0.0	0.0	62.8
BaO	c	−548.0	−520.3	72.1
$BaCl_2$	c	−855.0	−806.7	123.7
$BaCO_3$	c	−1216.3	−1137.6	112.1
$BaSO_4$	c	−1473.2	−1362.2	132.2
Br_2	g	30.9	3.1	245.5

Appendix C (Continued)

Substances	State	$\Delta_f H_m^\ominus$/(kJ/mol)	$\Delta_f G_m^\ominus$/(kJ/mol)	$S_m^\ominus$/(J·K^{-1}/mol)
Br$_2$	l	0.0	0.0	152.2
HBr	g	−36.3	−53.4	198.7
C(diamond)	c	1.9	2.9	2.4
C(graphite)	c	0.0	0.0	5.7
CO	g	−110.5	−137.2	197.7
CO$_2$	g	−393.5	−394.4	213.8
Ca	c	0.0	0.0	41.6
CaCl$_2$	c	−795.4	−748.8	108.4
CaO	c	−634.9	−603.3	38.1
CaCO$_3$ (calcite)	c	−1207.6	−1129.1	91.7
CaSO$_4$	c	−1434.5	−1322.0	106.5
Cl$_2$	g	0.0	0.0	223.1
HCl	g	−92.3	−95.3	186.9
Co	c	0.0	0.0	30.0
CoCl$_2$	c	−312.5	−269.8	109.2
Cu	c	0.0	0.0	33.2
CuS	c	−53.1	−53.6	66.5
Cu$_2$O	c	−168.6	−146.0	93.1
CuO	c	−157.3	−129.7	42.6
CuSO$_4$	c	−771.4	−662.2	109.2
F$_2$	g	0.0	0.0	202.8
HF	g	−273.3	−275.4	173.8
Fe	c	0.0	0.0	27.3
Fe$_2$O$_3$	c	−824.2	−742.2	87.4
Fe$_3$O$_4$	c	−1118.4	−1015.4	146.4
H$_2$	g	0.0	0.0	130.7
H$^+$	aq	0.0	0.0	0.0
H$_2$O	g	−241.8	−228.6	188.8
H$_2$O	l	−285.8	−237.1	70.0
H$_2$O$_2$	l	−187.8	−120.4	109.6
Hg	l	0.0	0.0	75.9
HgCl$_2$	c	−224.3	−178.6	146.0
HgO(red)	c	−90.8	−58.5	70.3
HgI$_2$(red)	c	−105.4	−101.7	180.0
HgS	c	−58.2	−50.6	82.4
I$_2$	c	0.0	0.0	116.1
I$_2$	g	62.4	19.3	260.7
HI	g	26.5	1.7	206.6
K	c	0.0	0.0	64.7
KCl	c	−436.5	−408.5	82.6
KBr	c	−393.8	−380.7	95.9

Appendix C (*Continued*)

Substances	State	$\Delta_f H_m^\ominus$/(kJ/mol)	$\Delta_f G_m^\ominus$/(kJ/mol)	$S_m^\ominus$/(J·K^{-1}/mol)
KI	c	−327.9	−324.9	106.3
KMnO$_4$	c	−837.2	−737.6	171.7
KOH	c	−424.6	−378.7	78.9
Mg	c	0.0	0.0	32.7
MgO	c	−601.6	−569.3	27.0
MgCO$_3$	c	−1095.8	−1012.1	65.7
MgSO$_4$	c	−1284.9	−1170.6	91.6
Mn	c	0.0	0.0	32.0
MnO$_2$	c	−520.0	−465.1	53.1
N$_2$	g	0.0	0.0	191.6
NH$_3$	g	−45.9	−16.4	192.8
N$_2$H$_4$	l	50.6	149.3	121.2
N$_2$H$_4$	g	95.4	159.4	238.5
HN$_3$	l	264.0	327.3	140.6
HN$_3$	g	294.1	328.1	239.0
NH$_4$Cl	c	−314.4	−202.9	94.6
NH$_4$NO$_3$	c	−365.6	−183.9	151.1
NO	g	91.3	87.6	210.8
NO$_2$	g	33.2	51.3	240.1
N$_2$O$_4$	l	−19.5	97.5	209.2
N$_2$O$_4$	g	11.1	99.8	304.4
HNO$_3$	l	−174.1	−80.7	155.6
Na	c	0.0	0.0	51.3
NaCl	c	−411.2	−384.1	72.1
Na$_2$CO$_3$	c	−1130.7	−1044.4	135.0
NaNO$_3$	c	−467.9	−367.0	116.5
NaOH	c	−425.6	−379.5	64.5
O$_2$	g	0.0	0.0	205.2
O$_3$	g	142.7	163.2	238.9
P(white)	c	0.0	0.0	41.1
P(red)	c	−17.6	—	22.8
PCl$_3$	l	−319.7	−272.3	217.1
PCl$_5$	c	−443.5	—	—
Pb	c	0.0	0.0	64.8
PbCl$_2$	c	−359.4	−314.1	136.0
PbO(yellow)	c	−217.3	−187.9	68.7
PbSO$_4$	c	−920.0	−813.0	148.5
Pb$_3$O$_4$	c	−718.4	−601.2	211.3
PbO$_2$	c	−277.4	−217.3	68.6
PbS	c	−100.4	−98.7	91.2
S(trimetric)	c	0.0	0.0	32.1

Appendix C (Continued)

Substances	State	$\Delta_f H_m^\ominus/(kJ/mol)$	$\Delta_f G_m^\ominus/(kJ/mol)$	$S_m^\ominus/(J \cdot K^{-1}/mol)$
S(homocline)	c	0.3	—	—
H_2S	g	−20.6	−33.4	205.8
SO_2	g	−296.8	−300.1	248.2
SO_3	g	−395.7	−371.1	256.8
SiO_2 (quartz)	c	−910.7	−856.3	41.5
$SnCl_2$	c	−325.1	—	—
SnO(tetragonal)	c	−280.7	−251.9	57.2
SnO_2(tetragonal)	c	−577.6	−515.8	49.0
$SbCl_3$	c	−382.2	−323.7	184.1
Zn	c	0.0	0.0	41.6
$ZnSO_4$(S)	c	−982.8	−817.5	−110.5
ZnS(blende)	c	−206.0	−201.3	57.7
CH_4	g	−74.6	−50.5	186.3
C_2H_4	g	52.4	68.4	219.3
C_2H_6	g	−84.0	−32.0	229.2
C_2H_5OH	i	−277.6	−174.8	160.7

* Reference: Weast RC. CRC Handbook of Chemistry and Physics, 80th ed. CRC Press, 1999~2000.

Appendix D Values of K_a/K_b for Some Common Acids/Bases

Name (Formula)	K_a/K_b	Name (Formula)	K_a/K_b
Hydrofluoric acid (HF)	7.2×10^{-4}	Selenious acid (H_2SeO_4)	$K_1 = 3.5 \times 10^{-3}$
Hydrocyanic acid (HCN)	6.2×10^{-10}		$K_2 = 5 \times 10^{-8}$
Sulfocyanic acid (HSCN)	1.41×10^{-1}	Chromic acid (H_2CrO_4)	$K_1 = 1.8 \times 10^{-1}$
Hypochloric acid (HClO)	2.95×10^{-8}		$K_2 = 3.20 \times 10^{-7}$
Hypobromous acid (HBrO)	2.06×10^{-9}	Oxalic acid ($H_2C_2O_4$)	$K_1 = 5.9 \times 10^{-2}$
Hypoiodous acid (HIO)	2.3×10^{-11}		$K_2 = 6.4 \times 10^{-5}$
Hydrogen peroxide (H_2O_2)	2.4×10^{-12}	Metasilici acid (H_3SiO_3)	$K_1 = 1.7 \times 10^{-10}$
Nitrous acid (HNO_2)	4.0×10^{-4}		$K_2 = 1.58 \times 10^{-12}$
Arsenous acid (H_3AsO_3)	5.62×10^{-3}		$K_3 = 1.0 \times 10^{-12}$
Boric acid (H_3BO_3)	7.3×10^{-10}	Arsenic acid (H_3AsO_4)	$K_1 = 5.62 \times 10^{-3}$
Iodic acid (HIO_3)	1.69×10^{-1}		$K_2 = 1.70 \times 10^{-7}$
Periodic acid (H_5IO_6)	2.3×10^{-2}		$K_3 = 2.95 \times 10^{-12}$
Formic acid (HCOOH)	1.8×10^{-4}	Phosphoric acid (H_3PO_4)	$K_1 = 7.52 \times 10^{-3}$
Acetic acid (CH_3COOH)	1.76×10^{-5}		$K_2 = 6.23 \times 10^{-8}$
Ortho-hydroxybenzonic acid ($HC_7H_5O_3$)	1.05×10^{-3}		$K_3 = 2.2 \times 10^{-13}$
Sulfuric acid (H_2SO_4)	K_1: large	Citric acid ($H_3C_6H_5O_7$)	$K_1 = 7.1 \times 10^{-4}$
	$K_2 = 1.20 \times 10^{-2}$		$K_2 = 1.68 \times 10^{-5}$
Sulfurous acid (H_2SO_3)	$K_1 = 1.5 \times 10^{-2}$		$K_3 = 4.1 \times 10^{-7}$
	$K_2 = 1.0 \times 10^{-7}$	Ammonia water ($NH_3 \cdot H_2O$)	1.76×10^{-5}
Hydrosulfuric acid (H_2S)	$K_1 = 8.9 \times 10^{-8}$	Hydroxylamine (NH_2OH)	1.07×10^{-8}
	$K_2 = 1.2 \times 10^{-13}$	Carbamide (H_2NCONH_2)	1.3×10^{-14}
Carbonic acid (H_2CO_3)	$K_1 = 4.30 \times 10^{-7}$	Calcium hydroxide [$Ca(OH)_2$]	3.74×10^{-3}
	$K_2 = 5.61 \times 10^{-11}$		4.0×10^{-2}
Ortho-phosphorous acid (H_3PO_3)	$K_1 = 1.0 \times 10^{-2}$	Ethylene diamine ($H_2NCH_2CH_2NH_2$)	$K_1 = 8.5 \times 10^{-5}$
	$K_2 = 2.6 \times 10^{-7}$		$K_2 = 7.1 \times 10^{-8}$

Appendix E Values of K_{sp}, at 298.15 K for Common Ionic Solids*

Ionic Solid	K_{sp}	Ionic Solid	K_{sp}
AgAc	1.94×10^{-3}	$BaSO_4$	1.08×10^{-10}
AgBr	5.35×10^{-13}	BaP_2O_7	3.2×10^{-11}
$AgBrO_3$	5.38×10^{-5}	$Ba_3(AsO_4)_2$	8.0×10^{-51}
$AgCH_3COO$	1.94×10^{-3}	$BiAsO_4$	4.43×10^{-10}
AgCN	5.97×10^{-17}	BiOBr	3.0×10^{-7}
AgCl	1.77×10^{-10}	BiOCl	1.8×10^{-31}
AgI	8.52×10^{-17}	$Bi(OH)_3$	4×10^{-31}
$AgIO_3$	3.17×10^{-8}	$BiO(NO_2)$	4.9×10^{-7}
AgN_3	2.8×10^{-9}	$BiO(NO_3)$	2.82×10^{-3}
$AgNO_2$	3.22×10^{-4}	BiOOH	4×10^{-10}
AgOH	2.0×10^{-8}	BiOSCN	1.6×10^{-7}
AgSCN	1.03×10^{-12}	$BiPO_4$	1.3×10^{-23}
AgSeCN	4.0×10^{-16}	Bi_2S_3	1.82×10^{-99}
Ag_2CO_3	8.46×10^{-12}	$CaCO_3$	3.36×10^{-9}
$Ag_2C_2O_4$	5.40×10^{-12}	CaC_2O_4	1.46×10^{-10}
$Ag_2[Co(NO_2)_6]$	8.5×10^{-21}	$CaC_2O_4 \cdot H_2O$	2.32×10^{-9}
Ag_2CrO_4	1.12×10^{-12}	$CaCrO_4$	7.1×10^{-4}
$Ag_2Cr_2O_7$	2.0×10^{-7}	CaF_2	3.45×10^{-11}
Ag_2S	6.3×10^{-50}	$CaHPO_4$	1.0×10^{-7}
Ag_2SO_3	1.50×10^{-14}	$Ca(IO_3)_2$	6.47×10^{-6}
Ag_2SO_4	1.20×10^{-5}	$Ca(IO_3)_2 \cdot 6H_2O$	7.10×10^{-7}
Ag_3AsO_3	1.0×10^{-17}	$Ca(OH)_2$	5.02×10^{-6}
Ag_3AsO_4	1.03×10^{-22}	$CaSO_3$	6.8×10^{-8}
Ag_3PO_4	8.89×10^{-17}	$CaSO_4$	4.93×10^{-5}
$Ag_4[Fe(CN)_6]$	1.6×10^{-41}	$CaSO_4 \cdot 0.5H_2O$	3.1×10^{-7}
$Al(OH)_3$	1.1×10^{-33}	$CaSO_4 \cdot 2H_2O$	3.14×10^{-5}
$AlPO_4$	9.84×10^{-21}	$CaSiO_3$	2.5×10^{-8}
As_2S_3	2.1×10^{-22}	$Ca_3(PO_4)_2$	2.07×10^{-33}
$BaCO_3$	2.58×10^{-9}	$Cd(CN)_2$	1.0×10^{-8}
BaC_2O_4	1.6×10^{-7}	$CdCO_3$	1.0×10^{-12}
$BaCrO_4$	1.17×10^{-10}	$CdC_2O_4 \cdot 3H_2O$	1.42×10^{-8}
BaF_2	1.84×10^{-7}	CdF_2	6.44×10^{-3}
$BaHPO_4$	3.2×10^{-7}	$Cd(IO_3)_2$	2.5×10^{-8}
$Ba(IO_3)_2$	4.01×10^{-9}	$Cd(OH)_2$	7.2×10^{-15}
$Ba(IO_3)_2 \cdot 2H_2O$	1.5×10^{-9}	CdS	1.40×10^{-29}
$Ba(IO_3)_2 \cdot H_2O$	1.67×10^{-9}	$Cd_2[Fe(CN)_6]$	3.2×10^{-17}
$Ba(MnO_4)_2$	2.5×10^{-10}	$Cd_3(AsO_4)_2$	2.2×10^{-33}
$Ba(NO_3)_2$	4.64×10^{-3}	$Cd_3(PO_4)_2$	2.53×10^{-33}
$Ba(OH)_2$	5×10^{-3}	$CoCO_3$	1.4×10^{-13}
$Ba(OH)_2 \cdot 8H_2O$	2.55×10^{-4}	CoC_2O_4	6.3×10^{-8}
$BaSO_3$	5.0×10^{-10}	$Co(IO_3)_2 \cdot 2H_2O$	1.21×10^{-2}
$Co(OH)_2$[pink]	1.09×10^{-15}	Hg_2Br_2	6.40×10^{-23}
$Co(OH)_2$[blue]	5.92×10^{-15}	$Hg_2(CN)_2$	5×10^{-40}

Appendix E (*Continued*)

Ionic Solid	K_{sp}	Ionic Solid	K_{sp}
$Co(OH)_3$	1.6×10^{-44}	Hg_2CO_3	3.6×10^{-17}
$\alpha\text{-}CoS$	4.0×10^{-21}	$Hg_2C_2O_4$	1.75×10^{-13}
$\beta\text{-}CoS$	2.0×10^{-25}	Hg_2Cl_2	1.43×10^{-18}
$\gamma\text{-}CoS$	3.0×10^{-26}	Hg_2CrO_4	2.0×10^{-9}
$Co_2[Fe(CN)_6]$	1.8×10^{-15}	Hg_2F_2	3.10×10^{-6}
$Co_3(AsO_4)_2$	6.80×10^{-29}	Hg_2HPO_4	4.0×10^{-13}
$Co_3(PO_4)_2$	2.05×10^{-35}	Hg_2I_2	5.2×10^{-29}
$CrAsO_4$	7.7×10^{-21}	$Hg_2(IO_3)_2$	2.0×10^{-14}
CrF_3	6.6×10^{-11}	$Hg_2(OH)_2$	2.0×10^{-24}
$Cr(OH)_3$	6.3×10^{-31}	Hg_2S	1.0×10^{-47}
$CuBr$	6.27×10^{-9}	$Hg_2(SCN)_2$	3.2×10^{-20}
$CuCN$	3.47×10^{-20}	Hg_2SO_3	1.0×10^{-27}
$CuCO_3$	1.4×10^{-10}	Hg_2SO_4	6.5×10^{-7}
CuC_2O_4	4.43×10^{-10}	$KClO_4$	1.05×10^{-2}
$CuCl$	1.72×10^{-7}	$KHC_4H_4O_6$	3×10^{-4}
$CuCrO_4$	3.6×10^{-6}	KIO_4	3.71×10^{-4}
CuI	1.27×10^{-12}	$K_2Na[Co(NO_2)_6] \cdot H_2O$	2.2×10^{-11}
$Cu(IO_3)_2$	7.4×10^{-8}	$K_2[PdCl_6]$	6.0×10^{-6}
$Cu(IO_3)_2 \cdot H_2O$	6.94×10^{-8}	$K_2[PtBr_6]$	6.3×10^{-5}
$CuOH$	1×10^{-14}	$K_2[PtCl_6]$	7.48×10^{-6}
CuS	1.27×10^{-36}	Li_2CO_3	8.15×10^{-4}
$CuSCN$	1.77×10^{-13}	LiF	1.84×10^{-3}
$Cu_2[Fe(CN)_6]$	1.3×10^{-16}	$MgCO_3$	6.82×10^{-6}
$Cu_2P_2O_7$	8.3×10^{-16}	$MgCO_3 \cdot 3H_2O$	2.38×10^{-6}
Cu_2S	2.26×10^{-48}	$MgCO_3 \cdot 5H_2O$	3.79×10^{-6}
$Cu_3(AsO_4)_2$	7.95×10^{-36}	MgF_2	5.16×10^{-11}
$Cu_3(PO_4)_2$	1.40×10^{-37}	$MgHPO_4 \cdot 3H_2O$	1.5×10^{-6}
$FeAsO_4$	5.7×10^{-21}	$Mg(IO_3)_2 \cdot 4H_2O$	3.2×10^{-3}
$FeCO_3$	3.13×10^{-11}	$Mg(OH)_2$	5.61×10^{-12}
FeF_2	2.36×10^{-6}	$Mg_3(PO_4)_2$	1.04×10^{-24}
$Fe(OH)_2$	4.87×10^{-17}	$MnCO_3$	2.24×10^{-11}
$Fe(OH)_3$	2.79×10^{-39}	$MnC_2O_4 \cdot 2H_2O$	1.70×10^{-7}
$FePO_4$	1.3×10^{-22}	$Mn(IO_3)_2$	4.37×10^{-7}
$FePO_4 \cdot 2H_2O$	9.92×10^{-29}	$Mn(OH)_2$	2.06×10^{-13}
$Fe(P_2O_7)_3$	3×10^{-23}	MnS	4.65×10^{-14}
FeS	1.3×10^{-18}	$Mn_2[Fe(CN)_6]$	8.0×10^{-13}
Fe_2S_3	1×10^{-88}	$Mn_3(AsO_4)_2$	1.9×10^{-29}
HgC_2O_4	1.0×10^{-7}	$(NH_4)_2PtCl_6$	9.0×10^{-6}
HgI_2	2.9×10^{-29}	$NiCO_3$	1.42×10^{-7}
$Hg(OH)_2$	3.13×10^{-26}	NiC_2O_4	4×10^{-10}
HgS	6.44×10^{-53}	$Ni(IO_3)_2$	4.71×10^{-5}
$Ni(OH)_2$	5.48×10^{-16}	SnS	1.0×10^{-25}
NiS	1.07×10^{-21}	SnS_2	2.5×10^{-27}

Appendix E (*Continued*)

Ionic Solid	K_{sp}	Ionic Solid	K_{sp}
α-NiS	3×10^{-19}	$SrCO_3$	5.60×10^{-10}
β-NiS	1×10^{-24}	SrC_2O_4	5.61×10^{-7}
γ-NiS	2×10^{-26}	$SrC_2O_4 \cdot H_2O$	1.6×10^{-7}
$Ni_2[Fe(CN)_6]$	1.3×10^{-15}	SrF_2	4.33×10^{-9}
$Ni_3(AsO_4)_2$	3.1×10^{-26}	$Sr(IO_3)_2$	1.14×10^{-7}
$Ni_3(PO_4)_2$	4.74×10^{-32}	$Sr(IO_3)_2 \cdot 6H_2O$	4.65×10^{-7}
$Pb(Ac)_2$	1.8×10^{-3}	$Sr(IO_3)_2 \cdot H_2O$	3.58×10^{-7}
$PbBr_2$	6.60×10^{-6}	$Sr(OH)_2$	3.2×10^{-4}
$Pb(BrO_3)_2$	2.0×10^{-2}	$SrSO_3$	4×10^{-8}
$PbCO_3$	7.4×10^{-14}	$SrSO_4$	3.44×10^{-7}
PbC_2O_4	8.51×10^{-10}	$Sr_3(AsO_4)_2$	4.29×10^{-19}
$PbCl_2$	1.70×10^{-5}	$Sr_3(PO_4)_2$	4.0×10^{-28}
$PbCrO_4$	2.8×10^{-13}	$ZnCO_3$	1.46×10^{-10}
PbF_2	3.3×10^{-8}	$ZnCO_3 \cdot H_2O$	5.41×10^{-11}
$PbHPO_4$	1.3×10^{-10}	ZnC_2O_4	2.7×10^{-8}
PbI_2	9.8×10^{-9}	$ZnC_2O_4 \cdot 2H_2O$	1.38×10^{-9}
$Pb(IO_3)_2$	3.69×10^{-13}	ZnF_2	3.04×10^{-2}
$Pb(OH)_2$	1.42×10^{-20}	$Zn[Hg(SCN)_4]$	2.2×10^{-7}
$PbOHCl$	2×10^{-14}	$Zn(IO_3)_2$	4.29×10^{-6}
PbS	9.04×10^{-29}	γ-$Zn(OH)_2$	6.86×10^{-17}
$Pb(SCN)_2$	2.11×10^{-5}	β-$Zn(OH)_2$	7.71×10^{-17}
PbS_2O_3	4.0×10^{-7}	ϵ-$Zn(OH)_2$	4.12×10^{-17}
$PbSO_4$	2.53×10^{-8}	ZnS	2.93×10^{-25}
$Pb_3(PO_4)_2$	8.0×10^{-43}	α-ZnS	1.6×10^{-24}
$Pd(SCN)_2$	4.39×10^{-23}	β-ZnS	2.5×10^{-22}
PdS	2×10^{-37}	$ZnSeO_3$	2.6×10^{-7}
PtS	1×10^{-52}	$Zn_2[Fe(CN)_6]$	4.0×10^{-16}
$Sb(OH)_3$	4.0×10^{-42}	$Zn_3(AsO_4)_2$	3.12×10^{-28}
Sb_2S_3	1.5×10^{-93}	$Zn_3(PO_4)_2$	9.0×10^{-33}
$Sn(OH)_2$	5.45×10^{-27}		

* Reference: Weast RC. CRC Handbook of Chemistry and Physics, 80th ed. CRC Press, 1999~2000.

Appendix F Standard Reduction Potentials (V) at 298.15K*

1. In acidic solution

Half reaction	$E_A^{\ominus}$/V
$Ag^+ + e \rightleftharpoons Ag$	$+0.7996$
$Ag^{2+} + e \rightleftharpoons Ag^+$	$+1.980$
$AgBr + e \rightleftharpoons Ag + Br^-$	$+0.07133$
$AgBrO_3 + e \rightleftharpoons Ag + BrO_3^-$	$+0.546$
$AgCl + e \rightleftharpoons Ag + Cl^-$	$+0.22233$
$AgI + e \rightleftharpoons Ag + I^-$	-0.15224
$Ag_2S + 2e \rightleftharpoons 2Ag + S^{2-}$	-0.691
$Ag_2S + 2H^+ + 2e \rightleftharpoons 2Ag + H_2S$	-0.0366
$AgSCN + e \rightleftharpoons Ag + SCN^-$	$+0.08951$
$Al^{3+} + 3e \rightleftharpoons Al$	-1.662
$As + 3H^+ + 3e \rightleftharpoons AsH_3$	-0.608
$H_3AsO_4 + 2H^+ + 2e \rightleftharpoons HAsO_2 + 2H_2O$	$+0.560$

Appendix F (*Continued*)

Half reaction	$E_A^\ominus/V$
$Au^+ + e \rightleftharpoons Au$	+1.692
$Au^{3+} + 3e \rightleftharpoons Au$	+1.498
$AuBr_4^- + 3e \rightleftharpoons Au + 4Br^-$	+0.854
$AuCl_4^- + 3e \rightleftharpoons Au + 4Cl^-$	+1.002
$B(OH)_3 + 7H^+ + 8e \rightleftharpoons BH_4^- + 3H_2O$	−0.481
$H_3BO_3 + 3H^+ + 3e \rightleftharpoons B + 3H_2O$	−0.8698
$Ba^{2+} + 2e \rightleftharpoons Ba$	−2.912
$Be^{2+} + 2e \rightleftharpoons Be$	−1.847
$Bi^+ + e \rightleftharpoons Bi$	+0.5
$Bi^{3+} + 3e \rightleftharpoons Bi$	+0.308
$BiO^+ + 2H^+ + 3e \rightleftharpoons Bi + H_2O$	+0.320
$BiOCl + 2H^+ + 3e \rightleftharpoons Bi + Cl^- + H_2O$	+0.1583
$Br_2(aq) + 2e \rightleftharpoons 2Br^-$	+1.0873
$Br_2(l) + 2e \rightleftharpoons 2Br^-$	+1.066
$BrO_3^- + 6H^+ + 6e \rightleftharpoons Br^- + 3H_2O$	+1.423
$HBrO + H^+ + e \rightleftharpoons 1/2Br_2(aq) + H_2O$	+1.574
$HBrO + H^+ + e \rightleftharpoons 1/2Br_2(l) + H_2O$	+1.596
$(CN)_2 + 2H^+ + 2e \rightleftharpoons 2HCN$	+0.373
$2CO_2 + 2H^+ + 2e \rightleftharpoons H_2C_2O_4$	−0.49
$CO_2 + 2H^+ + 2e \rightleftharpoons HCOOH$	−0.199
$Ca^+ + e \rightleftharpoons Ca$	−3.80
$Ca^{2+} + 2e \rightleftharpoons Ca$	−2.868
$Cd^{2+} + 2e \rightleftharpoons Cd$	−0.4030
$Ce^{3+} + 3e \rightleftharpoons Ce$	−2.336
$Cl_2 + 2e \rightleftharpoons 2Cl^-$	+1.35827
$ClO_2 + H^+ + 2e \rightleftharpoons HClO_2$	+1.277
$ClO_3^- + 2H^+ + e \rightleftharpoons ClO_2 + H_2O$	+1.152
$ClO_3^- + 3H^+ + 2e \rightleftharpoons HClO_2 + H_2O$	+1.214
$ClO_3^- + 6H^+ + 5e \rightleftharpoons 1/2Cl_2 + 3H_2O$	+1.47
$ClO_3^- + 6H^+ + 6e \rightleftharpoons Cl^- + 3H_2O$	+1.451
$ClO_4^- + 2H^+ + 2e \rightleftharpoons ClO_3^- + H_2O$	+1.189
$ClO_4^- + 8H^+ + 7e \rightleftharpoons 1/2Cl_2 + 4H_2O$	+1.39
$ClO_4^- + 8H^+ + 8e \rightleftharpoons Cl^- + 4H_2O$	+1.389
$HClO + H^+ + 2e \rightleftharpoons Cl^- + H_2O$	+1.482
$HClO + H^+ + e \rightleftharpoons 1/2Cl_2 + H_2O$	+1.611
$HClO_2 + 2H^+ + 2e \rightleftharpoons HClO + H_2O$	+1.645
$HClO_2 + 3H^+ + 3e \rightleftharpoons 1/2Cl_2 + 2H_2O$	+1.628
$Co^{2+} + 2e \rightleftharpoons Co$	−0.28
$Co^{3+} + e \rightleftharpoons Co^{2+}$	+1.92
$Cr^{2+} + 2e \rightleftharpoons Cr$	−0.913
$Cr^{3+} + 3e \rightleftharpoons Cr$	−0.744
$Cr^{3+} + e \rightleftharpoons Cr^{2+}$	−0.407
$HCrO_4^- + 7H^+ + 3e \rightleftharpoons Cr^{3+} + 4H_2O$	+1.350
$Cr_2O_7^{2-} + 14H^+ + 6e \rightleftharpoons 2Cr^{3+} + 7H_2O$	+1.232
$Cs^+ + e \rightleftharpoons Cs$	−3.026
$Cu^+ + e \rightleftharpoons Cu$	+0.521
$Cu^{2+} + 2e \rightleftharpoons Cu$	+0.3419
$Cu^{2+} + e \rightleftharpoons Cu^+$	+0.153
$CuI_2^- + e \rightleftharpoons Cu + 2I^-$	0.00
$F_2 + 2e \rightleftharpoons 2F^-$	+2.866

Appendix F (*Continued*)

Half reaction	$E_A^\ominus/V$
$F_2 + 2H^+ + 2e \rightleftharpoons 2HF$	$+3.053$
$Fe^{2+} + 2e \rightleftharpoons Fe$	-0.447
$Fe^{3+} + e \rightleftharpoons Fe^{2+}$	$+0.771$
$Fe^{3+} + 3e \rightleftharpoons Fe$	-0.037
$FeO_4^{2-} + 8H^+ + 3e \rightleftharpoons Fe^{3+} + 4H_2O$	$+2.200$
$2HFeO_4^- + 8H^+ + 6e \rightleftharpoons Fe_2O_3 + 5H_2O$	$+2.09$
$Ga^{3+} + 3e \rightleftharpoons Ga$	-0.549
$Ge^{2+} + 2e \rightleftharpoons Ge$	$+0.24$
$Ge^{4+} + 4e \rightleftharpoons Ge$	$+0.124$
$2H^+ + 2e \rightleftharpoons H_2$	0.00000
$2Hg^{2+} + 2e \rightleftharpoons Hg_2^{2+}$	$+0.920$
$Hg^{2+} + 2e \rightleftharpoons Hg$	$+0.851$
$Hg_2Cl_2 + 2e \rightleftharpoons 2Hg + 2Cl^-$	$+0.26808$
$I_2 + 2e \rightleftharpoons 2I^-$	$+0.5355$
$I_3^- + 2e \rightleftharpoons 3I^-$	$+0.536$
$IO_3^- + 6H^+ + 6e \rightleftharpoons I^- + 3H_2O$	$+1.085$
$H_5IO_6 + H^+ + 2e \rightleftharpoons IO_3^- + 3H_2O$	$+1.601$
$In^{3+} + 3e \rightleftharpoons In$	-0.3382
$Ir^{3+} + 3e \rightleftharpoons Ir$	$+1.156$
$K^+ + e \rightleftharpoons K$	-2.931
$La^{3+} + 3e \rightleftharpoons La$	-2.379
$Li^+ + e \rightleftharpoons Li$	-3.0401
$Lu^{3+} + 3e \rightleftharpoons Lu$	-2.28
$Md^{3+} + 3e \rightleftharpoons Md$	-1.65
$Mg^+ + e \rightleftharpoons Mg$	-2.70
$Mg^{2+} + 2e \rightleftharpoons Mg$	-2.372
$Mn^{2+} + 2e \rightleftharpoons Mn$	-1.185
$MnO_2 + 4H^+ + 2e \rightleftharpoons Mn^{2+} + 2H_2O$	$+1.224$
$MnO_4^- + 8H^+ + 5e \rightleftharpoons Mn^{2+} + 4H_2O$	$+1.507$
$Mo^{3+} + 3e \rightleftharpoons Mo$	-0.200
$N_2 + 2H_2O + 6H^+ + 6e \rightleftharpoons 2NH_4OH$	$+0.092$
$2NH_3OH^+ + H^+ + 2e \rightleftharpoons N_2H_5^+ + 2H_2O$	$+1.42$
$N_2O + 2H^+ + 2e \rightleftharpoons N_2 + H_2O$	$+1.766$
$N_2O_4 + 2H^+ + 2e \rightleftharpoons 2HNO_2$	$+1.065$
$NO_3^- + 3H^+ + 2e \rightleftharpoons HNO_2 + H_2O$	$+0.934$
$NO_3^- + 4H^+ + 3e \rightleftharpoons NO + 2H_2O$	$+0.957$
$Na^+ + e \rightleftharpoons Na$	-2.71
$Nb^{3+} + 3e \rightleftharpoons Nb$	-1.099
$Nd^{3+} + 3e \rightleftharpoons Nd$	-2.323
$Ni^{2+} + 2e \rightleftharpoons Ni$	-0.257
$No^{2+} + 2e \rightleftharpoons No$	-2.50
$No^{3+} + e \rightleftharpoons No^{2+}$	$+1.4$
$Np^{3+} + 3e \rightleftharpoons Np$	-1.856
$O(g) + 2H^+ + 2e \rightleftharpoons H_2O$	$+2.421$
$O_2 + 2H^+ + 2e \rightleftharpoons H_2O_2$	$+0.695$
$O_2 + 4H^+ + 4e \rightleftharpoons 2H_2O$	$+1.229$
$O_3 + 2H^+ + 2e \rightleftharpoons O_2 + H_2O$	$+2.076$
$H_2O_2 + 2H^+ + 2e \rightleftharpoons 2H_2O$	$+1.776$
$P(red) + 3H^+ + 3e \rightleftharpoons PH_3(g)$	-0.111
$P(white) + 3H^+ + 3e \rightleftharpoons PH_3(g)$	-0.063

Appendix F (*Continued*)

Half reaction	$E_A^\ominus/V$
$H_3PO_2 + H^+ + e \rightleftharpoons P + 2H_2O$	-0.508
$H_3PO_3 + 2H^+ + 2e \rightleftharpoons H_3PO_2 + H_2O$	-0.499
$H_3PO_3 + 3H^+ + 3e \rightleftharpoons P + 3H_2O$	-0.454
$H_3PO_4 + 2H^+ + 2e \rightleftharpoons H_3PO_3 + H_2O$	-0.276
$Pa^{3+} + 3e \rightleftharpoons Pa$	-1.34
$Pb^{2+} + 2e \rightleftharpoons Pb$	-0.1262
$PbCl_2 + 2e \rightleftharpoons Pb + 2Cl^-$	-0.2675
$PbO_2 + 4H^+ + 2e \rightleftharpoons Pb^{2+} + 2H_2O$	$+1.455$
$PbO_2 + SO_4^{2-} + 4H^+ + 2e \rightleftharpoons PbSO_4 + 2H_2O$	$+1.6913$
$PbSO_4 + 2e \rightleftharpoons Pb + SO_4^{2-}$	-0.3588
$Pd^{2+} + 2e \rightleftharpoons Pd$	$+0.951$
$Pm^{2+} + 2e \rightleftharpoons Pm$	-2.2
$Pt^{2+} + 2e \rightleftharpoons Pt$	$+1.18$
$[PtCl_6]^{2-} + 2e \rightleftharpoons [PtCl_4]^{2-} + Cl^-$	$+0.68$
$Ra^{2+} + 2e \rightleftharpoons Ra$	-2.8
$Re^{2+} + 2e \rightleftharpoons Re$	$+0.300$
$Rh^{3+} + 3e \rightleftharpoons Rh$	$+0.758$
$S + 2e \rightleftharpoons S^{2-}$	-0.47627
$S + 2H^+ + 2e \rightleftharpoons H_2S(aq)$	$+0.142$
$SO_4^{2-} + 4H^+ + 2e \rightleftharpoons H_2SO_3 + H_2O$	$+0.172$
$S_2O_8^{2-} + 2e \rightleftharpoons 2SO_4^{2-}$	$+2.010$
$S_2O_8^{2-} + 2H^+ + 2e \rightleftharpoons 2HSO_4^-$	$+2.123$
$S_4O_6^{2-} + 2e \rightleftharpoons 2S_2O_3^{2-}$	$+0.08$
$Sb_2O_5 + 6H^+ + 4e \rightleftharpoons 2SbO^+ + 3H_2O$	$+0.581$
$Sc^{3+} + 3e \rightleftharpoons Sc$	-2.077
$Se + 2e \rightleftharpoons Se^{2-}$	-0.924
$Se + 2H^+ + 2e \rightleftharpoons H_2Se(aq)$	-0.399
$H_2SeO_3 + 4H^+ + 4e \rightleftharpoons Se + 3H_2O$	$+0.74$
$SiF_6^{2-} + 3e \rightleftharpoons Si + 6F^-$	-1.24
$SiO_2(quartz) + 6H^+ + 4e \rightleftharpoons Si + 2H_2O$	$+0.857$
$Sn^{2+} + 2e \rightleftharpoons Sn$	-0.1375
$Sn^{4+} + 2e \rightleftharpoons Sn^{2+}$	$+0.151$
$Sr^{2+} + 2e \rightleftharpoons Sr$	-2.899
$TcO_4^- + 4H^+ + 3e \rightleftharpoons TcO_2 + 2H_2O$	$+0.782$
$TcO_4^- + 8H^+ + 7e \rightleftharpoons Tc + 4H_2O$	$+0.472$
$Ti^{2+} + 2e \rightleftharpoons Ti$	-1.630
$TiO_2 + 4H^+ + 2e \rightleftharpoons Ti^{2+} + 2H_2O$	-0.502
$Tl^+ + e \rightleftharpoons Tl$	-0.336
$Tl^{3+} + 2e \rightleftharpoons Tl^+$	$+1.252$
$UO_2^{2+} + 4H^+ + 6e \rightleftharpoons U + 2H_2O$	-1.444
$UO_2^{2+} + 4H^+ + 2e \rightleftharpoons U^{4+} + 2H_2O$	$+0.327$
$V_2O_5 + 6H^+ + 2e \rightleftharpoons 2VO^{2+} + 3H_2O$	$+0.957$
$W^{3+} + 3e \rightleftharpoons W$	$+0.1$
$XeO_3 + 6H^+ + 6e \rightleftharpoons Xe + 3H_2O$	$+2.10$
$Zn^{2+} + 2e \rightleftharpoons Zn$	-0.7618
$Zr^{4+} + 4e \rightleftharpoons Zr$	-1.45

2. In basic solution

Half reaction	$E_B^{\ominus}/V$
$Ag_2CO_3 + 2e \rightleftharpoons 2Ag + CO_3^{2-}$	+0.47
$Ag_2O + H_2O + 2e \rightleftharpoons 2Ag + 2OH^-$	+0.342
$Al(OH)_3 + 3e \rightleftharpoons Al + 3OH^-$	-2.31
$Al(OH)_4^- + 3e \rightleftharpoons Al + 4OH^-$	-2.328
$H_2AlO_3^- + H_2O + 3e \rightleftharpoons Al + 4OH^-$	-2.33
$AsO_2^- + 2H_2O + 3e \rightleftharpoons As + 4OH^-$	-0.68
$Ba(OH)_2 + 2e \rightleftharpoons Ba + 2OH^-$	-2.99
$Bi_2O_3 + 3H_2O + 6e \rightleftharpoons 2Bi + 6OH^-$	-0.64
$BrO^- + H_2O + 2e \rightleftharpoons Br^- + 2OH^-$	+0.761
$BrO_3^- + 3H_2O + 6e \rightleftharpoons Br^- + 6OH^-$	+0.61
$[Co(NH_3)_6]^{3+} + 2e \rightleftharpoons [Co(NH_3)_6]^{2+}$	+0.108
$Ca(OH)_2 + 2e \rightleftharpoons Ca + 2OH^-$	-3.02
$Cd(OH)_2 + 2e \rightleftharpoons Cd(Hg) + 2OH^-$	-0.809
$ClO^- + H_2O + 2e \rightleftharpoons Cl^- + 2OH^-$	+0.81
$ClO_2^- + 2H_2O + 4e \rightleftharpoons Cl^- + 4OH^-$	+0.76
$ClO_2^- + H_2O + 2e \rightleftharpoons ClO^- + 2OH^-$	+0.66
$ClO_3^- + H_2O + 2e \rightleftharpoons ClO_2^- + 2OH^-$	+0.33
$ClO_4^- + H_2O + 2e \rightleftharpoons ClO_3^- + 2OH^-$	+0.36
$Co(OH)_2 + 2e \rightleftharpoons Co + 2OH^-$	-0.73
$Co(OH)_3 + e \rightleftharpoons Co(OH)_2 + OH^-$	+0.17
$CrO_2^- + 2H_2O + 3e \rightleftharpoons Cr + 4OH^-$	-1.2
$CrO_4^{2-} + 4H_2O + 3e \rightleftharpoons Cr(OH)_3 + 5OH^-$	-0.13
$Cu_2O + H_2O + 2e \rightleftharpoons 2Cu + 2OH^-$	-0.360
$2Cu(OH)_2 + 2e \rightleftharpoons Cu_2O + 2OH^- + H_2O$	-0.080
$2H_2O + 2e \rightleftharpoons H_2 + 2OH^-$	-0.8277
$H_2BO_3^- + 5H_2O + 8e \rightleftharpoons BH_4^- + 8OH^-$	-1.24
$Hg_2O + H_2O + 2e \rightleftharpoons 2Hg + 2OH^-$	+0.123
$H_3IO_6^{2-} + 2e \rightleftharpoons IO_3^- + 3OH^-$	+0.7
$Mg(OH)_2 + 2e \rightleftharpoons Mg + 2OH^-$	-2.690
$Mn(OH)_2 + 2e \rightleftharpoons Mn + 2OH^-$	-1.56
$MnO_4^- + 2H_2O + 3e \rightleftharpoons MnO_2 + 4OH^-$	+0.595
$MnO_4^{2-} + 2H_2O + 2e \rightleftharpoons MnO_2 + 4OH^-$	+0.60
$NO_2^- + H_2O + e \rightleftharpoons NO + 2OH^-$	-0.46
$2NO_3^- + 2H_2O + 2e \rightleftharpoons N_2O_4 + 4OH^-$	-0.85
$Ni(OH)_2 + 2e \rightleftharpoons Ni + 2OH^-$	-0.72
$O_2 + 2H_2O + 4e \rightleftharpoons 4OH^-$	+0.401
$O_2 + H_2O + 2e \rightleftharpoons H_2O_2 + 2OH^-$	-0.146
$O_3 + H_2O + 2e \rightleftharpoons O_2 + 2OH^-$	+1.24
$HPO_3^{2-} + 2H_2O + 2e \rightleftharpoons H_2PO_2^- + 3OH^-$	-1.65
$PO_4^{3-} + 2H_2O + 2e \rightleftharpoons HPO_3^{2-} + 3OH^-$	-1.05
$S + H_2O + 2e \rightleftharpoons HS^- + OH^-$	-0.478
$2SO_3^{2-} + 3H_2O + 4e \rightleftharpoons S_2O_3^{2-} + 6OH^-$	-0.571
$SO_4^{2-} + H_2O + 2e \rightleftharpoons SO_3^{2-} + 2OH^-$	-0.93
$SbO_3^- + H_2O + 2e \rightleftharpoons SbO_2^- + 2OH^-$	-0.59
$SiO_3^{2-} + 3H_2O + 4e \rightleftharpoons Si + 6OH^-$	-1.697
$Zn(OH)_2 + 2e \rightleftharpoons Zn + 2OH^-$	-1.249
$ZnO + H_2O + 2e \rightleftharpoons Zn + 2OH^-$	-1.260
$ZnO_2^{2-} + 2H_2O + 2e \rightleftharpoons Zn + 4OH^-$	-1.215

* Reference: Weast RC. CRC Handbook of Chemistry and Physics. 80th ed. CRC Press, 1999~2000.

Reference

[1] 张天蓝. 无机化学. 第 5 版. 北京：人民卫生出版社，2007.
[2] 张英珊. 化学概论 General Chemistry. 北京：化学工业出版社，2005.
[3] 许善锦. 无机化学. 第 4 版. 北京：人民卫生出版社，2003.
[4] Fred Basolo，Ronald C. Johnson. 配位化学. 宋银柱，王耕霖等译. 北京：北京大学出版社，1982.
[5] D. F. Shriver，P. W. Atkins，C. H. Langford. 无机化学. 第 2 版. 高忆慈，史启贞，曾克慰，李丙瑞等译. 北京：高等教育出版社，1997.
[6] F. Basolo，R. Pearson. 无机反应机理. 陈荣悌，姚允斌译. 北京：科学出版社，1987.
[7] CRC Handbook of Chemistry and Physics. 80th edition. CRC Press，1999~2000.
[8] P. A. Cox. Instant Notes in Inorganic Chemistry. BIOS Scientific Publishers Limited. 北京：科学出版社，2000.
[9] Mark J. Winter. d-Block Chemistry. Oxford：Oxford University Press，1995.
[10] Jerry A. Bell. Chemistry：a project of the American Chemical Society. New York：W. H. Freeman and Company，2005.
[11] J. N. Spencer，G. M. Bodner，L. H. Rickard. CHEMISTRY：Structure and Dynamics. 3rd edn. New Jersey：Wiley，2006.
[12] Darrell D. Ebbing，Steven D. Gammon，Ronald O. Ragsdale. Essentials of General Chemistry. 2nd edn. New York：Houghton Mifflin Company，2006.
[13] Catherine E. Housecroft，Alan G. Sharpe. Inorganic Chemistry. 2nd edn. Essex：PEARSON，2005.
[14] John C. Kotz，Paul M. Treichel，Patrick A. Harman. Chemistry and Chemical Reactivity，5th edn. Stamford：Thomson Learning，Inc.，2003.
[15] Peter Atkins，Loretta Jones，Roy Tasker. Chemical Principles. 2nd edn. New York：W. H. Freeman and Company，2002.
[16] Raymond Chang. Chemistry，7th Edn. New York：The McGraw-Hill Companies，Inc.，2002.
[17] John Olmsted Ⅲ，Gregory M. Williams. Chemistry，3rd Edn. New Jersey：John Wiley & Sons，Inc.，2002.
[18] John W. Moore，Courad L. Stanitski，Peter C. Jurs. Chemistry：The Molecular Science. 2nd edn. Belmont：Brooks/Cole，2005.
[19] John McMurry，Robert C. Fay. Chemistry. 3rd edn. New Jersey：Prentice-Hall，Inc.，2001.